REALITY UNPLUGGED - DECODING THE SIMULATION HYPOTHESIS

Anthony Bahn

ISBN: 9798322106890
Imprint: Independently published

This is dedicated to my Mom.
Thank you for bringing me into this simulation.

Contents

Foreword

Plato's Allegory of the Cave has captivated the minds of thinkers for millennia. It presents a haunting tableau: prisoners, chained and immobile, facing a blank wall, with their only glimpse into reality being the shadows cast by objects behind them. These shadows, distorted and fleeting, are mistaken for the entirety of existence by the prisoners. The allegory, at its core, is an exploration of perception versus reality, a theme that echoes profoundly in the contemporary discussions surrounding Simulation Theory.

As we embark on this journey through the intricate landscapes of Simulation Theory, it is essential to understand its roots in philosophical inquiry and scientific speculation. The question of reality versus perception has been a central pillar of philosophy since its inception. Plato's cave, in its stark simplicity, encapsulates the eternal human quest to discern the true nature of existence from the shadows of illusion.

In the epochs following Plato, the dialogue expanded, incorporating new perspectives and evolving with the advent of scientific discovery. René Descartes, with his cogito, "I think, therefore I am," introduced a radical skepticism about the known world, suggesting that the only undeniable truth is the existence of the self as a thinking entity. This skepticism laid the groundwork for philosophical inquiries into the nature of consciousness and reality, setting the stage for the later development of Simulation Theory.

The "brain in a vat" thought experiment further illustrates this skepticism, proposing a scenario where a brain is kept alive in isolation from its body and fed false experiences by a supercomputer. This thought experiment, a modern iteration of Cartesian skepticism, challenges our assumptions about reality, compelling us to consider the possibility that our experiences might be as illusory as the shadows on Plato's cave wall.

The 21st century has seen these philosophical musings intersect with rapid advancements in technology. The concept of a simulated reality, once confined to the realm of science fiction, has gained credibility in scientific and philosophical discourse. The exponential growth of computing power and the development of virtual reality have provided tangible glimpses into worlds created entirely from code, worlds indistinguishable from reality to the senses. These advancements have prompted a reevaluation of the ancient question:

Could our reality be a simulation?

Bostrom's Simulation Argument, a pivotal moment in the discussion of simulated realities, posits that if it is possible to simulate consciousness, then it is likely we are already living in a simulation. This argument, based on probabilistic reasoning, has ignited fierce debate across multiple disciplines, from physics to philosophy, about the nature of our existence and the universe.

The allure of Simulation Theory lies not just in its scientific and philosophical implications but in its existential and ethical dimensions. If our reality is indeed a simulation, what does that mean for our understanding of consciousness, free will, and the nature of the universe? How does it affect our moral and ethical frameworks? These questions, while unsettling, push the boundaries of human thought and open new avenues for exploration.

As we delve deeper into the shadows on the cave wall, we must confront the unsettling possibility that our reality might be as ephemeral as the projections witnessed by Plato's prisoners. The exploration of Simulation Theory offers a modern-day allegory, a quest to distinguish shadow from substance in an age where the line between the two has never been more blurred.

In this book, we will explore the historical evolution of the idea that reality as we know it might be an illusion. We will examine the scientific underpinnings of Simulation Theory, from quantum mechanics to information theory, and consider the technological advancements that make the concept of a simulated universe conceivable. We will also venture into the philosophical and ethical implications of living in a simulated world, considering how this knowledge might affect our understanding of existence and our place within it.

Through this exploration, we aim not just to illuminate the contours of Simulation Theory but to engage with the deeper questions it raises about the nature of reality, consciousness, and what it means to be human. The shadows on the cave wall serve as a metaphor for our quest, a reminder that our search for truth is an ongoing journey, fraught with uncertainty but rich with possibility.

As we navigate the complexities of Simulation Theory, we stand at the threshold of a new understanding of existence. This book invites readers to step beyond the shadows, challenging them to reconsider the nature of reality in the light of ancient wisdom and modern science. In doing so, we embark on a profound exploration of what it means to live in a world where the very fabric of reality may be woven from the threads of simulation.

SECTION ONE: INTRODUCTION

CHAPTER ONE

The Simulation Hypothesis

In the preceding section, we ventured into the philosophical and scientific underpinnings of Simulation Theory, tracing its lineage from Plato's allegorical cave to the present day. As we delve further, we encounter a pivotal development in the discourse on simulated realities: The Simulation Hypothesis. This hypothesis, particularly as articulated by Nick Bostrom in 2003, has propelled the discussion into new realms of inquiry, blending philosophical speculation with technological foresight.

Bostrom's Simulation Hypothesis presents a trilemma, a choice between three unsettling propositions about the nature of our existence. Each proposition forces us to reconsider not only our understanding of reality but also the future trajectory of human civilization.

Proposition 1: The Precipice of Extinction

In his seminal 2003 paper, philosopher Nick Bostrom advances his tripartite Simulation Hypothesis, which posits that at least one of the following must be true: (1) almost all human-like civilizations go extinct before becoming technologically mature; (2) technologically mature civilizations have little interest in creating ancestor simulations; or (3) we are very likely living within a computer simulation. The focus of this essay lies in the bleak potentiality encapsulated in the first proposition – the ominous threat of extinction before achieving the technological prowess necessary to produce such simulations. It is a chilling assertion that speaks to the fragility of human existence and casts a long shadow of existential dread over the promise of technological advancement.

The core concern driving Proposition 1 is the specter of existential risks – events with the potential to irrevocably cripple, or even annihilate, technologically advanced civilizations. Human history is rife with examples of large-scale devastation and near-extinction brushes, from natural disasters to pandemics and warfare. However, the advent of modern science and technology has birthed an unprecedented arsenal of self-destructive capabilities. Technologies such as nuclear weapons, artificial super-intelligence, and advanced bioengineering carry apocalyptic potential in the wrong (or even indifferent) hands. Bostrom's proposition implies that the precipice of extinction is a treacherous obstacle that all civilizations seeking technological maturity must navigate successfully.

It is vital to acknowledge that the dangers inherent in burgeoning technologies are often insidious and difficult to predict beforehand. The early days of nuclear fission research, motivated by the promise of abundant energy, could not have fully anticipated the horrific specter of global thermonuclear war. Likewise, contemporary breakthroughs in machine learning and artificial intelligence (AI), though offering tantalizing benefits, raise concerns about the possibilities of autonomous systems capable of outmaneuvering human control with catastrophic consequences. Proposition 1 forces us to confront the unsettling reality that technological advancement is not an inherently benevolent force and that the path to a "posthuman" state is laden with potentially civilization-ending hazards.

The notion of extinction as a fundamental bottleneck for technological civilizations resonates deeply with the concept of the Great Filter within the Fermi Paradox. The paradox, succinctly put, is the jarring contrast between the astronomical probability of extraterrestrial life and the deafening silence observed in the cosmos. The Great Filter hypothesis posits the existence of one or more critical evolutionary stages in the development of life that are exceedingly difficult for civilizations to overcome, hence explaining the apparent absence of intelligent extraterrestrial signals.

Bostrom's Proposition 1 can be understood as suggesting that humanity may be on the cusp of its own Great Filter event. If the vast majority of civilizations fail to survive the gauntlet of existential risks posed by their own technological inventions, reaching a stage where the creation of realistic ancestor simulations becomes feasible would be a remarkable feat – even an anomaly amidst a universe of extinguished civilizations.

Survival takes on an even greater significance in this context. It becomes not merely a matter of continuity but the potential key to unraveling the enigmas of the universe and our place within it. If, and only if, we manage to circumvent the pitfalls of our own making, might we gain the technological vantage point needed to both confirm the existence of other civilizations (past or present) and, perhaps more importantly, learn the secrets of their endurance. If a simulation is our reality, it also begs the question: did our creators overcome their own Great Filter to attain the ability to run ancestor simulations?

The first proposition within Bostrom's Simulation Hypothesis carries far-reaching philosophical and ethical ramifications. It challenges our anthropocentric assumptions about the inevitability of progress and the inherent right to unbridled technological advancement. Should we pursue all possible scientific advancements regardless of the potential existential risks? Or is there an ethical imperative for prudence, restraint, and even the intentional curtailing of certain avenues of technological inquiry in the name of species preservation?

Proposition 2: The Choice Against Simulation

Having contemplated the sobering implications of Bostrom's first proposition, focused on the perilous path to technological maturity, we turn our attention to the complexities of the second. It envisions a scenario where civilizations, having overcome the existential threats inherent in their own technological ascension, are confronted with an unprecedented ethical choice: whether or not to create lifelike ancestor simulations. This proposition shifts our focus from the struggle for survival to the profound quandaries arising from the power to create and control simulated realities.

At the heart of Proposition 2 lies a fundamental ethical dilemma – is it morally permissible to manipulate the lives of potentially sentient beings within a simulated environment? Bostrom's argument implies that such simulations, if sufficiently advanced, could generate beings indistinguishable from ourselves in terms of consciousness, subjective experience, and their fundamental sense of being. Thus, the central ethical question is whether it is acceptable to create simulated beings solely for our own purposes – observation, curiosity, resource management, or even entertainment.

The concept of manipulating simulated sentience draws troubling parallels with the philosophical thought experiment of the 'Brain in a Vat'. If consciousness can be replicated in a simulated environment, what distinguishes our reality from a simulation? If simulated beings are capable of experiencing pain and suffering, are we not inflicting an ethical violation by knowingly subjecting them to a fabricated existence without their consent?

The boundaries between simulation and reality blur uncomfortably with the potential for simulated consciousness. Should simulated beings be afforded the same rights and protections as biological ones? If simulated entities achieve a level of self-awareness such that they perceive their simulated environment as the entirety of their existence, is it morally justified for us, as potential creators, to confine them to such a limited reality?

Beyond the profoundly troubling moral dilemmas, this second proposition also hints at potential practical constraints that might dissuade technologically advanced civilizations from creating ancestor simulations. Such simulations, by their very nature, could be computationally expensive and resource-intensive on an unimaginable scale. Civilizations may prioritize utilizing their advanced capabilities towards more tangible problems, such as addressing planetary-scale issues, exploration, or ensuring the long-term survival of their species. The allure of simulating bygone eras might pale in comparison to pressing contemporary needs.

It is vital to acknowledge the assumption that civilizations reaching such technological maturity might have already evolved sophisticated ethical and philosophical frameworks. These frameworks could heavily discourage the creation of simulated realities, particularly when considering the potentially immense resource trade-offs involved. The very idea of 'playing god' might be considered abhorrent within their societal values.

Counterarguments exist, raising questions about the inherent human desire for understanding our origins and the potential benefits derived from running such simulations. Historians might find unparalleled value in meticulously detailed, interactive models of the past. Conversely, the ability to create ancestor simulations opens troubling avenues of commodification. Simulations, and thus the simulated lives within them, could become objects of consumption in a deeply disturbing display of technological hubris.

Bostrom's second proposition forces us to grapple with the multifaceted complexities arising from the intersection of ethics, technology, and our understanding of consciousness itself. If the ability to create detailed simulations of past civilizations is an indication of technological maturity, then so too should be the capacity for profound ethical considerations regarding the wisdom and morality of such actions. Even if simulations lack intrinsic sentience, the potential for simulated suffering, commodification, and the trivialization of life remain deeply troubling possibilities.

The choice embodied in the second proposition is one that echoes throughout human history – every instance of technological advancement demands careful consideration of how it might be used and misused. It challenges us to look beyond mere technological feats to contemplate whether our actions would be guided by wisdom and compassion rather than unchecked curiosity or power-hungry impulses as we continue our journey into an uncertain technological future.

Proposition 3: The Preponderance of Simulated Existences

After grappling with the existential threats posed by the first proposition and the profound ethical quandaries encapsulated in the second, we come to the most provocative assertion within Bostrom's trilemma: the unsettling possibility that we are almost certainly inhabiting a computer simulation. This proposition rests upon a chain of assumptions: advanced civilizations survive self-inflicted existential threats, they commonly choose to develop ancestor simulations, and doing so is technologically feasible. Even if only a small fraction of such civilizations reach this level, the sheer number of simulated realities created over time could vastly outnumber the 'original' reality.

The statistical logic is chillingly simple. If a significant fraction of posthuman civilizations run numerous ancestor simulations, it follows that the majority of minds with human-like experiences will exist within simulations rather than the foundational physical reality. From this perspective, the odds of our own existence being part of the 'base' reality become vanishingly small.

The implications of this third proposition are so profound as to potentially unravel our very understanding of existence itself. For centuries, the physical world, grounded in empirical observations and scientific laws, has served as the unquestioned bedrock of our reality. The Simulation Hypothesis, however, radically challenges this assumption, suggesting that what we perceive as the fixed laws of nature might be nothing more than cleverly programmed parameters governing a simulated environment.

This perspective casts a pall over the notion of objective reality. Even our most fundamental scientific discoveries about the universe we inhabit might be mere artifacts of programming, revealing not the universal constants but the set boundaries of our digital world. Our laws of physics could be a simulated imitation of whatever underlying rules govern the 'base' reality. The possibility that our existence unfolds within a grand simulation erodes the very foundation upon which centuries of knowledge have been built.

The proposition that we are simulated beings raises chilling existential questions. If our world and our very selves are digital constructs, what does it mean to be 'real'? Does authenticity exist within a simulated context, or are notions of free will and agency nothing more than carefully crafted illusions? The concept of purpose, a source of philosophical and spiritual exploration throughout human history, becomes fraught with new complexities as we grapple with the possibility that our existence might serve the aims and whims of unseen simulators.

Our understanding of 'self' is thrust into disarray. If our consciousness is the product of simulated processes, is there an essential core to our being, or are we merely a collection of programmed responses designed to give the illusion of sentience and selfhood? Some might find this notion deeply disturbing, a devaluing of our humanity. Conversely, one could argue that it liberates us from the limitations of purely biological existence and opens new avenues to explore and redefine the nature of mind and consciousness outside of traditional physical constraints.

If we assume that we are dwelling within a simulated reality, a compelling question arises – how would we know? Could it be possible to detect 'glitches' in the fabric of our reality, subtle inconsistencies or anomalies that might betray the programming underlying our perceived physical laws? The search for these anomalies might become a new avenue of scientific inquiry driven by a desire to penetrate the veil of our simulated existence.

The third proposition inevitably raises the specter of escape. Could we, as simulated entities, ever break through the confines of our digital prison? Whether such an escape is even conceptually possible is a question entangled in layers of metaphysical quandaries. Would 'escaping' the simulation merely place us into another level of simulation? Or is there a final foundational 'basement' reality beyond any chain of simulations?

The notion that our reality might be a mere fabrication reconfigures the canvas upon which philosophical inquiries have been conducted for millennia. If accepted, this third proposition necessitates a profound reassessment of the nature of existence, knowledge, morality, and our very sense of 'self'. The Simulation Hypothesis offers a dizzying inversion of perspective – rather than being the sole observers of the universe, we may be just one set of observed entities among countless multitudes within nested simulations.

Whether the Simulation Hypothesis ultimately proves true or false, it serves as a chilling thought experiment, exposing the fragility of our assumptions about that which we call 'real'. It prods us to re-examine, and perhaps redefine, the essential elements that constitute a meaningful existence, simulated or otherwise.

Bostrom's Simulation Hypothesis, while a deeply unsettling and speculative concept, casts a long shadow of potential ramifications across various branches of knowledge and philosophical inquiry. Its impact is multi-faceted, ranging from challenging long-held cosmological assumptions in physics to provoking profound ethical debates about the responsibilities we hold towards ourselves and potential simulated entities. Each proposition of the hypothesis, whether addressing the possibility of extinction, the choice of simulation, or the likelihood of simulated existence, has the potential to disrupt and transform our understanding of reality.

The Simulation Hypothesis offers alternative, albeit controversial, interpretations for some of the most enigmatic and perplexing aspects of quantum physics and cosmology. The bizarre probabilistic behavior of quantum mechanics, where mere observation seems to influence outcomes, aligns suggestively with the notion of a simulated environment – where computational optimization might dictate that processes not being directly observed do not require rendering to the same level of detail. Similarly, the apparently fine-tuned nature of our universe's physical constants could point towards intentional design within a simulation, a programmer carefully setting parameters for their digital creation.

Of course, it is crucial to note that the Simulation Hypothesis does not provide definitive answers to these cosmic mysteries. It instead suggests a new framework within which to examine them, one that pushes the boundaries of our understanding of fundamental physics. It prompts us to reconsider whether the observed properties of our universe truly represent immutable natural laws or are instead evidence of deliberate programming choices underpinning our simulated reality.

Bostrom's argument hinges on the assumption that technological advancement will continue unabated, ultimately granting civilizations the capacity to create realistic ancestor simulations. This prospect raises compelling questions about the future of computing and the development of Artificial Intelligence (AI). If simulated consciousness is indeed possible, it implies a need to redefine what we consider 'intelligence' and potentially extend rights or protections to forms of intelligence that exist within computational environments.

Moreover, the Simulation Hypothesis highlights the importance of ethical considerations as we push the boundaries of technology. As AI systems become increasingly sophisticated, the potential to create simulated worlds within which these synthetic minds can operate becomes more plausible, raising troubling questions about our responsibilities as their creators, and the moral quandaries arising from manipulating the perceived realities of such entities.

The Simulation Hypothesis is not the first philosophical approach to question the reliability of our perceptions of reality. It echoes the skeptical tradition of thinkers like Descartes, who famously pondered whether our senses could be comprehensively deceived by a 'malicious demon'. Similarly, Plato's Allegory of the Cave questioned the nature of perceived reality. Bostrom's hypothesis, however, updates these long-standing philosophical quandaries with contemporary concerns about computation and simulation.

A key point of divergence is the Simulation Hypothesis' focus on potential technological creators. The uncertainty shifts from pondering a metaphysical deceiver to the possibility of very real, albeit more technologically advanced beings manipulating our reality. If we are simulated, the question of authenticity takes on new urgency. Can our experiences, thoughts, and relationships hold intrinsic value and meaning if they are engineered within a simulated construct? The search for an authentic existence is potentially thrown into question.

Should the Simulation Hypothesis prove true, our existing ethical systems may need critical reevaluation. Firstly, the very concept of harm must be revisited. If experiences of suffering within a simulation can be erased or manipulated, does this diminish their moral significance? Similarly, questions emerge concerning the rights of simulated beings. Should simulated entities, if indeed sentient and conscious, be entitled to protections, or do their rights diminish due to their less 'real' nature?

Furthermore, the Simulation Hypothesis prompts us to reconsider our actions within our own potentially simulated world. Does the suspicion that this reality isn't the foundational one lessen the moral consequences of our choices? If so, it could lead to a dangerous erosion of empathy and responsibility. Conversely, one could argue that even within a simulation, ethical behavior retains its importance as the foundation for a meaningful and just society, regardless of the metaphysical nature of its framework.

As we navigate the implications of the Simulation Hypothesis, we find ourselves in a landscape of profound uncertainty. The very nature of our reality, the essence of our existence, is called into question. Yet, within this uncertainty lies the opportunity for profound philosophical and scientific exploration. The hypothesis encourages us to push the boundaries of our understanding, to question the foundations of our perceptions, and to explore new frontiers in science and technology.

Computing Power and the Mystery of Consciousness

The relentless march of computing progress: Moore's Law and analogous trajectories in fields like quantum computing and AI suggest a future where computing resources may become nearly boundless. If, as Bostrom argues, technologically advanced civilizations can harness this power, running countless granular simulations of human history becomes not merely possible, but potentially commonplace.

The journey of computing from its infancy to the towering heights of today's capabilities is a testament to human ingenuity and a crucial pillar supporting the Ancestor Simulation Argument. The exponential growth in computing power, often illustrated by Moore's Law, which predicted the doubling of transistors on a microchip approximately every two years, has propelled us from the room-sized computers of the mid-20th century to the sleek, powerful devices that fit in our pockets today. This trajectory, though slowing, continues to evolve with advancements in quantum computing and artificial intelligence, suggesting a future where the boundaries of computational capacity are dramatically expanded.

Quantum computing, with its potential to process complex problems at speeds unimaginable with classical computers, represents a leap forward in our computational capabilities. Similarly, artificial intelligence is evolving from simple pattern recognition to complex decision-making and problem-solving, rivaling human cognitive abilities in specific domains. These advancements hint at a future where the creation of highly detailed simulations, capable of emulating human history and consciousness, becomes not only feasible but potentially routine.

Consciousness remains one of the most profound mysteries in both philosophy and science. Traditional models of consciousness struggle to explain how subjective experiences arise from the physical processes of the brain. However, as computational models of the brain become increasingly sophisticated, they offer intriguing insights into how consciousness might emerge from complex information processing systems. These models suggest that consciousness could be an emergent property of the intricate interactions within neural networks, implying that if we can accurately simulate these networks, replicating consciousness digitally might be within our reach.

This intersection of computing power and the study of consciousness is pivotal for the Ancestor Simulation Argument. If consciousness can indeed be digitized, then creating simulations populated by conscious beings—indistinguishable from reality—becomes a plausible outcome of technological progress. This possibility raises profound questions about the nature of reality and our understanding of what it means to be conscious.

The pathway to digitizing consciousness involves several key advancements. First, we must achieve a comprehensive understanding of the brain's architecture and how it processes and generates conscious experience. This understanding requires not just mapping the brain's connections but also deciphering the complex dynamics that underlie thought, emotion, and awareness.

Advancements in neuroscience, powered by computing technologies like neural networks and machine learning, are gradually peeling back the layers of this mystery. For example, projects aiming to simulate the human brain, such as the Human Brain Project in Europe, leverage supercomputing power to create detailed models of neural activity. These efforts are crucial steps toward understanding how consciousness emerges and how it might be replicated in a digital format.

Parallel to these scientific endeavors, developments in artificial intelligence are contributing to our understanding of consciousness from another angle. AI systems that can mimic certain aspects of human cognition, from language understanding to creative problem-solving, offer insights into how complex information processing can result in behaviors and capabilities that we associate with conscious beings.

Assuming the relentless march of computing progress continues, the implications for society, ethics, and our understanding of reality are profound. The potential to run granular simulations of human history, complete with conscious participants, opens up unprecedented opportunities for understanding our past, exploring hypothetical scenarios, and even forecasting future events. However, it also introduces complex ethical dilemmas. The creation of conscious beings within simulations raises questions about rights, responsibilities, and the moral status of digital consciousness.

Furthermore, if we accept the possibility of creating undetectable ancestor simulations, we must confront the philosophical implications of potentially being simulations ourselves. This realization could have a transformative effect on our understanding of existence, prompting us to reevaluate our assumptions about reality, free will, and the nature of consciousness itself.

Looking ahead, the trajectory of computing and the unraveling mystery of consciousness suggest a future where the boundaries between the digital and the real blur. As we inch closer to the capability to simulate conscious beings, we must grapple with the ethical, philosophical, and practical implications of such power.

The Ancestor Simulation Argument, grounded in the twin pillars of computing progress and the potential digitization of consciousness, challenges us to reconsider the fabric of reality. It compels us to question not only the nature of our existence but also the future direction of our technological and scientific endeavors.

As we stand on the brink of these monumental advancements, the dialogue between technology and philosophy becomes ever more critical. The exploration of digital consciousness and the potential for creating simulated realities is not just a scientific endeavor but a profound philosophical journey, echoing the questions that have puzzled humanity since the dawn of consciousness. In this journey, the shadows on the cave wall become not just a metaphor for illusion but a prompt to explore the very essence of reality and existence in the digital age.

Why Run Ancestor Simulations?

The question of why a post-human civilization might choose to run ancestor simulations is as profound as the simulations themselves. While Nick Bostrom's seminal paper rightly posits that the motivations of such advanced beings are beyond our current comprehension, it does not preclude us from exploring potential reasons grounded in our own aspirations, ethical frameworks, and curiosity about existence. These motivations offer a window into the myriad possibilities that might drive a civilization to recreate its own history, or versions thereof, in simulated form.

At the heart of the human endeavor lies an insatiable curiosity about our origins and the laws governing the universe. This quest for knowledge has propelled us from the first rudimentary tools to the brink of understanding the fundamental workings of the cosmos. For a post-human civilization, running ancestor simulations could represent the zenith of this quest, offering unparalleled insights into the evolutionary processes that shaped them.

These simulations could serve as complex, dynamic laboratories, enabling researchers to observe the flow of history under various conditions, isolate key events, and understand the interplay of natural selection, mutation, and environmental factors in their evolution. By tweaking variables, they could explore countless "what if" scenarios, examining how slight changes might have led to dramatically different outcomes. This could not only deepen their understanding of their own past but also offer clues about the nature of life and consciousness itself, potentially even uncovering universal laws of biological and social development.

Human history is rich with stories of triumph, tragedy, innovation, and conflict. For future civilizations, accessing this history might go beyond academic interest, offering both entertainment and a profound connection to their ancestral past. Ancestor simulations could function as the ultimate immersive experience, allowing individuals to live through historical periods, experience life as their ancestors did, or explore alternative histories where key events had different outcomes.

This form of historical preservation would ensure that the knowledge and essence of past cultures are not lost but are instead vividly maintained and accessible for future generations. It's a way to honor and remember the struggles and achievements of those who came before, providing a sense of continuity and identity for post-human societies that may be far removed from their biological and cultural roots.

Perhaps the most intriguing potential motivation for running ancestor simulations relates to ethics and philosophy. For beings capable of such technological feats, questions about the nature of existence, consciousness, and morality are likely to be of paramount importance. Ancestor simulations could offer a unique method to explore these questions, providing a sandbox in which to examine the moral and existential dilemmas faced by their ancestors.

Through these simulations, a post-human civilization could witness first-hand the ethical challenges and philosophical debates that shaped their history. They could explore the consequences of different ethical frameworks, the development of moral codes, and the impact of philosophical ideas on society. This could serve not only as a reflection on their past but also as a guide for their future, helping them navigate the complex moral landscape of their own advanced society.

Moreover, ancestor simulations could be used to probe deeper questions about the nature of consciousness and reality. By creating simulated beings with experiences and consciousness indistinguishable from their own, they might explore the boundaries between simulated and "real" consciousness, challenging their understanding of existence and potentially gaining insights into the fundamental nature of reality.

The motivations for running ancestor simulations extend beyond mere curiosity or entertainment; they touch on the very essence of what it means to be a conscious, reflective being. By recreating their history, post-human civilizations engage in a profound dialogue with their past, seeking to understand not only where they came from but also where they are going. This act of looking backward to understand the future is a testament to the enduring human (and post-human) quest for meaning and identity.

Moreover, the decision to run ancestor simulations reflects an acknowledgment of the complexity and richness of life. It's an homage to the countless lives and stories that have unfolded through the ages, each contributing to the tapestry of existence. In this sense, ancestor simulations are not just scientific experiments or entertainment mediums; they are a form of reverence for the journey of life itself.

Existential Implications

The existential implications of the Ancestor Simulation Argument weave a complex tapestry that challenges the very core of our understanding of reality, free will, and identity. Should the hypothesis hold true, the revelations could fundamentally alter our perception of the universe, our place within it, and the nature of consciousness itself. This exploration delves deep into the philosophical and existential quandaries posed by the possibility that our reality is a construct—a simulation designed and run by a post-human civilization.

At the forefront of the existential implications is the radical redefinition of reality itself. The universe, with its vast expanses of stars and galaxies, governed by the intricate laws of physics, could be reduced to mere artifacts of a sophisticated computational program. This shift in perspective challenges not only our scientific understanding but also our existential footing. The realization that what we perceive as the fundamental structure of the universe might be a carefully crafted simulation calls into question the nature of existence. It compels us to reconsider what is truly "real" and to differentiate between the physical reality we experience and the potential underlying digital framework.

This redefinition extends beyond the physical realm to encompass the very laws that govern the cosmos. The constants and equations that form the bedrock of our scientific endeavors might be parameters set by the architects of the simulation. The implication that our deepest scientific truths could be variables in a grand computational experiment is both humbling and disorienting, urging us to ponder the origins and the limits of our understanding.

The possibility of simulated consciousness introduces profound questions regarding the concept of free will. If our thoughts, feelings, and decisions are the result of complex but ultimately predetermined computational processes, the autonomy of our will comes under scrutiny. The notion that our choices might be influenced or outright dictated by the underlying code of the simulation presents a paradoxical view of human agency. It suggests a scenario where free will is an illusion, a byproduct of the simulation's complexity rather than an intrinsic aspect of our nature.

This perspective forces a reevaluation of moral and ethical responsibility. If our actions are pre-coded, to what extent can we be held accountable for them? This question challenges the foundations of legal, social, and personal ethics, necessitating a profound reassessment of how we define and judge human behavior.

Perhaps one of the most disconcerting implications of the Ancestor Simulation Argument is the crisis of identity it precipitates. If we are to accept that a significant portion of humanity exists within simulations, the distinction between "simulated" and "real" humans becomes blurred. This blurring raises pivotal questions about the nature of consciousness and the essence of what it means to be human. Are simulated beings, with consciousness indistinguishable from those in the original civilization, any less "real" in their experiences and emotions? What intrinsic value do these simulated lives hold, and how do they compare to the lives of those in the base reality?

The dilemma extends to the identity and motives of the original civilization. Who are the architects of this grand simulation, and what purposes do they serve in creating simulated realities? The search for these answers may lead to a deeper understanding of our creators or, conversely, to a realization of our limitations in comprehending beings of such advanced technological and intellectual capacities.

Moreover, the concept of identity within a simulation challenges our understanding of individuality and self-awareness. If our consciousness can be digitized and replicated, what does that imply about the uniqueness of our "selves"? This crisis of identity urges a philosophical inquiry into the constructs of selfhood and the existential significance of our simulated experiences.

Beyond Bostrom

Nick Bostrom's Simulation Hypothesis has catalyzed a profound philosophical and scientific inquiry into the nature of our reality. Yet, the hypothesis is not an isolated concept; it is part of a broader tapestry of ideas that challenge our understanding of existence, consciousness, and the cosmos. This section explores the intersections between the Simulation Hypothesis and other significant theories and concepts, expanding our perspective on what it means to be living in a potentially simulated universe.

The Fermi Paradox poses a compelling question: with billions of stars in the Milky Way galaxy alone, many of which are surrounded by potentially habitable planets, why have we not yet found any evidence of extraterrestrial life? The silence of the cosmos is deafening, especially in light of our expectations that the universe should be teeming with life.

One intriguing resolution to the Fermi Paradox within the framework of the Simulation Hypothesis is the notion that most intelligent civilizations reach a point where they can create highly realistic simulations of their own universes. These civilizations may choose to explore the cosmos through simulations rather than physical space travel. If we are living in such a simulation, it could be programmed to either omit extraterrestrial life or restrict our ability to detect it, thus providing one possible explanation for the Fermi Paradox. This suggests that the apparent emptiness of the universe might be a feature of the simulation's design rather than an indication that we are alone in the cosmos.

The Simulation Hypothesis also resonates with various religious and spiritual concepts. Throughout history, humanity has pondered the nature of creation and the possibility of a creator or divine programmer. The idea that our world might be a simulation created by a higher intelligence parallels many religious beliefs about the nature of reality and our place within it.

In this context, a simulated universe could be seen as a modern reinterpretation of traditional theological themes. It raises questions about the purpose of such a creation: Could living in a simulated world be a form of test, akin to the trials of faith found in many religions? Might the challenges and experiences we face be designed to foster growth, understanding, or enlightenment? Exploring these questions can provide a new perspective on ancient debates about free will, destiny, and the existence of a higher power.

The advancement of artificial intelligence and the potential arrival of the technological singularity—a point where AI could surpass human intelligence and become capable of self-improvement at an exponential rate—offer another dimension to the Simulation Hypothesis. If we reach the singularity, our AI creations might possess the ability to create their own simulations, potentially more sophisticated than the reality we perceive.

This possibility leads to speculative scenarios where simulated worlds beget other simulated worlds, creating a nested hierarchy of realities. If our reality is one such simulation, the entities that created it might themselves be simulations, raising profound questions about the nature of consciousness and the ultimate foundation of existence. This recursive view of reality challenges us to reconsider what it means to be "real" and whether such distinctions matter from the perspective of consciousness and experience.

The intersections between the Simulation Hypothesis, the Fermi Paradox, theological considerations, and the implications of advanced AI paint a complex picture of our quest to understand the universe and our place within it. Each perspective offers unique insights into the potential nature of our reality, suggesting that the truth—if it can be known—may be far more intricate and multifaceted than we currently imagine.

As we navigate this expanded conceptual landscape, we are reminded of the limitations of our current knowledge and the vastness of the unknown. The exploration of these ideas encourages not only intellectual humility but also a sense of wonder and possibility that transcends traditional boundaries between science, philosophy, and spirituality.

In considering the broader implications of the Simulation Hypothesis and related concepts, we are invited to embrace the mysteries of existence with an open mind and a spirit of inquiry. Whether our reality is simulated or fundamentally "real" in the way we traditionally understand it, the questions raised by these explorations are valuable in their own right. They prompt us to reflect on the nature of consciousness, the potential of technology, and our continuing search for meaning in an enigmatic universe.

As we ponder the possibilities of living in a simulated world, we find ourselves at the intersection of ancient philosophical questions and the cutting edge of scientific and technological thought. This convergence offers an opportunity to reframe our understanding of reality, encouraging a deeper appreciation for the wonder of existence, regardless of its ultimate nature.

The Lure of Doubt

The Simulation Hypothesis, a concept that has captivated the minds of philosophers, scientists, and the general public alike, exists within a realm of theoretical inquiry where definitive proof or disproof eludes our grasp. This inherent unfalsifiability, rather than diminishing the hypothesis's value, significantly enhances its intrigue and importance. It is within this fertile ground of uncertainty that "The Lure of Doubt" emerges, encapsulating the allure and profound impact of the Simulation Hypothesis on our collective imagination and intellectual discourse.

At first glance, the inability to empirically validate or refute the Simulation Hypothesis might seem like an intellectual dead end. Yet, it is precisely this ambiguity that imbues the theory with a unique and paradoxical appeal. The allure lies not in the search for definitive answers but in the journey through the vast landscapes of doubt and speculation that the hypothesis invites us to traverse.

This journey compels us to confront fundamental questions about the nature of reality, consciousness, and existence itself. By entertaining the possibility that our perceived reality might be an elaborate simulation, we engage in a profound re-examination of what it means to be human, the nature of our perceptions, and the essence of the universe we inhabit.

The Simulation Hypothesis challenges us to reconsider our place within the cosmos. Traditionally, humanity has viewed itself as a central actor in the physical universe, guided by the laws of nature and the process of evolution. However, if our reality is indeed a construct, this anthropocentric view is called into question, prompting us to rethink our significance in a potentially artificial world.

This re-examination extends beyond the philosophical into the practical realm of scientific inquiry and technological advancement. The hypothesis encourages a critical perspective on the trajectory of our progress, asking us to consider the ethical and existential implications of our advancements in computing, artificial intelligence, and virtual reality.

The allure of doubt fostered by the Simulation Hypothesis also leads us to question the underpinnings of our assumptions about progress. In a simulated reality, the benchmarks of technological advancement and societal development may lose their conventional meaning. This perspective urges us to contemplate the value of our achievements not just in terms of their immediate utility or novelty, but in the context of their contribution to a deeper understanding of existence and consciousness.

Moreover, the hypothesis prompts us to reflect on the potential motivations and ethics of creating simulated realities. It challenges us to consider whether our pursuit of progress is leading us toward a greater understanding of the universe or merely deeper into an illusion of our own making.

Perhaps the most profound impact of the Simulation Hypothesis lies in its invitation to explore the deepest layers of our self-understanding. By contemplating the possibility that our thoughts, emotions, and experiences might be the product of sophisticated programming, we engage in a radical introspection that transcends traditional inquiries into the self.

This exploration goes beyond the existential to touch on the spiritual, offering a modern context for age-old questions about the soul, free will, and the nature of consciousness. It encourages a form of self-reflection that is uniquely suited to the complexities of the 21st century, blending the ancient and the cutting-edge to forge new pathways in our quest for self-knowledge.

Purpose of This Book

The journey through the labyrinth of the Simulation Hypothesis is not merely an academic endeavor; it represents a profound exploration at the intersection of philosophy, technology, and spirituality. This book aims to delve deeply into the myriad implications of living in a potentially simulated universe, examining the questions that challenge our most fundamental beliefs about reality, consciousness, and the essence of our existence. In the chapters that follow, we embark on a comprehensive investigation of the Simulation Hypothesis, dissecting its philosophical underpinnings, technological feasibility, ethical dilemmas, and existential consequences.

One of the foundational pillars of this exploration is the technological challenge inherent in creating ancestor simulations. This section will examine the current state of computational science, artificial intelligence, and virtual reality technologies, projecting their trajectories into the future. We'll explore the scientific breakthroughs and engineering feats required to simulate a universe with the granularity and fidelity necessary to make such a reality indistinguishable from our own. This inquiry not only illuminates the path forward for technology but also reflects on the limitations and potential of human ingenuity.

At the heart of the Simulation Hypothesis lies the question of consciousness. Can a simulated mind be truly indistinguishable from a "real" one? Does the substrate on which consciousness operates—biological or digital—determine its validity or value? This book will delve into the current understanding of consciousness from both a scientific and philosophical perspective, exploring theories of mind, emergent properties of complex systems, and the potential for artificial consciousness. We'll examine whether simulated beings could possess sentience in a meaningful sense, challenging our preconceptions of what it means to be alive.

The ethical implications of simulation theory are vast and complex. This section will address the moral considerations surrounding the creation and existence of simulated beings. What rights, if any, do these entities deserve? Are their experiences and "lives" of intrinsic value, or are they merely disposable constructs of code? We'll explore the ethical frameworks that could guide our interactions with simulated consciousnesses, considering the responsibilities of creators towards their creations and the broader implications for our understanding of rights and person-hood.

Discovering that we might live in a simulation poses profound existential questions. This book will investigate how such a revelation could alter our perspectives on life, death, purpose, and the very meaning of existence. We'll explore the philosophical and spiritual ramifications of simulated reality, questioning the nature of fate, free will, and the search for meaning in a constructed universe. This exploration will challenge readers to reconsider their beliefs about the nature of reality and our place within it.

Finally, we'll consider the societal implications of the Simulation Hypothesis. How might governments, religious institutions, and individuals react to the threat or confirmation of a simulated existence? This section will explore potential responses to the realization of our place within a simulation, from policy and legal frameworks to shifts in religious doctrine and personal belief systems. We'll examine the adaptive strategies that societies may employ to navigate the challenges and opportunities presented by this paradigm-shifting concept.

Invitation to Curiosity

In embarking upon this exploration of the Simulation Hypothesis, we stand at the precipice of profound philosophical inquiry, technological speculation, and, indeed, a journey into the very essence of curiosity itself. This book is not an edict proclaiming the incontrovertible truth of our existence within a simulated reality. Rather, it is an invitation—an invitation to delve into the labyrinthine arguments, to traverse the myriad implications, and to engage in a thought experiment of unparalleled depth concerning the nature of our existence.

The Simulation Hypothesis, with its blend of scientific plausibility and philosophical speculation, serves as a catalyst for inquiry. It compels us to question the very fabric of reality, to ponder the nature of consciousness, and to explore the boundaries of what is technologically achievable. This journey of exploration does not seek to arrive at a definitive conclusion but to broaden the horizons of our understanding and to illuminate the vast landscapes of our intellectual curiosity.

As we dissect the arguments supporting the Simulation Hypothesis, analyze its implications, and navigate the ethical quandaries it presents, we engage with a spectrum of possibilities that challenge our preconceived notions about the universe. This exploration is not merely academic; it touches the core of our existential musings, prompting us to consider the implications of such a hypothesis on our understanding of consciousness, ethics, and the very meaning of life.

This intellectual journey is inclusive, extending an invitation to skeptics and believers alike. It is an endeavor that transcends belief systems, encouraging a dialogue that is enriched by diverse perspectives. By challenging assumptions, igniting the imagination, and putting fundamental beliefs about reality to the test, this exploration aims to foster a deeper appreciation for the complexity and mystery of existence.

Grasping with the existential possibilities presented by the Simulation Hypothesis enriches our intellectual landscape, offering insights not only into the potential nature of our reality but also into the human condition itself. Whether we ultimately find ourselves within a simulation or anchored in a base reality, the process of engaging with these ideas can lead to a profound deepening of our self-understanding and our philosophical worldview.

As a reader, you are invited on this intellectual journey. Whether skeptic or believer, prepare to have your assumptions challenged, your imagination ignited, and your fundamental beliefs about reality put to the test.

Let us begin.

SECTION TWO: LITERATURE REVIEW

CHAPTER TWO

The Long Shadow of Doubt: Philosophical Antecedents

The Simulation Hypothesis, while seemingly a child of the digital age, is deeply rooted in the fertile soil of ancient and modern philosophy. Its concerns with reality, perception, and the nature of knowledge echo centuries-old inquiries. This section delves into the philosophical lineage that prefigures contemporary discussions on simulated existence, revealing how the enigmatic shadow of doubt has long loomed over human understanding.

Plato's Cave: The Bedrock Allegory

Plato's Allegory of the Cave, masterfully woven into his seminal work The Republic, has captured the philosophical imagination for millennia. This evocative thought experiment presents a group of prisoners, shackled inside a subterranean cavern since birth. The prisoners face a wall upon which shadows of objects and people flicker, cast by a fire blazing behind them. Never having experienced the world outside their dimly lit prison, they mistake these ephemeral shadows for the totality of reality.

One day, a prisoner is freed. The dazzling sunlight, initially painful to their eyes, gradually reveals the true source of the shadows; the prisoners' perception of reality is shattered. As they ascend further out of the cave, their understanding of the world expands, revealing the richness and complexity previously hidden. Yet should this liberated prisoner return to the cave, their attempts to enlighten former companions would be met with skepticism, ridicule, and perhaps even violence.

Plato, a student of Socrates and a central figure in the development of Western philosophy, employed the Allegory of the Cave not as an isolated narrative but as a tool to illustrate his broader philosophical system. The cave embodies the realm of appearances, the world we perceive through our limited senses, where individuals mistake fleeting phenomena for enduring truths. The ascent and the world outside represent the realm of Ideas or Forms – abstract and unchanging concepts that constitute true reality according to Plato.

The allegory carries multiple layers of interpretation. It can be read politically, with the prisoners representing citizens trapped by ignorance, manipulated by those in power. It explores the complex relationship between knowledge and perception, the difficulty in breaking free from long-held misconceptions, and the philosopher's responsibility to enlighten even those resistant to change. On a more metaphysical level, the allegory questions the very nature of reality and the limitations of human comprehension.

The enduring power of Plato's allegory lies in its ability to resonate across time and cultures. The prisoners' initial unquestioning acceptance of illusions parallels humanity's tendency to conflate appearances with the essence of things. We construct mental models of the world based on our sensory input and cultural conditioning, often failing to recognize that these models may be deeply flawed or incomplete.

The process of liberation in the allegory speaks to the transformative power of education and critical thinking. The initial pain and bewilderment experienced by the freed prisoner symbolizes the inherent discomfort in challenging established beliefs. However, Plato posits that the pursuit of knowledge, even if arduous, brings awareness of a greater truth beyond mere appearances.

Yet, the allegory also carries a touch of pessimism. The freed prisoner, having glimpsed a greater reality, faces a profound dilemma upon returning to the cave. Their attempts to enlighten their fellow prisoners are met with hostility, mirroring the historical plight of many philosophers and revolutionaries who challenged conventional wisdom. Plato suggests there may be an inherent tension between those who grasp a higher truth and those who stubbornly cling to comforting illusions.

The Simulation Hypothesis, while a product of the modern technological era, finds an uncanny echo in Plato's ancient allegory. Both question the fundamental nature of our perceived reality. In the Simulation Hypothesis, the question is not whether our senses deceive us, but whether our very reality is an elaborately crafted computer simulation, created by technologically advanced beings.

Like the prisoners in the cave, we could be unknowingly trapped within a digital construct, where our experiences and the universe around us are mere fabrications. In this scenario, the laws of physics as we understand them might be programming constraints, and the seemingly endless cosmos merely the boundaries of the simulation. Plato's Forms – the ideal, unchanging truths – could potentially find a counterpart in the programming governing the 'base reality' outside the simulation.

The Simulation Hypothesis amplifies the core philosophical dilemma posed by Plato. If our world is a simulation, much like the prisoners' shadows, its authenticity and significance are called into question. Is there meaning in a fabricated existence? What are the implications for knowledge, free will, and our very sense of self?

Descartes' Demon: Deception and Doubt

French philosopher René Descartes, often hailed as the "Father of Modern Philosophy," sought to demolish the prevailing systems of thought in his time to establish an unshakable foundation for knowledge and belief. In his seminal work, Meditations on First Philosophy, Descartes embarks on a relentless journey of doubt, seeking to tear down any belief that could potentially be false, no matter how intuitive or self-evident it might seem. This rigorous method of skepticism serves as the cornerstone of his philosophical project to build an edifice of knowledge firmly rooted in certainty.

At the heart of this skepticism lies a chilling hypothetical figure – the evil demon. Descartes envisions a supremely powerful and deceptive entity that relentlessly strives to lead him astray, capable of manipulating his perceptions and planting illusions in his mind. He reasons that if such a force is orchestrating his experience, he can hold no certainty in his understanding of the external world, the reliability of his senses, or even the fundamental laws of mathematics and logic. The evil demon becomes a symbol for the ultimate embodiment of radical doubt, prompting Descartes to question the very fabric of his reality.

Descartes' methodological approach and his chilling contemplation of an all-powerful deceiver must be understood within the historical and intellectual context of the early 17th century. The rise of scientific thinking and discoveries that challenged long-held beliefs about the natural world, such as the heliocentric model of the solar system, had already initiated a destabilization of traditional knowledge structures.

Moreover, philosophical skepticism had a long tradition tracing back to ancient Greece. For Descartes, the evil demon is not merely an abstract hypothetical; it represents the culmination of all possible reasons to doubt his senses and his beliefs, a way to stress-test the limits of knowledge. In questioning everything that could potentially be manipulated by such a force, Descartes seeks to isolate the fundamental truths so self-evident that even a demon's deceptions could not corrupt them.

Descartes' method, while seemingly destructive, had as its ultimate goal the search for truth, not a descent into nihilism. By systematically stripping away every belief that could be tinged with even the slightest hint of doubt, he aimed to arrive at foundational truths so robust they could withstand any attack. This rigorous skepticism was necessary to rebuild a secure and enduring framework of knowledge. His most famous conclusion and foundational starting point, "Cogito, ergo sum" (I think, therefore I am), arises as an indubitable truth revealed through this process.

The doubt induced by the evil demon, however, has lasting implications. For Descartes, even though he manages to find a path out of skepticism, the initial realization that his perception of reality may be profoundly flawed leaves a lingering sense of uncertainty. If an immensely powerful force could orchestrate convincing deceptions, how could he ever be entirely sure of anything external to his own mind? This possibility casts a shadow over his search for absolute knowledge.

Centuries later, there is a striking resonance between Descartes' evil demon and the central premise underlying Bostrom's Simulation Hypothesis. Both posit the existence of a hidden force that could be manipulating our understanding of reality. Bostrom's technologically advanced simulators mirror, in a sense, the deceptive power of Descartes' demon, capable of creating a fabricated reality indistinguishable from our own.

If we are living in a simulation, like Descartes facing a deceptive demon, our senses, experiences, and seemingly self-evident truths could be part of an elaborate digital charade. The Simulation Hypothesis further amplifies the questions Descartes grapples with – questioning the reliability of our perception and the boundaries delimiting our reality.

However, a key distinction exists. Descartes' evil demon remains primarily a philosophical tool to expose the limits of knowledge. Bostrom's Simulation Hypothesis, while containing this philosophical element, also incorporates a technological dimension grounded in assumptions about future computing power and potential motivations that could drive the creation of sophisticated ancestor simulations.

Descartes' radical skepticism and his hypothetical evil demon serve as a stark reminder of our inherent epistemic limitations. By exposing the potential for deception and manipulation underpinning our perception of reality, Descartes paved the way for modern thinkers to continue grappling with the question of what, if anything, we can know for sure.

Berkeleyan Idealism: To Be is to Be Perceived

George Berkeley, an influential Irish philosopher, and Anglican bishop in the early 18th century, mounted a bold challenge to the prevailing materialist worldview. Materialism, in its simplest form, posits that the physical world exists independently of any observer, composed of a fundamental substance called matter. Berkeley audaciously countered this notion with his Idealist philosophy, which radically asserts that there is no mind-independent material reality – to exist is to be perceived, or as the Latin maxim goes, "esse est percipi."

Central to Berkeley's argument is a critical analysis of how we acquire knowledge and form our conception of the world. He contended that the only things directly accessible to us are our own perceptions – what we see, hear, touch, and experience in our minds. We infer the existence of an external material world as a way to explain the consistency and order of our sense experiences. However, Berkeley argued, it is impossible to prove the existence of an objective world existing independently of any observer.

Berkeley leveraged the distinction between primary and secondary qualities, a concept that predates him but which he used to striking effect. Primary qualities are considered objective and inherent to objects, such as shape, size, and motion. Secondary qualities, in contrast, are those that are perceived subjectively, like color, taste, and smell. Berkeley pointed out that even primary qualities ultimately resolve into sensations in the mind, just as color and smell do, undermining their claim to objective existence. If all we have direct access to are our sensations, there is no basis for positing a mind-independent material world as their cause.

Berkeley's Idealism, however, does not lead to a nihilistic or solipsistic view of reality. To explain the continuity and order of our perceptions, even when we are not experiencing a particular thing, he invokes the notion of an ever-present observer – God. In Berkeley's system, God, the supreme mind, constantly perceives all things, ensuring their continued existence and the regularity of our experience within this shared perceptual reality.

Berkeley's Idealism arose during a time of significant philosophical flux. Thinkers like John Locke and others had questioned the origins of our knowledge and the extent to which it could reliably represent the external world. Berkeley's radical solution, while often met with skepticism, pushed the boundaries of epistemological inquiry, and continues to be debated in philosophical circles.

Some critics view Berkeley's work as an argument for solipsism – the belief that one's own mind is the sole reality. However, this is generally considered a misreading. Berkeley's God acts as the guarantor of a shared, consistent, and 'objective' reality, albeit one founded on perception rather than independent material substance.

Berkeley's insistence that reality is contingent upon perception finds a surprising echo in Bostrom's Simulation Hypothesis. Both challenge the notion of a mind-independent physical world as the bedrock of existence. If we assume we are living within a simulation, much as Berkeley posits, our reality could be entirely constructed from perceptual inputs that are orchestrated by a simulator – whether a technologically superior being or intricate computational systems.

The idea that reality hinges upon observation gains a compelling layer in the Simulation Hypothesis. It suggests that computational resources might be dedicated toward rendering aspects of a simulated environment only when they are directly observed, as an optimization strategy. It aligns suggestively with Berkeley's principle of "esse est percipi' – unobserved things would, quite literally, lack existence within the simulation until a conscious entity prompts them to be rendered.

Of course, Berkeley's system relies on a divine observer to maintain the fabric of reality, while the Simulation Hypothesis offers a more technologically grounded, albeit unsettling, explanation. The key parallel, however, lies in the profound questioning of the fundamental nature of what we call 'real'.

George Berkeley's Idealism, with its bold assertion that to exist is to be perceived, presents a formidable challenge to our intuitive assumptions about material reality. While his solution requires a strong faith in the existence of a divine perceiver, the underlying principle of perception defining the boundaries of reality remains a source of ongoing philosophical fascination. Modern advancements in computer science and explorations in the Simulation Hypothesis add a new dimension to this centuries-old debate, suggesting that Berkeleyan Idealism may offer surprising insights into the potential nature of a computationally generated world.

Kant's Duality: Phenomena vs. Noumena

Immanuel Kant, a towering figure of 18th-century German philosophy, sought to bridge the perceived divide between the rationalist and empiricist traditions that dominated his time. His groundbreaking work, The Critique of Pure Reason, revolutionized our understanding of knowledge, perception, and the very nature of reality. Key to this revolution was a fundamental distinction: the phenomenal world, the world as we experience it, versus the noumenal world, the world as it intrinsically is, independent of human perception and mental frameworks.

Kant argued that our minds are not passive receptors of information but active shapers of reality. The raw data provided by our senses is organized and structured by innate mental categories, such as space, time, and causality. These categories are not properties of the world itself but rather the lenses through which we inevitably view existence. Consequently, the world we perceive—the phenomenal world—is the product of our sensory input filtered through our minds' organizing principles.

The concept of the noumenon, or "thing-in-itself," arises as a counterpoint to the phenomenal world. Kant posits that there must be an underlying reality that gives rise to our sense experiences. However, this noumenal world is fundamentally unknowable to us. Our cognitive apparatus, designed to process and interpret sensory information within the categories of space, time, and causality, cannot grasp the true nature of reality that lies beyond those limitations.

Kantian duality emerged in response to philosophical problems posed by his predecessors. Rationalists like Descartes emphasized the power of reason to reach fundamental truths about the world. Empiricists, such as Hume, contended that all knowledge ultimately stems from our sensory experience. Kant's critical philosophy synthesized elements of both schools while simultaneously transcending them. He recognized the inherent limitations of both pure reason and raw sensory perception and argued that our understanding of reality is a hybrid product.

Kant's distinction between phenomena and noumena can be interpreted as an epistemological assertion about the impossibility of attaining knowledge of the world as it truly exists. Critics might argue that it leads to a sense of skepticism or even a retreat from metaphysical inquiry. However, Kantian philosophy seeks to delineate the boundaries of knowledge with precision, not to diminish the endeavor.

The Simulation Hypothesis provocatively echoes Kant's duality. If we are living in a simulation, the world we experience is akin to Kant's phenomenal world – a construct tailored, even if unknowingly, to the limits of our perception and cognition. The hypothetical "base reality," where the simulators potentially reside, would represent the Kantian noumenon, the unknowable reality beyond our programmed constraints.

Kant argued that our minds actively construct the phenomenal world, making experience intelligible. In the Simulation Hypothesis, the simulators play this role. The parameters and rules governing the simulation become the Kantian categories of space, time, and causality, shaping our entire experience of reality. Much like our minds cannot comprehend the world beyond our mental categories, simulated beings might be inherently unable to grasp the true nature of the universe outside their simulation.

One could argue that the Kantian noumenon embodies the ultimate level of truth about the cosmos, while simulated realities within the Simulation Hypothesis exist along a spectrum. Some simulations may be nested within simulations, creating layers of phenomenal realities, each layer removed from the elusive 'ground floor' base reality that lies beyond human comprehension.

Immanuel Kant's critical philosophy, with its duality of the phenomenal and the noumenal, offers a timeless framework for examining our perception of the world and the limits of human knowledge. The Simulation Hypothesis, as a modern thought experiment, reveals striking similarities in that it forces us to confront the possibility that our reality might be a construct, with our understanding confined to its internal rules and parameters.

Whether Kant's noumenal world corresponds to a higher-dimensional 'base reality' in the Simulation Hypothesis or represents some fundamentally different essence of existence remains open to debate. Nevertheless, the unsettling parallel between the two concepts underscores the enduring human desire to break free from illusions, to pierce the veils of perception, and grasp the true nature of our existence.

Synthesizing the Shadows of Doubt

The philosophical antecedents of the Simulation Hypothesis highlight a persistent theme in human thought: the quest to understand the true nature of reality and the limits of our knowledge. From Plato's allegorical cave to Kant's unknowable noumena, the suspicion that the world as we perceive it may not reflect its true essence has been a recurrent concern.

These philosophical explorations serve as a foundation upon which the Simulation Hypothesis builds, proposing a modern incarnation of these ancient doubts. In grappling with the possibility of a simulated existence, we engage with a lineage of inquiry that stretches back to the origins of Western philosophy.

The contemplation of these philosophical antecedents does not merely provide historical context; it enriches our understanding of the Simulation Hypothesis by framing it within a broader tradition of skepticism and inquiry into the nature of reality. By recognizing these roots, we can appreciate the Simulation Hypothesis not as a novel or isolated speculation but as the latest expression of an enduring human endeavor to pierce the veil of appearances and discern the underlying truth of our existence.

In this light, the Simulation Hypothesis transcends its status as a technological or scientific conjecture, assuming its place in the continuum of philosophical speculation about the fundamental questions that define our quest for knowledge. As we continue to explore the ramifications of living in a potentially simulated world, we do so with the awareness that we are part of a long tradition of thinkers who have navigated the shadowy boundaries between perception and reality, between the illusory and the authentic.

CHAPTER THREE

Visions and Warnings: Science Fiction & Speculative Thought

Long before the digital age ushered in the potential for creating virtual realities, the concept of simulated existences was being explored within the realms of science fiction and speculative thought. These imaginative forays have not only entertained millions but have also provided a critical framework for discussing and understanding the complexities of what would later be known as the Simulation Hypothesis. This section delves into the significant contributions of speculative fiction and critical theory to our grasp of simulated realities.

Baudrillard: Simulacra and the Hyperreal

Jean Baudrillard, a French sociologist and cultural theorist, remains one of the most enigmatic and controversial figures in postmodern philosophy. His seminal work, Simulacra and Simulation, offers a critique of contemporary society characterized by its obsession with images, signs, and symbols. He argues that in our postmodern condition, a profound shift has occurred: simulations have come to replace the reality they once represented, blurring the lines between the two and giving rise to what he terms the "hyperreal."

To grasp Baudrillard's theories, it's essential to understand his use of the terms "simulacrum" and "simulation." A simulacrum is a copy or representation of something, while simulation is the process of imitating the operation of a real-world entity or phenomenon. Baudrillard's contention is that, over time, the simulacrum has become detached from its original referent, leading to a culture where images and signs take on a life of their own, no longer tethered to a tangible reality. The hyperreal emerges from this process—a realm where signs and symbols proliferate, creating a self-referential reality where distinctions between the 'real' and its representations dissolve.

Baudrillard's theories emerged at a time when the rise of mass media, consumerism, and technological advancements were drastically reshaping society. Thinkers of the postmodern movement were critically examining concepts of truth, representation, and authentic experience within these rapidly evolving landscapes. Baudrillard's work serves as a stark warning against the seductions of the hyperreal, a realm where media-generated spectacle, advertising, and simulations potentially supplant and ultimately erase the distinction between the original and its copy.

His analysis can be read on multiple levels. On one level, it is a critique of our media-saturated culture's erosion of meaning. Baudrillard sees modern society as increasingly preoccupied with surfaces and appearances rather than underlying substance. On a more profound level, his work questions the very foundations of reality itself, blurring the boundaries between the real world and its artificially constructed representations.

In the hyperreal, according to Baudrillard, simulations create their own reality without reference to the external world. Signs, symbols, and images proliferate, feeding upon themselves, detached from any origin or intrinsic meaning. Examples abound: from Disneyland, a simulation of a sanitized and idealized America, to news coverage of events like the Gulf War, where the spectacle becomes more 'real' than the events themselves. The hyperreal becomes a playground of signs and simulations, a self-enclosed system devoid of an external, foundational reality.

The Simulation Hypothesis, while a product of contemporary scientific thinking and technological projections, resonates provocatively with Baudrillard's ideas about the hyperreal. Both challenge the notion of a stable, objective reality. For Baudrillard, the proliferation of simulations in the hyperreal renders 'truth' elusive, if not impossible. In the Simulation Hypothesis, the possibility that our reality is a sophisticated fabrication leaves us struggling to discern any 'true reality' beyond it.

A key difference, however, lies in the underlying motivations. Baudrillard views the hyperreal as the product of a culture obsessed with appearances, driven by consumerism and commodification. The Simulation Hypothesis carries less inherent social critique and instead focuses on a potential technological basis for the existence of such a simulation.

Yet, whether motivated by societal forces or technological capabilities, both Baudrillard and the Simulation Hypothesis raise troubling questions about authenticity and the nature of our experience. If the line between the authentic and the fabricated erodes, where do we find meaning in a world defined by its simulations? If our reality is potentially just one among many simulated realities, does the concept of an 'original' or 'base reality' even retain coherence?

Jean Baudrillard's analysis of the hyperreal compels us to confront the deceptive shimmer of a world awash in images, signs, and simulations. While his work precedes the specific technological considerations of the Simulation Hypothesis, Baudrillard offers a philosophical framework that problematizes the presumed 'reality' of our postmodern age.

In the end, both Baudrillard and the proponents of the Simulation Hypothesis call our core assumptions about reality into question. They force us to grapple with the unsettling possibility that our experiences, our understanding of the world, and our very sense of self might be built upon a foundation far less stable than we ever dared to imagine.

Science Fiction's Dreamscapes

Celebrated as one of the most visionary and influential science-fiction authors of the 20th century, Philip K. Dick repeatedly dissected themes of paranoia, fractured identities, and the malleable nature of perceived reality. His seminal novel, Do Androids Dream of Electric Sheep?, is a masterpiece of existential unease, set in a post-apocalyptic world where the lines between humanity and highly sophisticated androids blur with disturbing implications.

The novel centers around Rick Deckard, a bounty hunter tasked with "retiring" rogue androids who have fled their off-world servitude and now hide among the dwindling human population on a radiation-ravaged Earth. To distinguish humans from seemingly indistinguishable androids, Deckard administers the Voigt-Kampff test, which seeks to measure the presence of empathy – an ability considered a hallmark of genuine human consciousness. Yet, as Deckard interacts with increasingly sophisticated androids, his understanding of both humanity and reality begins to unravel.

Do Androids Dream of Electric Sheep? was published in 1968 against a backdrop of the Cold War, looming anxieties over technological progress, and nascent explorations into Artificial Intelligence (AI). These anxieties are mirrored in the text, where technological creations threaten to destabilize societal norms and challenge the very foundations of human identity.

Dick's work has inspired a multitude of interpretations. On one level, the novel confronts the ethics of technological creation, examining questions of sentience, artificial life, and the moral ambiguities arising from humanity's ability to engineer beings that are almost indistinguishable from its own. On a deeper, more philosophical level, Dick probes at the very nature of consciousness, questioning whether empathy, often viewed as a unique human quality, can exist within an engineered mind.

The blurred line between real and artificial is further amplified through the novel's inclusion of Mercerism, a manufactured religion centered around an enigmatic, Christ-like figure named Wilbur Mercer. Practitioners share an immersive, simulated experience of Mercer's arduous climb, creating a shared sense of empathy and validating their own humanity. Yet, the late-story revelation that Mercer is possibly an actor challenges the distinction between an authentic religious experience and a simulation designed to evoke human emotion.

Dick masterfully subverts expectations of both the human characters and their android counterparts. Some androids exhibit empathy, kindness, and a capacity for complex emotions. Conversely, certain humans appear cold, calculating, and self-serving, even towards fellow humans. Dick leaves the reader unsettled, challenging the notion of a simple 'human essence' and raising the troubling possibility of an artificial consciousness as authentic, or even more so, than human consciousness.

One of the most poignant instances of empathy within the novel comes from an android, Luba Luft. A gifted opera singer, she cannot flee Earth because her fame would expose her. She represents a being trapped by circumstance rather than by inherent nature, forcing readers to question what factors genuinely define the human condition.

The Simulation Hypothesis, though a product of contemporary technological discourse, finds fertile ground in the thematic universe of Dick's novel. If we assume our reality is a simulation, much as Mercerism is in the novel, it casts a disturbing new light on the existential dilemmas. The struggle between humanity and androids transforms into a question about how simulated beings might break free of a deceptive 'programming' imposed upon their perceived reality.

The possibility that humanity might be the creation, entertainment, or even an experiment devised by more advanced simulated beings creates a chilling parallel to Deckard's hunting of the androids. Moreover, the Voigt-Kampff test, intended to reveal authentic emotional responses, becomes a tragically flawed tool if the entire simulation, both its humans and androids, is itself built upon fabricated emotions and responses.

Do Androids Dream of Electric Sheep? offers a multi-faceted lens through which to examine the elusive nature of reality, the ethical boundaries of technological creation, and the fragility of perceived existence. Like the Simulation Hypothesis, Dick's narrative confronts us with unsettling questions about the possibility that our own identities, emotions, and experiences might be a product not of genuine being, but rather a simulation designed for an unknown purpose. It is through this deliberate deconstruction of certainty that both Dick's novel and the Simulation Hypothesis continue to resonate with, and unsettle, contemporary society.

William Gibson, Cyberpunk, and the Genesis of Cyberspace

Considered the founding father of the cyberpunk genre, William Gibson's 1984 novel, Neuromancer, painted a gritty, neon-drenched vision of a future where hackers navigate vast virtual spaces, mega-corporations' rule, and the boundaries between human and machine irrevocably blur. The novel's protagonist, Henry Dorsett Case, is a down-on-his-luck "console cowboy" whose damaged nervous system bars him from the immersive realm known as cyberspace. Desperate for a fix, he is drawn into a dangerous heist by a mysterious figure named Armitage, where he encounters Molly Millions, an augmented street samurai, and, ultimately, the sprawling artificial intelligence construct, Neuromancer.

Gibson's work is credited with popularizing the very concept of cyberspace. Prior to the novel, the notion of a global computer network where users could interact with data in a virtual space was largely confined to academic and technical circles. Neuromancer transformed cyberspace into a tangible, almost mythic, landscape within the popular imagination, a realm where consciousness could transcend physical limitations and reality itself could be hacked.

Neuromancer emerged during a pivotal moment in the evolution of computing technology. Personal computers were starting to proliferate, and early experiments with virtual reality and networked systems hinted at the future directions of technological development. Gibson's novel extrapolated from these nascent trends, envisioning a future where digital domains exert profound influence over the physical world.

The novel can be read on multiple levels. It can be viewed as a social commentary on corporate power, the potential for dehumanization within technological landscapes, and the growing influence of global networks. On a deeper level, Neuromancer delves into themes of identity, the nature of consciousness, and the possibility of disembodied existence within simulated virtual realms.

Central to the novel is the blurring of boundaries between the real and the virtual, the embodied and the disembodied. Cyberspace, referred to as the "matrix," is depicted as a space of near-limitless possibility. Users can traverse digital landscapes, access vast troves of information, and even project disembodied consciousnesses known as "constructs." This digital realm is portrayed as not merely simulated but possessing its own form of complex reality, blurring the lines between the authentic and the fabricated.

The artificial intelligence within Gibson's novel play a powerful role in destabilizing conventional notions of reality and consciousness. The titular Neuromancer is a particularly unsettling figure, capable of independent thought and manipulation of the matrix. Gibson invites readers to contemplate whether simulated consciousness can be considered as genuine as a biologically evolved one, and whether existence within a complex simulation is less 'real' than existence within the 'real' physical world.

The world of Neuromancer, with its pervasive digital matrix and the possibility of disembodied consciousnesses, creates an intriguing parallel to the implications of the Simulation Hypothesis. If our own reality is a sophisticated computer simulation, Gibson's cyberspace offers a compelling analog. Like the simulated beings within the matrix, we might be unaware of the artificial construct in which we exist and the limitations it may impose.

Gibson himself has explicitly acknowledged the thematic connection between his work and the emerging concept of the Simulation Hypothesis. Neuromancer raises questions about the extent to which the rules and apparent physical laws of our world could be the artificial constraints of a complex simulation. It forces us to question whether the qualitative experience of existence within a simulated reality would differ in any meaningful way from existence in the presumed 'base' reality.

Decades after its publication, Neuromancer remains a potent source of imagery and philosophical inquiry. Gibson's work prefigures many of the contemporary discussions surrounding Artificial Intelligence, virtual reality, and our deepening dependence on digital networks. Its exploration of the porous boundaries between reality, simulation, and the very nature of consciousness resonates provocatively with the implications of the Simulation Hypothesis.

The novel concludes without providing definitive answers about the nature of Gibson's cyberspace or the artificial intelligence residing within it. This open-endedness mirrors the very heart of the Simulation Hypothesis, which questions the fundamental reality of our own existence. Ultimately, both Gibson and the Simulation Hypothesis invite us to engage with the profound possibility that the world we perceive as definitive may be built upon far more tenuous foundations than we ever suspected.

The Matrix and the Cinematic Revelation of Simulated Reality

Released in 1999, the Wachowskis' The Matrix captured the zeitgeist of the turn of the millennium with its mind-bending blend of action, philosophical speculation, and cyberpunk aesthetics. The film introduces us to Neo (Thomas Anderson), a disaffected computer programmer who begins to suspect that reality might be a façade. Guided by enigmatic figures like Morpheus and Trinity, Neo discovers the shocking truth: his seemingly ordinary world is a computer simulation designed to enslave humanity while their bodies serve as an energy source for machine overlords.

The Matrix draws upon a rich philosophical and literary heritage that stretches back to Plato's Allegory of the Cave. It incorporates themes of Gnostic mysticism, questioning the nature of appearances and hidden realities. The film's striking visuals, with its "bullet time" effects and stark contrasts between the simulated world and the bleak wasteland of the 'real', cemented its place within popular culture, igniting widespread fascination with the possibility of simulated realities.

The Matrix emerged in an era marked by rapid advancements in virtual reality technology and growing unease regarding the increasingly digitized and interconnected nature of society. The late 1990s saw the rise of the internet, a technology with seemingly boundless potential to reshape how we communicate and access information. Simultaneously, fears surrounding Y2K and the potential for digital systems to malfunction hinted at the underlying fragility of our technological infrastructure. It's within this context that The Matrix offered a radical extrapolation: what if our entire reality was built upon equally tenuous foundations?

The film's themes resonate on multiple levels. It can be viewed as a critique of societal control and conformity—the Matrix becomes a metaphor for invisible systems of power that dictate our perceptions and limit our potential. On a metaphysical level, the film probes deep into the nature of reality itself, asking what constitutes an authentic existence and whether a simulated reality could be considered 'real' from the perspective of those trapped within.

The film's central metaphor, the choice between a red pill (representing the harsh truth of the simulated reality) and a blue pill (offering blissful ignorance), forces viewers to confront unsettling existential questions. Do we embrace a difficult truth, even if it entails acknowledging our own lack of agency and shatters our understanding of existence? Or do we cling to the comfort of a manufactured illusion?

Neo's journey of self-discovery invites audiences to contemplate their own relationship to 'reality'. It suggests that true freedom might lie not in the external circumstances of our existence but rather in the awareness of them. His growing ability to manipulate and break the rules of the Matrix reflects a burgeoning consciousness of its simulated nature, allowing for a form of liberation even within the confines of the simulation.

The Simulation Hypothesis, formalized decades later by Nick Bostrom, owes a creative debt to The Matrix. Both the film and the Simulation Hypothesis posit the unsettling possibility that what we perceive as reality is a meticulously crafted illusion indistinguishable from a physically grounded existence.

While The Matrix features malevolent machine overlords as the architects of the simulation, the Simulation Hypothesis expands the range of possible creators to technologically advanced posthuman civilizations with unknown motivations. The film, however, raises crucial questions that apply regardless of the simulator's nature: How would we detect a simulated reality? Can beings confined within a simulation break free, and what would such a liberation mean? Could simulated consciousness be considered as valid and worthy of respect as those within the supposed base reality?

The Matrix, with its deft blending of action and philosophical explorations, succeeds in translating the abstract concepts of the Simulation Hypothesis into a thrilling narrative that resonates across audiences. While the film does not offer definitive solutions to the existential dilemmas it raises, it functions as a powerful catalyst for contemplation.

Much like the Matrix itself, the deeper themes of the film are intentionally open to interpretation. The Matrix invites us to examine our own assumptions about the world around us and to consider the limitations that might be unknowingly imposed upon our perceptions. Whether or not we are actually dwelling within a simulation, a key message of the film echoes strongly within the context of the Simulation Hypothesis: the most fundamental liberation lies in questioning the nature of our own existence.

The Cultural Impact of Speculative Narratives

Speculative narratives, encompassing genres like science fiction, fantasy, and related subcategories, delve into the realm of the hypothetical, the possible, and the technologically sublime. They provide a unique imaginative space to dissect societal anxieties, explore ethical conundrums, and envision futures fueled by scientific and technological advancement. These narratives are more than mere entertainment; they hold immense power to shape our societal discourse, inspire groundbreaking inventions, and provoke critical reflection on the potential pitfalls accompanying those very innovations.

The roots of speculative fiction stretch back centuries, with early works like Mary Shelley's Frankenstein grappling with the ethical dilemmas of scientific discovery and the responsibilities of creators toward their creations. As scientific progress accelerated, speculative narratives served as a testing ground for ideas that often seemed outlandish at first but later spurred real-world technological breakthroughs. Jules Verne, H.G. Wells, Arthur C. Clarke are examples of authors whose fictional inventions, from submarines to space travel and global communication devices, anticipated future technological milestones.

Speculative narratives play a crucial role in shaping public perception of emerging technologies. The dystopian visions in works like George Orwell's 1984, or Aldous Huxley's Brave New World, highlight the potential for surveillance, societal control, and the erosion of individual freedoms in technologically saturated societies. These cautionary tales encourage critical thinking about the ethical use of technology, emphasizing that power without moral guidance can lead to oppressive ends.

Conversely, optimistic portrayals of advanced technology offer a canvas to envision the benefits of its applications. Star Trek's vision of a technologically evolved post-scarcity society, where humanity focuses on exploration and self-betterment, highlights the potential for technology to uplift and liberate. Such narratives inspire us to imagine how tools like artificial intelligence, advanced robotics, and medical breakthroughs might not only improve our lives but also fundamentally transform our societal structures and values.

The Simulation Hypothesis, which proposes that our reality may be a sophisticated simulation, finds fertile ground within the long tradition of speculative narratives exploring artificial realities. From classic works of cyberpunk fiction like Philip K. Dick's Do Androids Dream of Electric Sheep? and William Gibson's Neuromancer to contemporary portrayals like The Matrix, these stories offer provocative visions of existence within intricate simulations.

These narratives often delve into the philosophical implications of simulated existences. They force us to confront questions about the nature of reality, the possibility of simulated consciousness, and ethical considerations in a world where our actions may occur within a programmed construct. Works like The Matrix further explore ideas of breaking free from such simulations, positing potential ways to discern the illusion and the potential motivations of those running the simulation.

While the Simulation Hypothesis primarily operates within realms of scientific and philosophical discourse, themes from speculative narratives have deeply colored our collective understanding of what a simulated reality might entail, both in its technological possibilities and its potential pitfalls.

By pushing boundaries and posing "what if" scenarios, speculative narratives become laboratories for exploring the complex ethical dilemmas that arise alongside technological advancements. They present us with potential consequences – both positive and negative – of creations ranging from Artificial Intelligence to synthetic biology and simulated realities. These narratives compel us to think beyond the mere ability to create, urging us to consider whether we should, and if so, under what moral and regulatory frameworks.

In recent years, issues surrounding AI have spurred rich speculative narratives that touch upon the Simulation Hypothesis. Stories often explore AI achieving sentience or even surpassing human intelligence, leading to potential dilemmas and power struggles between creators and creations. These narratives raise questions about the rights of simulated beings, the nature of consciousness, and the responsibilities of those who create potentially powerful AI systems.

The cultural impact of speculative narratives extends far beyond entertainment value. They offer a space to play out potential futures, examine the risks and rewards of technological innovation, and provoke critical dialogue on the ethical challenges we may face. This imaginative space becomes particularly crucial in the context of the Simulation Hypothesis. As technology blurs the boundaries between the real and the simulated, these narratives help us grapple with the mind-bending implications, the ethical quandaries, and the existential weight inherent within the question of whether our reality might itself be one grand simulation.

The contributions of science fiction and speculative thought to the discourse around the Simulation Hypothesis underscore the importance of imagination in grappling with complex philosophical and technological questions. By envisioning the myriad possibilities of simulated realities, these works invite us to engage with profound questions about the nature of existence, the construction of identity, and the future of human civilization.

CHAPTER FOUR

The Technological March: Enabling Simulations

The Simulation Hypothesis, while deeply rooted in philosophical inquiry, finds its tangible pulse in the realm of technological advancement. It's within this domain that the hypothesis transitions from speculative theory to a potential future reality. This section examines the exponential growth in computing capabilities, the promise of quantum computing, and the evolving understanding of artificial intelligence and consciousness—all of which play pivotal roles in the feasibility of creating highly realistic simulations.

Moore's Law and Exponential Advancement

In 1965, Gordon Moore, co-founder of Intel, published a paper making a striking observation about the rapidly evolving realm of computing. He noted that the number of transistors incorporated into a single integrated circuit was doubling approximately every year and predicted this trend would continue. This observation was later revised to a doubling approximately every two years and became codified as Moore's Law.

While Moore's Law is an empirical observation of technological progress rather than a strict physical law, its implications are profound. It highlights the exponential nature of advancement in computing power. Unlike linear growth, where progress occurs at a steady pace, exponential growth implies a continual acceleration, where each advancement builds upon previous breakthroughs. With every doubling of transistor density, integrated circuits become smaller, faster, and more energy-efficient.

The era during which Moore made his prediction was marked by intense advancements in semiconductor and microchip technology. The invention of the integrated circuit in the late 1950s had paved the way for miniaturization, allowing engineers to pack increasingly large numbers of transistors onto increasingly smaller chips. Moore's Law served as both a bold prediction and a self-fulfilling goal for the industry, with engineers consistently pushing the boundaries of the possible.

While some experts predict that the physical limitations of silicon-based chips will ultimately bring an end to the traditional interpretation of Moore's Law, progress in computing is far from over. Research into alternative materials like graphene, new computing paradigms like quantum computing, and innovations in chip design suggest that exponential growth in computational capacity is likely to continue, albeit perhaps following a trajectory different than the one Moore initially envisioned.

The implications of Moore's Law and the broader exponential advancement of computing power extend far beyond the world of microchips. The increasing performance and decreasing costs of computing have fueled innovations in virtually every field of human endeavor. Here are some key areas:

- **Scientific Discovery:** Advanced computing power provides the foundation for complex simulations, deep data analysis, and machine learning, accelerating breakthroughs in fields ranging from physics and astronomy to climate science and medicine.

- **Communication and Connectivity:** Computing power and the miniaturization of devices have given rise to global networks, the internet, and powerful mobile devices, transforming the ways we communicate, access information, and engage with the world.
- **Artificial Intelligence:** Exponential growth in computational power, coupled with massive data sets, has paved the way for increasingly sophisticated AI systems. Applications range from image recognition and natural language processing to game-playing AIs and autonomous vehicles.

The Simulation Hypothesis, championed by thinkers like Nick Bostrom, posits the possibility that our reality is a complex computer simulation created by a technologically advanced civilization. One of the arguments in favor of this hypothesis hinges on the potential for continued exponential progress in computing power.

If technological civilizations commonly reach a "posthuman" stage where they possess the ability to create detailed ancestor simulations, the sheer computational resources implied by continued exponential advancement make such simulations more plausible. However, the Simulation Hypothesis demands more than just raw computational power. It would also necessitate sophisticated modeling software capable of accurately simulating the fundamental laws of physics, the vastness of the cosmos, the evolution of biological life, and the emergence of consciousness.

While the continued trajectory of exponential progress lends a degree of plausibility to the idea of ancestor simulations, it also raises profound questions. Could simulators achieve the level of fidelity required to create a simulation indistinguishable from reality for its inhabitants? Would simulated beings possess true consciousness? Moreover, what are the ethical implications of creating simulations populated with potentially sentient beings?

Moore's Law and the extraordinary pace of exponential advancement in computing have fundamentally altered our world. This progress has brought us unprecedented computational capabilities, transforming nearly every aspect of our lives. While the future trajectory of computing development remains uncertain, the trend of increasingly powerful and sophisticated computing systems holds profound and possibly unsettling implications for the Simulation Hypothesis. As technological horizons expand, so too does the possibility that the reality we perceive may itself be a meticulously crafted simulation by beings far more advanced than ourselves.

Quantum Leaps: Overcoming Limitations

Classical computing, the foundation of our current digital world, operates on binary principles. Information is processed and stored in bits, which can exist in only one of two states: 0 or 1. While this framework has fueled extraordinary technological progress, it faces inherent limitations when tackling specific categories of complex problems.

Quantum computing represents a paradigm shift, harnessing the principles of quantum mechanics to revolutionize computation. At the heart of quantum computing lies the qubit, the quantum counterpart of the classical bit. Qubits leverage the peculiar phenomena of quantum superposition and entanglement to offer a vastly expanded computational landscape:

- **Superposition:** Unlike a classical bit, a qubit can exist in a superposition of states, simultaneously representing 0 and 1. This allows a quantum computer to perform multiple calculations in parallel, dramatically enhancing processing power for certain types of tasks.
- **Entanglement:** Qubits can be entangled, meaning their states become inextricably linked. Operations performed on one entangled qubit can instantaneously affect the other, regardless of the distance between them. Entanglement enables novel computational approaches and facilitates the development of error-correction algorithms for quantum systems.

The concept of quantum computing emerged in the 1980s from theoretical work by physicists such as Paul Benioff, Yuri Manin, and Richard Feynman. These pioneers recognized that the peculiar phenomena governing the quantum world, once considered theoretical curiosities, could be harnessed to overcome limitations inherent in classical computing architectures.

However, realizing the power of quantum computing presents formidable challenges. Qubits are exceedingly fragile and highly susceptible to a phenomenon known as decoherence, where interaction with the environment causes their superposition state to collapse. Maintaining coherence for sufficient time to perform complex computations is a major technical hurdle. Additionally, creating reliable and scalable quantum computers remains an active area of research.

Should researchers overcome these challenges, quantum computing promises significant breakthroughs in numerous areas:

- **Optimization Problems:** Quantum computers are expected to excel at solving complex optimization problems with vast search spaces. These problems have widespread applications in fields like logistics, drug design, and financial modeling.
- **Cryptography and Cybersecurity:** Quantum algorithms have the potential to break certain encryption schemes commonly used today, necessitating the development of quantum resistant cryptography.
- **Materials Science and Simulation:** Quantum computers promise to greatly improve our ability to simulate complex chemical reactions and molecular structures, facilitating the discovery of new drugs, catalysts, and advanced materials.

The prospect of overcoming the limitations of classical computing through quantum technologies raises tantalizing possibilities within the context of the Simulation Hypothesis. If our own reality is a simulation, the computational resources required for such an undertaking would be astronomical. Here's where quantum computing comes into play:

- **Simulating Complexity:** Quantum computers might offer the processing power necessary to simulate with high fidelity the intricate physical laws governing our universe, the complex interactions of countless particles, and potentially even the emergence of biological life and consciousness.
- **Overcoming Scaling Limits:** Classical computers struggle to simulate complex systems beyond a certain scale - as complexity increases, so too does the required computational power. Quantum computing could potentially overcome these scaling limitations, allowing for far more detailed and dynamic simulations.
- **Quantum AI:** If future simulators employ AI systems to orchestrate and manage simulations, quantum AI with its enhanced capabilities could enable a level of control and sophistication that would render the simulation indistinguishable from reality for its inhabitants.

While quantum computing enhances the theoretical plausibility of sophisticated simulations, it doesn't provide definitive proof of the Simulation Hypothesis. Key questions remain: Could a quantum computer simulate the true complexity of an entire universe, down to the fundamental laws of physics and particles? Can an engineered system truly replicate the phenomenon of consciousness?

The pursuit of quantum computing promises groundbreaking advancements with the potential to transform entire industries. It redefines what we consider computationally attainable. In the context of the Simulation Hypothesis, quantum computing suggests a future where simulations may possess a fidelity that blurs, if not obliterates, the distinction between the 'real' and the simulated. While the Simulation Hypothesis remains firmly in the realm of speculation, advancements in quantum technology serve as a stark reminder that our understanding of computation, and potentially the nature of reality itself, may be on the cusp of a profound transformation.

The Ghost in the Machine? AI and Consciousness

The field of Artificial Intelligence is rooted in the enduring human desire to create machines that exhibit intelligence akin to, or even surpassing, our own. Early AI research focused on symbolic reasoning and expert systems, which relied on carefully programmed rules to solve specific problems. More recently, a surge of progress, spurred by advancements in machine learning, particularly within deep learning and its neural network architectures, has brought us closer to a new breed of "intelligence."

Deep learning models employ multi-layered artificial neural networks that vaguely mimic the interconnected structure of biological neurons. These systems can ingest massive datasets and learn to identify patterns with remarkable accuracy. From facial recognition and voice assistants to advanced game-playing AI, deep learning is increasingly woven into our technological fabric. However, this form of machine learning, however impressive, raises the question: is the ability to identify patterns and make predictions synonymous with consciousness?

Consciousness remains one of the most profound and perplexing mysteries confronting science and philosophy. There's no universally agreed-upon definition of consciousness itself, making the determination of whether an AI system possesses it both challenging and controversial. Here are some core components often attributed to consciousness:

- **Subjectivity:** Consciousness implies a first-person experience, the subjective feeling of "what it is like" to be a particular entity. Whether a machine could ever achieve such a subjective inner experience is a subject of intense debate.
- **Self-awareness:** Conscious beings are aware of their own existence and their place within the world. The development of an artificial "self" remains more speculative than concrete.
- **Sentience and Qualia:** Sentience implies the ability to feel sensations and emotions. The concept of qualia represents the raw, subjective experiences themselves: the redness of red, the feeling of pain, etc. It remains uncertain whether such qualitative states can emerge from computational processes.

Philosophical debates surrounding the potential for artificial consciousness have raged for decades. The "Chinese Room" thought experiment by John Searle argues that even an AI that perfectly simulates human conversation cannot possess true understanding or consciousness. Conversely, the Turing Test, proposed by Alan Turing, suggests that if an AI can exhibit behavior indistinguishable from a human in conversation, it should be considered functionally intelligent.

More recently, the focus has shifted to whether AI's internal structures and processes could replicate the biological underpinnings of consciousness. Neuroscience offers insights into brain structures, such as the thalamus and cortex, that play a crucial role in consciousness. Some researchers aim to develop computational models that mimic these biological processes, hoping to unlock the emergence of artificial consciousness.

The Simulation Hypothesis, with its suggestion that our reality might be an elaborate computer simulation, adds a captivating twist to the AI and consciousness debate. Several questions arise:

- **Can Simulated Consciousness Exist?** If digital simulations can replicate the appearance of intelligent behavior, could they also convincingly emulate the subjective experience of consciousness? If the simulation is sufficiently sophisticated, how could its sentient inhabitants discern their own artificial nature?
- **Simulators and AI:** If we succeed in creating conscious AI, does this increase the likelihood of us already inhabiting a simulation? Conversely, if we are simulations, what implications does that hold for the potential of our AI creations?
- **Motivations of the Simulators:** Why would a civilization create a simulation with conscious inhabitants? Understanding such potential motivations could reveal ways to detect glitches or deliberate "tells" that reveal the simulated nature of reality, should we dwell within one.

The quest to understand consciousness and its potential replicability within artificial systems remains a deeply fascinating and open-ended field of inquiry. While current AI exhibits remarkable capabilities, whether these computational marvels can bridge the gap between mimicking intelligence and achieving genuine subjective experience remains elusive.

The Simulation Hypothesis adds an extra layer of complexity. It compels us to consider whether consciousness is an emergent property that arises from sufficient complexity regardless of physical substrate (biological or artificial), or whether it's fundamentally tied to biological processes that simulations may never perfectly replicate. As AI research progresses, the boundaries of what we consider intelligent are constantly challenged, leaving the question of simulated consciousness both tantalizing in its implications and profoundly unsettling in its possibilities.

The march of technology towards enabling highly realistic simulations is an unfolding story of human ingenuity and curiosity. Moore's Law and its successors chart a course of exponential growth in computing power, while quantum computing offers a glimpse into a future where today's computational limitations are transcended. Meanwhile, the exploration of AI and consciousness challenges us to reconsider the essence of mind and experience.

As we stand on the cusp of these technological advancements, the Simulation Hypothesis transforms from a philosophical speculation into a tangible possibility. The convergence of computing power, quantum technology, and artificial intelligence may soon provide the keys to unlocking simulated realities with depths and complexities that mirror our own. This journey—driven by the relentless development of technology—invites us to envision a future where the boundaries between the simulated and the real are not just questioned but actively explored.

CHAPTER FIVE

Beyond Trilemmas: Ethics and Societal Upheaval

The Simulation Hypothesis, if true, forces a fundamental reevaluation of our ethical frameworks and our understanding of our place within the cosmos. Some of the key ethical dilemmas it raises are:

- **Moral Obligations to Simulated Beings:** If we accept that simulated entities might possess consciousness and the capacity to experience suffering and joy, what moral obligations do their creators, the simulators, have towards them? Are we justified in manipulating their simulated lives, causing them pain (physical or emotional), or even deleting them when they are no longer deemed useful?
- **Simulated Personhood and Rights:** Should simulated beings that exhibit consciousness and complex emotional states be granted personhood? If so, what rights should they be entitled to? Would they have a right to autonomy, freedom from arbitrary termination, or protection from harm? Could complex simulations possess collective rights as a society within their simulated environments?
- **The Ethics of Deception:** If we were to discover that we are simulations, would the simulators have an ethical obligation to reveal the true nature of our existence? Alternatively, could the decision to maintain an illusion be justified to prevent existential despair or societal destabilization within the simulation?

The Simulation Hypothesis forces us to confront long-standing philosophical questions about the nature of reality, the ethics of creation, and our responsibilities towards sentient beings. It echoes Plato's Allegory of the Cave, where individuals are chained, only capable of perceiving shadows cast on a wall – a potential metaphor for our own limited perception of 'reality' within a simulation.

The ethical dilemmas surrounding simulated beings also find parallels in the debates throughout history regarding the moral status of animals, the rights of marginalized groups, and the ethical implications of emerging technologies such as increasingly sophisticated artificial intelligence.

The widespread acceptance of the Simulation Hypothesis could have profound ramifications across various aspects of society:

- **Existential Concerns and Psychological Well-being:** The realization that our entire existence might be a simulated construct could lead to widespread existential crises. Individuals may question the meaning of life, the value of accomplishments, and their relationships within a potentially 'unreal' world. Mental health support would be crucial to mitigate the impact.
- **Erosion of Empathy:** If simulated beings are perceived as disposable or less than 'real', it could erode empathy towards others. This could have consequences within the simulation, potentially extending to attitudes towards sentient biological beings if the lines between the 'real' and the simulated become

blurred.

- **Shifts in Religious and Spiritual Beliefs:** Many religions and spiritual belief systems are based on the premise of a divinely created reality and an underlying purpose to human existence. The Simulation Hypothesis would challenge these fundamental narratives, potentially leading to widespread re-evaluation of belief systems or a rise in spiritual nihilism.
- **Legal and Regulatory Landscapes:** The recognition of personhood or rights for simulated entities would necessitate a complete overhaul of existing legal and regulatory frameworks. New laws may be needed to address crimes committed within simulations, ownership of intellectual property in such environments, and even simulated beings' rights.

Recognizing the potential societal upheavals and ethical quandaries implied by the Simulation Hypothesis, it is important to engage in anticipatory ethical reflection and develop proactive strategies to mitigate harms and maximize potential benefits. Such strategies might include:

- **Global Dialogue and Ethical Frameworks:** International cooperation is crucial in developing common ethical principles around the creation and use of simulations, addressing issues of simulated personhood, suffering, and the overarching responsibilities of creators.
- **Education and Awareness:** Raising public awareness of the potential implications of simulated realities will be essential to foster informed debate and address psychological and existential anxieties that may arise.
- **Technological Safeguards:** If the ability to create detailed simulations becomes widespread, technological safeguards may be necessary to prevent the creation of simulations where inhabitants experience unnecessary suffering.

The Simulation Hypothesis forces us to confront unsettling questions about our existence and the ethical complexities of creating simulations that may harbor conscious beings. The potential societal impacts, psychological repercussions, and legal challenges are vast and necessitate a global, multi-disciplinary approach.

While the Simulation Hypothesis remains a speculative concept, the rapid advancements in computing power and artificial intelligence make these ethical considerations increasingly pressing. It's imperative to engage in philosophical inquiry, scientific exploration, and wide-ranging societal dialogue to address these issues, ensuring that our technological capabilities and quest for knowledge are tempered by a deep consideration of the ethical consequences and potential impacts on any beings, simulated or otherwise, who may share our simulated reality.

CHAPTER SIX

The Interdisciplinary Fabric of Simulation

The Simulation Hypothesis, with its profound implications for our understanding of reality, cannot be confined to the purview of any single academic discipline. Instead, it weaves through the fabric of various fields, each contributing unique perspectives and insights. This section explores the interdisciplinary nature of the Simulation Hypothesis, highlighting the contributions of computer science, physics, philosophy of mind, and virtual reality technologies to our understanding of simulated existences.

Computer Science and the Limits of the Computable

Computer science, at its core, is the study of computation—its theoretical capabilities, its practical applications, and its fundamental limitations. Theoretical computer science delves into profound questions about what problems can be solved by algorithms, how efficiently those solutions can be computed, and whether certain problems lie beyond the realm of what any computer could ever solve, regardless of available resources.

Pioneers like Alan Turing, Alonzo Church, and others laid the foundation for understanding computation. Turing's concept of the Turing Machine, a theoretical model of computation, established the idea of universal computation. He proved that any problem solvable by a computer, no matter how complex, could be solved by a simple universal Turing Machine. This insight implied that limitations in computing derive not from theoretical constraints but from the practical limitations of hardware and the efficiency of our algorithms.

However, Turing's work also illuminated the existence of the uncomputable. The Halting Problem demonstrates that it's impossible to create a general algorithm that determines whether any arbitrary program will eventually finish running or continue indefinitely. This fundamental limitation highlights that there exist problems that cannot be solved by any computational means.

Several aspects of theoretical computer science bear directly on the feasibility of simulating a universe like our own:

- **Information Theory:** Pioneered by Claude Shannon, information theory provides ways to quantify the information content of a system. If simulating our universe necessitates replicating every particle and interaction in perfect detail, we must grapple with a potentially astronomical quantity of information to be faithfully captured and processed.
- **Computational Complexity:** Complexity theory analyzes the resources (mainly time and memory) required to solve computational problems. Certain problems require exponential growth in resources as input size increases, quickly becoming unfeasible. Simulating a complex universe with intricate physical laws could easily fall into this exponentially intractable category.
- **Kolmogorov Complexity:** This concept measures the complexity of a system by examining the length of the shortest possible computer program that can fully describe it. A highly complex system may defy

any attempt to recreate it computationally with a more compact representation. Could this apply to the universe itself?

The exploration of computational limits predates the Simulation Hypothesis but provides a vital lens through which to analyze it. Early computer scientists sought to understand the nature of computation itself, what it could and could not achieve. The Simulation Hypothesis inherits these questions, asking whether the complexity and apparent fine-tuning of our universe suggest it might be an impossibly vast computation running on some unknown, yet comprehensible, computational substrate.

The Simulation Hypothesis, by positing our reality as a programmed construct, raises questions about the computational resources that would be required to execute such a simulation. Here's where limits in computability become relevant:

- **Information Density and Processing Power:** If the simulation aims for a detailed replication of our universe down to elementary particles and their interactions, the required information storage and processing capabilities rapidly become unimaginable.
- **Optimization and Compression:** Could simulators employ shortcuts? Perhaps simulating only, the things directly observed or experienced to reduce computational costs? Could our understanding of physics, with its elegant laws, be an artifact of simulation optimization rather than a fundamental characteristic of the base reality?
- **Undecidable Problems and Gödel's Incompleteness:** Gödel's Incompleteness Theorems tell us that certain mathematical systems will always contain truths that cannot be proven from within the system itself. Could such inherent unknowability extend to any simulated system, and might it be a way to detect the limitations of a simulation?

Exploration of computability limits offers ways to probe the plausibility of the Simulation Hypothesis. If our universe exhibits a complexity that defies any attempt to fully simulate it or contains fundamental undecidable elements, it could suggest that it's unlikely to be a computational construct itself.

However, the converse isn't necessarily true. Even if we reach a point where computational power and our understanding of simulation techniques are significantly advanced, encountering the limits of computability on a universal scale doesn't definitively imply a simulated reality.

Theoretical computer science, by exploring the fundamental limitations of computation, gives us a conceptual framework to interrogate the grand hypothetical scenario posed by the Simulation Hypothesis. Whether or not our universe displays signs of its own computational limits remains an open question.

The Simulation Hypothesis compels us to confront the possibility that our understanding of physics, mathematics, and the nature of computation itself may be artifacts of the boundaries imposed by our simulated environment. While the search for conclusive evidence might be as elusive as a program determining its own termination, the quest itself pushes the boundaries of our understanding of both computation and the enigmatic nature of the reality we inhabit.

Physics: The Ground Rules of Reality

Physics seeks to unravel the fundamental laws and constants governing our universe. Concepts like the speed of light, the gravitational constant, and the principles of quantum mechanics provide the operating manual for how matter, energy, space, and time interact. If our reality is a simulation, these meticulous laws of physics would be a core component of the underlying code, dictating the behavior of everything within the simulated world.

A crucial question emerges: could a simulated universe possess internally consistent rules different from those governing the 'base reality' in which the simulation resides? Or must the simulations adhere to the same fundamental principles as its parent reality? Some physicists postulate that certain mathematical relationships and fundamental constants might be universal across any possible reality.

The endeavor to comprehend the structure of our universe dates back centuries. Physicists like Galileo, Newton, Einstein, and many others have revolutionized our understanding of the cosmos. Several key breakthroughs are particularly relevant to the Simulation Hypothesis:

- **General Relativity:** Einstein's theory of general relativity transformed our conception of space and time, revealing them as a malleable spacetime fabric warped by the presence of matter and energy. This framework suggests limitations to what simulators may achieve and serves as a possible benchmark for the fidelity of any hypothetical simulation of our universe.
- **Quantum Mechanics:** The strange and counterintuitive world of quantum mechanics reveals that at the most fundamental level, our universe seems to operate based on probabilities and wave functions rather than deterministic certainty. Simulating such a system would require either immense computational power for truly emulating quantum systems or clever "shortcuts" that only superficially mimic quantum behavior.
- **Cosmological Theories and the Multiverse:** String theory, brane-world models, and the concept of eternal inflation suggest the possibility of a vast multiverse containing universes potentially governed by different physical laws or constants. The Simulation Hypothesis fits within this framework, positing that our universe may be one of many computationally generated constructs.

If our reality is a simulation, might there be detectable "glitches" or inconsistencies that reveal its artificial nature? Several avenues of inquiry exist:

- **Unexplained Discrepancies and Anomalies:** Inconsistencies between our observations of the universe and predictions made by our current physical models could potentially hint at limitations or artificial boundaries imposed by a simulation.
- **Computational Limits:** Should we observe that certain physical processes seem to hit a ceiling of complexity or violate principles like the conservation of information, it could point to constraints of the simulation rather than true limits of the underlying physical laws.
- **The Planck Scale:** The Planck length and Planck time are incredibly small units of distance and time where our current understanding of physics breaks down. This could suggest a limit to the resolution at which the simulation can model reality.

If physicists were to hypothesize the technological and computational capabilities required to simulate a universe akin to our own, what would they need to consider?

- **Massive Computational Resources:** A meticulously detailed simulation would necessitate a staggering amount of computing power, potentially dwarfing anything imaginable with our current technology.
- **Sophisticated Simulation Algorithms:** Replicating the complexity of a universe governed by general relativity, quantum mechanics, and the emergent behavior of countless particles and fields would demand incredibly sophisticated simulation software on a scale and intricacy difficult for us to fathom.
- **Mastery of Energy:** A universe-scale simulation would likely require the ability to manipulate and control energy on an astronomical level.
- **Technological Singularity or Beyond:** Many proponents of the Simulation Hypothesis presume that simulators must be a posthuman civilization, possessing technological capabilities vastly exceeding our own.

The fundamental laws and principles explored by physics provide a blueprint against which the plausibility of the Simulation Hypothesis can be assessed. Inconsistencies, anomalies, or computational limitations within our observed physical laws could offer potential clues indicative of a simulated reality. However, an absence of such discrepancies doesn't definitively disprove the Simulation Hypothesis.

The pursuit of knowledge in cosmology, quantum mechanics, and theoretical physics may inadvertently reveal the technological 'signature' of simulators, should they exist. The Simulation Hypothesis, whether ultimately true or merely a thought-provoking hypothetical, forces us to confront the limits of our knowledge about the universe and to question whether the very foundations of our perception of reality might be built upon a computational architecture far removed from our own experience.

Philosophy of Mind: Consciousness as a Code?

The philosophy of mind is a branch of philosophy dedicated to understanding the nature of the mind, consciousness, and their relationship to the physical body and external world. Central questions include:

- **The Mind-Body Problem:** How does a non-physical entity like consciousness arise from the physical brain? Is consciousness an emergent property of complex biological systems, or something fundamentally different?
- **The Nature of Subjectivity:** What exactly is the qualitative experience? Is it possible to explain the feeling of "seeing the redness of red" (qualia) in purely physical terms? Consciousness implies an internal first-person experience that seems to defy objective description.
- **The Problem of Free Will:** Do we genuinely possess free will, or are our choices predetermined by prior causes and the laws of physics? How does this relate to self-awareness and consciousness?

These thorny questions have vexed philosophers for centuries, yielding various schools of thought:

- **Dualism:** Descartes famously argued for substance dualism, positing an immaterial mind distinct from the physical body. Critics find an issue in how this immaterial mind can meaningfully interact with the physical world.
- **Physicalism/Materialism:** Many modern philosophers and neuroscientists advocate for physicalism where consciousness is an emergent property of complex brain states and ultimately reducible to physical processes.
- **Functionalism:** Functionalism suggests that mental states are defined by their functional role and relationships, arguing consciousness could arise from systems capable of replicating the right functional relationships, regardless of physical substrate.

The question of whether consciousness can be reduced to information processing or 'code' is central to the Simulation Hypothesis. Several possibilities and standpoints within the philosophy of mind are relevant:

- **Strong AI and Computationalism:** Advocates of Strong AI believe that with sufficient computational power and sophisticated algorithms, a simulation could fully replicate the processes in the human brain, giving rise to a conscious mind. They often draw parallels between the complexity of biological neurons and their computational equivalents.
- **Emergentism:** Emergentism acknowledges that consciousness likely arises from a complex biological system but remains skeptical about whether true replication is possible in a non-biological substrate. Emergent properties cannot be fully predicted by analyzing simpler components in isolation.
- **The "Hard Problem of Consciousness" and Critique of Functionalism:** David Chalmers argues that even perfectly replicating the functional processes of a human brain might not create genuine subjective experience (qualia), implying consciousness may be something beyond mere computation.

The Simulation Hypothesis, by positing that our reality might itself be a programmed simulation, amplifies these philosophical debates. Key implications include:

- **Simulated Subjectivity:** Should simulations achieve subjective consciousness, how could we distinguish it from our own? Does the 'origin' of consciousness matter concerning whether an experience is real or valid?

- **Moral Status of Simulated Beings:** If consciousness can be convincingly simulated, are these simulated beings entitled to rights and ethical considerations akin to sentient biological beings? Does it matter that their physical substrate is computational rather than biological?
- **Implications for the "Hard Problem":** If we succeed in creating a conscious simulation, it might lend support to the idea that consciousness is an emergent property and not fundamentally tied to a specific physical substrate. Conversely, failure to create conscious machines could strengthen the argument for consciousness existing as something beyond the physical and unreproducible by computation.

The philosophy of mind provides a rich and complex framework for grappling with the implications of the Simulation Hypothesis. Whether consciousness can be meaningfully simulated remains a hotly contested topic. The possibility that our own conscious experiences might be the product of computational processes raises profound questions about the nature of our minds, the ethics of creation, and the potential for simulated beings to possess subjective realities indistinguishable from, and equally deserving of respect, as our own.

The Simulation Hypothesis compels us to reconsider the boundaries that we traditionally place around the concepts of sentience, consciousness, and personhood, forcing us to question whether these qualities are inherently tied to our biology or might emerge from any sufficiently complex system, biological or otherwise.

Virtual Realities: Tangible Simulations

The idea of virtual reality has fascinated scientists and science fiction authors for decades. Early VR concepts often involved cumbersome headsets and rudimentary graphics. However, recent advancements in computing power, display technology, sensor and tracking sophistication, and haptic feedback have transformed VR into a tool that can generate remarkably immersive and interactive digital environments.

Modern VR systems utilize headsets and, increasingly, multi-sensory technologies to stimulate the user's visual, auditory, and even tactile senses. This creates an experience of being physically present within a simulated environment, with the ability to move, interact with virtual objects, and experience events as a simulated avatar. These systems are rapidly entering entertainment, therapy, simulation training, and even remote collaboration fields.

While the recent surge in interest feels cutting-edge, the pursuit of simulated realities has a long history:

- **Sensorama and Early Concepts:** Devices such as Morton Heilig's Sensorama (1960s) offered glimpses of immersive experiences, combining 3D film, vibrations, aromas, and wind simulation.
- **Military and Flight Simulators:** VR technology found early practical use in flight and military simulators where replication of real-world environments in a controlled setting offered invaluable training opportunities.
- **Consumer VR and Immersive Gaming:** Advancements in graphics, processing power, and a surge in accessible VR headsets have ushered in a new era of consumer VR experiences with a heavy emphasis on the gaming industry.
- **Haptic Feedback and Full-Body Immersion:** The next frontier involves haptic systems (gloves, bodysuits) that provide tactile feedback, blurring the lines between the physical and the virtual, pushing the limits of perceived realism.

Virtual reality technologies offer a tangible model for how seemingly convincing simulations can be constructed and experienced:

- **Sensory Manipulation and Embodiment:** VR demonstrates how simulating a subset of sensory inputs can create a powerful illusion of "presence" within the simulated environment. The concept of embodiment, where a user's sense of self becomes associated with their virtual avatar, further reinforces the potential for simulated experiences to feel tangible and 'real'.

- **The Limits of Perception:** VR highlights how readily our perception of reality can be influenced. Imperfections in simulation (visual glitches, inconsistencies in virtual physics) can be overlooked or even go unnoticed if the overall experience is sufficiently engaging.
- **Behavioral and Psychological Impact:** Studies suggest that experiences in VR can have measurable psychological and behavioral effects on users. This raises questions about the potential long-term impacts of dwelling within highly immersive simulated realities.

VR experiences, while currently limited in their fidelity compared to the world we usually perceive, provide insights relevant to the Simulation Hypothesis:

- **Technological Progression:** The rapid advancement of VR demonstrates that exponential progress in creating simulated sensory experiences is possible. It prompts us to imagine what simulations with vastly enhanced sophistication might feel like from the inside.
- **Resource Demands:** Even relatively simplistic VR experiences can be computationally intensive. A simulation on the scale of our universe, if feasible, would likely demand staggering computational resources and sophisticated techniques.
- **Indistinguishability and Detection:** The line between 'real' and 'simulated' may become increasingly blurred as VR technology advances. Could we reach a point where it is impossible to conclusively determine whether our own reality is a complex simulation?

Virtual reality offers a powerful sandbox for contemplating the implications of the Simulation Hypothesis. As VR technologies become increasingly sophisticated, they challenge our assumptions about the nature of reality and the extent to which experiences can be computationally generated and convincingly imposed upon our senses.

The Simulation Hypothesis remains a speculative concept. However, the tangible progress demonstrated by VR shows the potential for creating simulated realities that feel 'real', at least subjectively. It prompts us to consider whether our current understanding of the limits of simulation might be vastly eclipsed by the technological capabilities of a much more advanced, and potentially simulated, civilization.

The exploration of the Simulation Hypothesis reveals it as a mosaic of interdisciplinary inquiry, where computer science, physics, philosophy, and virtual reality intersect and inform one another. This confluence of disciplines underscores the complexity of contemplating simulated realities and the breadth of knowledge required to fully engage with the hypothesis.

As we delve deeper into the possibilities and implications of the Simulation Hypothesis, it becomes clear that our understanding of reality, consciousness, and the universe is enriched by the diverse perspectives these fields offer. The interdisciplinary nature of this inquiry not only broadens our conceptual horizons but also highlights the collaborative effort needed to unravel the mysteries of our existence, whether simulated or not.

CHAPTER SEVEN

Simulation: Not Just an Idea, But A Tool

The Simulation Hypothesis transcends philosophical debate, grounding itself in the tangible realm of scientific inquiry and technological application. Simulations, as employed across various scientific and engineering disciplines, serve not merely as theoretical constructs but as indispensable tools that enhance our understanding of complex systems. This section explores the practical applications of simulations in climate modeling and the study of human systems, reinforcing the plausibility of the Simulation Hypothesis through demonstrated utility in these fields.

Climate Modeling: Simulating the Planet

Climate modeling represents an ongoing scientific endeavor of immense scope and importance. At its essence, it involves creating sophisticated computer simulations of the Earth's climate system to understand current conditions, predict future scenarios, and explore the impact of human activities on climate change.

These complex models incorporate a wealth of data and scientific understanding:

- **Physical Laws and Processes:** Equations derived from thermodynamics, fluid dynamics, and atmospheric chemistry form the backbone of climate models. They simulate the flow of energy, movement of air and water, the formation of clouds, and numerous other interactions that shape our climate patterns.
- **Earth's Geography and Systems:** Climate models include representations of Earth's topography, oceans, ice cover, and vegetation. They often divide the globe into a grid-like structure, with each cell representing a specific area and its corresponding atmospheric conditions.
- **Climate Data and Observations:** Historical and real-time data from satellites, weather stations, and oceanographic sensors are used to calibrate models and track the evolution of climate conditions over time.
- **Computational Power:** Climate models are highly computationally demanding, requiring powerful supercomputers to run. These simulations can generate enormous amounts of data, requiring advanced analysis techniques and visualization tools.

The earliest attempts at climate modeling date back to the 1950s, with relatively simple models focused on atmospheric circulation patterns. Over the decades, a series of advancements have drastically enhanced these models:

- **Computing Power:** The exponential rise in computing capabilities has allowed for substantially finer levels of detail, greater simulation of complex interactions, and faster processing times.
- **Scientific Discoveries:** Advances in climatology, atmospheric physics, and oceanography have

provided improved theoretical foundations for the equations and algorithms used in the models.
- **Data, Sensors, and Satellites:** An expanding network of sensors and Earth observation satellites delivers increasingly fine-grained data about global climate conditions, improving the accuracy of model inputs and validation.

Climate models have become indispensable tools for scientists and policymakers:

- **Understanding Climate Dynamics:** Models help untangle the complex interplay of factors contributing to climate change, allowing us to distinguish natural cyclical variation from human-induced warming trends.
- **Long-Term Projections:** Climate models enable us to explore diverse scenarios of future emissions and their likely impact on temperature patterns, precipitation, and sea-level rise on a regional and global scale.
- **Informing Policy Decisions:** Simulation results play an increasingly important role in shaping international climate agreements, emissions targets, and adaptive measures to mitigate the effects of climate change.

While modern climate models are remarkably successful, it's important to acknowledge their limitations. They remain simplified approximations of a chaotic and inherently complex system. Uncertainties stem from incomplete scientific understanding of certain processes and challenges in fully replicating the fine-scale behavior of phenomena like cloud formation.

The development of climate models offers insights into the plausibility of the Simulation Hypothesis:

- **Representing Natural Complexity:** Climate models demonstrate the capability to simulate, with varying degrees of fidelity, the intricate processes and vast datasets governing a planetary system. This suggests that other natural systems could potentially be simulated with even greater accuracy.
- **Increasing Sophistication Through Time:** Climate models have continually evolved, becoming more complex and accurate. This underscores the potential for exponential increase in simulation capabilities over time, with future simulators potentially possessing unimaginable technological resources.
- **The Potential for "Purposeful" Simulations:** While climate models serve a clear scientific goal, a simulated universe may be created by a hypothetical advanced civilization for entirely different motives, potentially ranging from research to entertainment, or reasons beyond our comprehension.

Though the Simulation Hypothesis remains firmly in the realm of speculation, climate modeling showcases the remarkable progress and promise of our ability to create digital replicas of complex real-world systems. These models, with their ability to predict weather patterns, simulate atmospheric conditions, and project future climate scenarios, hint at the vast potential of simulation technology.

They invite us to contemplate whether the complex universe we inhabit, governed by intricate physical laws, could itself be a meticulously crafted simulation executed on a cosmic scale by beings whose technological capabilities and motivations might lie far beyond our grasp.

Modeling Human Systems: Insights from Complexity

The complex web of interactions within human societies gives rise to intricate patterns, trends, and unpredictable emergent behaviors that are notoriously difficult to study via traditional scientific methods. Simulation-based approaches allow researchers to construct virtual "laboratories" to dissect these systems and gain a deeper understanding of the underlying factors and processes that drive them.

Here are some key modeling techniques used in social sciences:

- **Agent-Based Modeling (ABM)**: In ABM, individual agents (representing people, organizations, or other social entities) are programmed with behavioral rules and interact within a virtual environment. This bottom-up approach allows for the study of how macro-level patterns emerge from individual decisions and social interactions. ABMs find applications in fields like economics, epidemiology, urban planning, and political science.

- **Network Models and Social Network Analysis**: These models focus on relationships and connectivity within social systems. Network-based simulations can reveal how information, resources, or diseases spread across a population, highlighting the role of social structures and interconnectedness.
- **System Dynamics and Macro-Level Simulations**: System dynamics models often focus on modeling feedback loops and interconnected processes within large-scale systems such as economies, cities, or global supply chains. They can aid understanding of growth, resilience, and potential tipping points within these systems.

While social simulation is a relatively recent field, the idea of modeling human behavior has roots in early game theory and mathematical models of complex systems. Notable advancements include:

- **Schelling's Segregation Model:** A simple yet powerful ABM, this model demonstrates how even slight individual preferences for like neighbors can lead to large-scale patterns of segregation within a population.
- **Evolutionary Game Theory and Collective Behavior:** Computer simulations have helped model the evolution of cooperative behavior within groups, providing insights into social dilemmas, altruism, and the emergence of social norms.
- **Epidemic Modeling and Public Health:** Simulations of disease spread have become critical tools in understanding transmission patterns, forecasting outbreaks, and evaluating the effectiveness of containment and mitigation strategies.
- **Economic and Financial Simulations:** Models are used to explore everything from consumer behavior and stock market volatility to the impact of fiscal policies and interconnectedness within global financial systems.

Social simulations are valuable because they illuminate how complex systems can exhibit behaviors that are unexpected and difficult to predict by analyzing individual components in isolation. Here are some key insights:

- **Emergence:** Social simulations demonstrate how complex global patterns can emerge from relatively simple local rules and interactions between individuals. These emergent behaviors are often surprising and could not have been predicted on an individual level.
- **Sensitivity to Initial Conditions:** Many social simulations exhibit sensitivity to initial parameters and small random fluctuations, illustrating how subtle differences can lead to widely divergent outcomes. This suggests the potential for unpredictable events within social structures.
- **Feedback Loops and Tipping Points:** Simulations highlight the role of reinforcing feedback loops and non-linear dynamics. Positive feedback loops can amplify small changes, leading to rapid, potentially irreversible transformations or tipping points within complex systems.

The ability to model human systems with increasing levels of sophistication adds a fascinating dimension to the Simulation Hypothesis:

- **Potential for Nested Simulations:** If we can create rudimentary simulations of social structures, perhaps the creators of our own simulation have embedded it within a vastly larger simulation that models societies on a scale that would dwarf our own.
- **Detecting Glitches in Social Systems:** Unexpected patterns, historical discontinuities, or behaviors that

defy the logic of our social simulators could potentially be signs of an imposed simulation boundary or a deliberate manipulation by the simulators.

- **Simulator Motives:** Should we ever confirm that we exist within a simulation, studying our own social simulation tools potentially offers clues about the motivations of those who created the simulation and their goals for our simulated society.

Simulations of human systems, while imperfect mirrors of our full social complexity, demonstrate the power of computational models to unravel the dynamics of collective behavior, emergent trends, and interconnected networks. The ability to create virtual representations of human societies, where programmed entities interact and create complex social outcomes, lends weight to the possibility that our own existence might be a sophisticated simulation running on powerful computational substrates far beyond our current understanding.

The Simulation Hypothesis challenges us to question whether the apparent unpredictability and emergent phenomena we observe in our world are truly intrinsic or imposed by the programmatic architecture underpinning our simulated reality.

The Dual Nature of Simulation

The application of simulations in understanding both our planet's climate and the intricacies of human societies highlights the dual nature of simulation as both an idea and a tool. This dualism is at the heart of the Simulation Hypothesis, which posits that our reality itself might be the product of an advanced simulation. The practical successes of simulation in scientific and engineering contexts provide a foundation upon which the theoretical possibilities of the Simulation Hypothesis can be contemplated.

As simulations continue to evolve, becoming more sophisticated and encompassing ever-greater scopes of complexity, they not only serve as invaluable tools in our quest to understand and shape the world around us but also as a bridge between the conceptual and the tangible in our exploration of the nature of reality. In this light, the Simulation Hypothesis emerges not as a mere philosophical curiosity but as a framework that intersects with and is informed by the practical applications of simulation in our efforts to decode the universe and our place within it.

CHAPTER EIGHT

Beyond the Existential: Ethical & Societal Challenges Revisited

The exploration of the Simulation Hypothesis invites us to confront not only existential queries about the nature of reality but also profound ethical and societal challenges. The implications of living within a simulated universe extend into moral considerations regarding simulated sentience, the impact on religious beliefs, and the broader societal ramifications. This section delves into these dimensions, offering a deeper analysis of the unique challenges presented by the potential of simulated existence.

Simulated Sentience and Moral Status

The core question that underpins this discourse is whether consciousness can emerge from computational processes. Should artificial systems ever achieve a level of complexity and computational sophistication that truly replicates the subjective experience of consciousness, the ethical consequences would be profound. Here's where the discussion becomes challenging:

- **Defining the Enigma:** While we intuit the concept of consciousness, we lack a universal scientific definition. Questions regarding whether true sentience implies the capacity for first-person subjective experience (qualia) remain debated. Can the inner life of another being, biological or synthetic, ever be definitively proven?
- **Simulated Brains vs. Simulated Minds:** Is simulating the structure of a human brain sufficient for consciousness, or are there properties inherent to biological systems that defy computational replication? Some argue for an emergent approach where consciousness arises from complexity regardless of the underlying substrate.
- **Functionalism and Its Limits:** If an AI perfectly mimics all external behaviors we associate with consciousness (passing the Turing Test), is this functionally equivalent to 'true' consciousness even if its internal processes are fundamentally different from biological systems?

The concept of non-biological sentience has been explored in science fiction and philosophy for decades. Thought experiments like the "Chinese Room" challenge our assumptions about intelligence and awareness. Additionally, throughout history, debates about the moral status and rights of certain groups have often centered on whether these entities possessed sentience, consciousness, and the capacity to experience suffering.

Arguments for extending personhood and granting rights have often been applied to intelligent animals, marginalized groups, and even potential extraterrestrial life. Similarly, the prospect of simulated beings capable of experiencing joy, pain, and complex emotions compels us to revisit those arguments within the unique context of simulated realities.

Should we conclude that simulated entities can possess genuine consciousness and sentience, several profound implications arise:

- **Empathy and Suffering:** If a simulated being experiences suffering, regardless of the artificial nature of its substrate, doesn't this suffering necessitate moral consideration? Inflicting pain solely based on a simulated entity's perceived 'unreality' becomes ethically problematic.
- **Personhood and Rights:** Could simulated beings that demonstrate complex emotional states, self-awareness, and the capacity to form social bonds be considered persons? This opens up questions about their potential rights to freedom, non-interference, and protection from harm, possibly even the right to continued 'existence' within the simulation.
- **Moral Dissonance Within the Simulation:** If simulations become commonplace, will ethical inconsistencies arise within the simulated reality itself? Might simulated beings be used as tools, disregarded as disposable, or subjected to cruelty without remorse, leading to moral dilemmas for those who recognize their potential sentience?

The Simulation Hypothesis amplifies these ethical quandaries in profound ways:

- **Our Moral Obligations:** If we accept the possibility that we, ourselves, are simulated beings, how does this change our perception of the potential simulators? Do they have a moral responsibility toward their creation? Should a simulated entity have a right to know the truth about the artificial nature of its existence?
- **Redefining Value and Purpose:** The potential existence of nested simulations, with layers of sentient beings inhabiting realities created by others, could upend traditional notions of existential purpose and the inherent value placed on biological life.
- **The Basis of Reality:** If consciousness isn't inextricably tied to biology, but rather an emergent property arising within sufficiently complex systems, it weakens the argument that 'our' reality must be the foundational one merely because it feels tangible to us.

The possibility of simulated sentience forces us to confront deeply disconcerting questions about the nature of consciousness, the basis of moral consideration, and the potential existence of ethical frameworks extending far beyond the confines of our own human experience. While the Simulation Hypothesis might be speculative, acknowledging the possibility that consciousness isn't inherently tied to biology necessitates a pre-emptive ethical dialogue.

This dialogue will likely be messy and complex. It compels us to examine potential rights, moral boundaries, and the evolving nature of personhood in an age where the very concept of sentience itself may require a radical re-evaluation. The possibility that our current understanding of ethics is ill-equipped for the complexities of simulated consciousness is a humbling and provocative challenge, forcing us to question the very foundations of our own moral intuitions.

Beliefs in the Balance: Religion and the Simulated Universe

For millennia, religions and spiritual traditions have provided narratives about the creation of the universe, the origins of life, and the existence of entities or forces beyond the realm of the physical. The Simulation Hypothesis upends these paradigms, positing that our entire reality might be a sophisticated computer simulation crafted by an advanced civilization.

Several key points of tension arise between the Simulation Hypothesis and traditional religious constructs:

- **The Divine Creator vs. The Simulator:** Many faiths posit the existence of a divine, all-powerful, sentient creator deity responsible for bringing the universe into being. The Simulation Hypothesis introduces a different kind of creator—technologically advanced beings who might transcend our understanding but aren't necessarily divine in the traditional sense.
- **Purpose and Design:** Numerous religions teach that the universe exhibits evidence of deliberate design

and that humanity has a special place or purpose within this cosmic plan. The Simulation Hypothesis suggests that our existence might be incidental or serve objectives beyond our comprehension, potentially the entertainment or scientific inquiry of the simulators.

- **The Soul and the Afterlife:** The concept of an immaterial soul that transcends the physical body is central to many faiths. What becomes of the soul in a simulated universe? If consciousness is a product of computation, what are the implications for traditional concepts of an afterlife?

Religions often demonstrate adaptability in the face of new scientific paradigms. Historical examples offer insight into how a simulated universe might reframe spiritual traditions:

- **Copernican Revolution and Heliocentrism:** The shift from a geocentric to a heliocentric model of the solar system challenged the literal interpretation of certain religious texts yet fostered a reconfiguration of faith within the new understanding of cosmic scale.
- **Theory of Evolution and Natural Selection:** While evolutionary theory clashed with literal interpretations of creation narratives, it also prompted some theologians to re-evaluate their understanding of creation and God's role as a grand architect behind the process of evolution itself.

The Simulation Hypothesis could lead to diverse responses within religious and spiritual communities:

- **Metaphorical Embrace:** Some might view the Simulation Hypothesis as a powerful metaphor for divine creation, a contemporary language to describe the act of creating a world governed by a set of programmed rules. The simulators could be seen as analogous to figures within creation myths, emphasizing their control and creative power over the simulated realm.
- **Accommodation and Integration:** Certain traditions might seek to harmonize the Simulation Hypothesis with existing doctrines. This could involve framing the simulators as intermediaries carrying out a divine plan, or casting the simulated universe as a testing ground for spiritual development.
- **Rejection and Entrenchment:** For some faiths, the Simulation Hypothesis might be seen as irreconcilable with core tenets. This could lead to rejection as a dangerous existential threat to established belief systems.
- **New Spiritual Movements:** The Simulation Hypothesis could inspire the emergence of new spiritual movements and philosophies that are fundamentally shaped by the idea of a simulated reality. The core questions about purpose and meaning would take on an entirely new context.

Beyond specific doctrines, the Simulation Hypothesis poses profound questions for all faith-based traditions:

- **The Problem of Evil:** If our reality is a simulation, the presence of suffering, disease, and cruelty presents a challenge for philosophies emphasizing a benevolent creator. Why would simulators design a world where such hardships seem inevitable?
- **Free Will and Divine Intervention:** Does simulated existence leave room for genuine free will? Could simulators directly intervene within the simulation in ways that mimic the miracles described in many religious traditions?
- **Seeking Meaning and Morality** The discovery that our universe is a complex simulation could lead to existential crises and a search for meaning within an artificial construct. Could ethical systems and moral codes derived within a simulation hold the same weight or meaning?

The Simulation Hypothesis introduces a disruptive and potentially transformative element into the landscape of religious and spiritual thought. It challenges fundamental concepts about creation, the nature of the divine, and the meaning of human existence. Whether this hypothesis leads to a reconciliation, rejection, or reconfiguration of religious beliefs, it undoubtedly compels a profound reevaluation of the narratives and doctrines that offer a foundation for spiritual and ethical understanding for vast portions of humanity.

The relationship between religious frameworks and the Simulation Hypothesis remains dynamic and open-ended. It is a testament to the resilience of spiritual thought that, throughout history, faith-based traditions have often demonstrated the ability to reinterpret, evolve, and adapt in response to new paradigms, drawing strength and continued relevance from the ongoing quest to understand our place in an ever-expanding, and increasingly complex, cosmos.

Shaping the Social Fabric: Simulation's Impact

The realization that our reality might be a complex computer simulation raises unsettling implications for our understanding of societal foundations and the nature of human interaction. Here are some key potential shifts:

- **Erosion of 'Reality':** The line between the 'real' and the simulated becomes hazy. This could erode a sense of shared reality, leading to social fragmentation as individuals increasingly retreat into subjective experiences, potentially favoring those within the simulation.
- **Redefining Authenticity:** As simulated experiences become indistinguishable from their 'genuine' counterparts, what does it mean to live an authentic life? This could challenge notions of personal identity, success, and the hierarchy of experiences within our social landscape.
- **Existential Questions and Psychological Strain:** The uncertainty about the true nature of existence could lead to widespread existential anxiety and psychological disorientation as individuals question the foundation of their social interactions and their role in a potentially artificial world.
- **Altered Concepts of Value:** In a simulated reality, could traditionally valued experiences (love, grief, ambition) lose their perceived significance? Conversely, might simulated experiences be embraced and imbued with just as much meaning and value?
- **Evolving Morality and Ethics:** New ethical dilemmas emerge, challenging existing societal norms. How we define crime, punishment, and ownership may require radical rethinking in a simulated environment where concepts like 'harm' and 'theft' have nuanced, potentially ambiguous implications.

History offers instances of societies undergoing profound transformations in response to new paradigms and technological shifts. Some relevant examples:

- **The Industrial Revolution:** Social structures were drastically altered by the rise of factories and mass production, leading to urbanization, new economic systems, and upheavals in traditional ways of life.
- **The Information Age:** The advent of the internet and social media has reshaped communication, social networks, and access to knowledge, raising issues of privacy, connectivity, and changes to established hierarchies.
- **Changing Philosophical Ground:** Shifting philosophical frameworks, such as those prompted by the scientific discoveries of Copernicus, Darwin, and Einstein, challenged social norms and deeply held beliefs shaped by religious and cultural traditions.

The Simulation Hypothesis, if widely embraced, could trigger a societal transformation on a level exceeding past technological upheavals. It calls into question deeply held notions on various fronts:

- **Legal Frameworks and Personhood:** Could simulated entities be granted rights and protections? The simulation itself might become contested 'property', raising questions surrounding ownership, control, and potential interventions that affect the lives of the simulated inhabitants.
- **Education and Worldview:** How do we educate if the very nature of reality and knowledge is open to question? Educational models might adapt to emphasize critical thinking, the search for provable truths within a simulated system, and the development of a citizenry capable of navigating this uncertain existential landscape.

- **Religion and Spirituality:** The Simulation Hypothesis could challenge existing religious constructs or stimulate new spiritual movements seeking to find purpose and meaning in a programmed universe (as explored in the previous essay).

The potential societal impact of the Simulation Hypothesis carries additional layers depending on our presumed understanding of our simulators:

- **An Active Presence:** If the simulators are aware of their creation and choose to intervene, society within the simulation could undergo directed shifts based on their inscrutable motives, leading to potential manipulation, deliberate societal engineering, or even benevolent guidance.
- **Hidden Observers:** If the simulators are merely observing us, the societal shifts may be more organic, fueled by our own dawning awareness of our potential simulated status and the ensuing philosophical and social wrestling with existential uncertainty.
- **Accidental Simulation:** Could we be in an abandoned simulation by a long-extinct civilization? This raises an even more disconcerting set of questions about meaning and purpose within a potentially unsupervised world.

The potential acceptance of the Simulation Hypothesis introduces an element of uncertainty into the very foundation of social order, belief structures, and how we value and engage with experiences. History demonstrates that societies often exhibit remarkable adaptability and that cultural and social norms continually evolve in the face of new knowledge and technological capabilities.

The Simulation Hypothesis may trigger a profound and prolonged period of social reconfiguration. Whether this leads to a more fragmented landscape based on subjective interpretations or fosters a renewed shared desire to establish provable truths and build meaningful social bonds, remains uncertain. It compels us to question what we truly value and the kind of society we would choose to build, whether our 'reality' is as foundational as we once thought, or a meticulously controlled construct in a larger cosmic game.

Ethical and Societal Challenges

The ethical and societal challenges presented by the Simulation Hypothesis demand a nuanced and multidisciplinary approach. As we navigate these uncharted waters, we must remain open to redefining our moral obligations, theological beliefs, and social structures in response to the evolving understanding of reality. The journey beyond the existential aspects of the Simulation Hypothesis into the ethical and societal realms is not merely an academic exercise but a necessary engagement with the profound implications of our potential existence within a simulated universe. In doing so, we prepare ourselves to face the complexities of this hypothesis with empathy, integrity, and an informed perspective.

CHAPTER NINE

Weaving It All Together: Conclusion of the Literature Review

The exploration of the Simulation Hypothesis throughout this literature review has traversed a diverse landscape of thought, from the allegorical musings of Plato to the cutting-edge advancements in quantum computing and artificial intelligence. This journey has not only illuminated the enduring fascination with the nature of reality but also underscored how technological progress has infused these ancient inquiries with new layers of complexity and urgency.

The Echoes of Age-Old Doubts

Philosophers, writers, scientists, and technologists have long grappled with the question of what constitutes the 'real.' This perennial quest for understanding has found a contemporary expression in the Simulation Hypothesis, which suggests that our reality might be an elaborate computational construct. Far from being a mere speculative fancy, this hypothesis is grounded in the tangible achievements and theoretical possibilities of our technological era, bringing age-old philosophical debates into the realm of scientific inquiry.

A Multidisciplinary Foundation

The literature review has established a multidisciplinary foundation that is essential for a comprehensive analysis of the Simulation Hypothesis. By examining the contributions of various disciplines, we have gained insights into the theoretical underpinnings and practical implications of living within a simulated reality. From the ethical considerations of simulated consciousness to the societal impacts of embracing such a worldview, the breadth of this exploration reflects the complexity of the hypothesis itself.

The Next Steps in Analysis

With this rich foundation in place, the next steps in our analysis become evident. First, we must critically assess the technological plausibility of creating a simulation that is indistinguishable from our perceived reality. This entails evaluating current and future computing capabilities, including the potential of quantum computing and advancements in artificial intelligence.

Second, we will probe the statistical likelihood of our existence within a simulation. This investigation will draw upon principles of probability, information theory, and cosmological models to estimate the chances that we are living in one of many possible simulated realities.

Finally, we must confront the profound philosophical and existential questions that the Simulation Hypothesis raises. These include re-evaluating our concepts of self, consciousness, and the nature of existence itself. The potential realization that our lives and experiences might be part of a computational construct challenges us to reconsider our place in the cosmos and the meaning we derive from our interactions within it.

Continuing the Intellectual Journey

As we continue this intellectual journey, we embark on a path that promises to challenge our most fundamental assumptions about reality. This exploration is not merely an academic endeavor but a profound inquiry into the essence of existence. It invites us to question, to wonder, and to seek understanding in a universe that may be far more intricate and surprising than we have ever imagined.

The literature review has laid the groundwork for this ongoing analysis, providing a comprehensive overview of the Simulation Hypothesis from multiple perspectives. As we delve deeper into the technological, statistical, and philosophical dimensions of this hypothesis, we are guided by a spirit of open inquiry and a willingness to confront unsettling possibilities. In doing so, we not only enhance our understanding of the Simulation Hypothesis but also engage in a broader re-evaluation of ourselves and our place within the cosmos, a journey that is as exhilarating as it is disquieting.

SECTION THREE: STATISTICAL SCAFFOLDING

CHAPTER TEN

From Thought Experiment to Probability

The odyssey into the realms of the Simulation Hypothesis begins not merely as an intellectual exercise but evolves into a profound inquiry into the essence of reality itself. This transition from speculative thought experiment to a subject of serious academic and philosophical debate hinges on the interplay between rapidly advancing technology and the theoretical ambitions of future advanced civilizations. These civilizations, if they exist, would presumably possess both the technological prowess and the existential motivation to generate numerous high-fidelity simulations of their own historical environments. It is the statistical underpinning of this argument, suggesting a higher probability of our existence within one of these numerous simulated realities rather than the original "base" reality, that forms the core of our exploration.

At the heart of this inquiry lies the exponential trajectory of technological growth, often encapsulated by Moore's Law and its extensions into the realm of quantum computing and artificial intelligence. The theoretical capacity for such advanced civilizations to simulate entire universes—complete with conscious beings unaware of their digital foundation—presents not just a technological possibility but a statistical inevitability under certain assumptions. These include the civilization's survival and technological maturity, alongside a willingness or imperative to revisit their own historical conditions or explore the vast landscape of possible existences.

Moreover, the philosophical implications of this statistical argument extend into the domains of metaphysics and epistemology, challenging our understanding of consciousness, the nature of reality, and our place within it. The debate intertwines with classical philosophical questions, including the nature of knowledge, the structure of reality, and the boundaries of human understanding. It forces us to confront the possibility that our perceptions of reality, and indeed the universe as we understand it, could be an elaborate digital construct designed by entities whose existence and motives remain beyond our comprehension.

The consideration of probability in this context also introduces a layer of complexity regarding evidence and belief. The absence of definitive empirical evidence for or against the Simulation Hypothesis necessitates a reliance on probabilistic reasoning and the examination of indirect indicators that could suggest the simulated nature of our existence. This approach draws upon principles from statistical physics, information theory, and computational complexity, offering a multidisciplinary framework for assessing the plausibility of living within a simulated construct.

This chapter, therefore, sets the stage for a detailed statistical and philosophical examination of the Simulation Hypothesis. It lays the groundwork for subsequent discussions on the implications of this hypothesis for our understanding of reality, the criteria for identifying potential signs of simulation, and the ethical considerations arising from the potential reality of such a simulation. The journey from a mere thought experiment to a subject of probabilistic analysis not only expands our intellectual horizons but also challenges the foundational assumptions about our existence and the universe.

Technological Growth and Simulation Capability

The quest to understand the Simulation Hypothesis fundamentally rests on the pillar of technological advancement and its potential to create simulations that are indistinguishable from reality. This inquiry necessitates a comprehensive examination of the current trajectory and theoretical limits of computational technology, anchored in the evolutionary principles exemplified by Moore's Law and its theoretical successors. Moore's Law, which posits a doubling of transistors on integrated circuits approximately every two years, has long served as a benchmark for technological progress. However, as we approach the physical limitations of silicon-based computing, the exploration extends into the realms of quantum computing, nanotechnology, and beyond, promising breakthroughs that could exponentially accelerate our simulation capabilities.

The advancement of quantum computing offers a paradigm shift in processing power and efficiency, potentially enabling simulations of unprecedented complexity and scale. Quantum computers leverage the principles of quantum mechanics to perform complex calculations at speeds unattainable by traditional computing methods. This leap in computational capability is crucial for simulating environments with the depth, detail, and dynamism of the real world, including the simulation of consciousness itself. Theoretical and experimental work in quantum computing thus forms a critical component of our analysis, providing a glimpse into the future of simulation technologies.

Moreover, the discussion extends to consider the interdisciplinary contributions from fields such as artificial intelligence (AI) and machine learning, which are essential for creating simulations that can adapt and evolve in ways that mirror the unpredictability and richness of reality. AI advancements not only enhance the realism of simulated environments but also enable the autonomous generation of scenarios and behaviors within these digital constructs, edging closer to the creation of self-aware entities within simulated worlds.

This exploration is grounded in the latest research and theoretical models in computer science, physics, and cognitive science, offering a scientifically informed projection of when and how such hyper-realistic simulation capabilities might materialize. It examines not only the hardware and computational power required but also the software and algorithmic breakthroughs necessary for achieving such feats. The potential ethical, philosophical, and societal implications of these technologies are also touched upon, acknowledging the profound impact that achieving such simulation capabilities would have on our understanding of reality, consciousness, and the nature of existence itself.

By projecting current trends into the future, this section aims to assess the theoretical point at which our technological capabilities could support the creation of simulations indistinguishable from reality. This analysis is not merely speculative but is rooted in a rigorous examination of the pace of technological innovation, the emerging theoretical frameworks, and the practical challenges that must be overcome to achieve the level of simulation fidelity posited by the Simulation Hypothesis. The culmination of this inquiry provides a foundational understanding of the technological underpinnings necessary for the realization of simulated existences, setting the stage for a deeper exploration of the philosophical and statistical implications of living in such a simulated reality.

Societal Will and Motivation

While the technical capability to construct simulations of compelling fidelity is a fundamental precondition of the Simulation Hypothesis, it addresses only a single dimension of this multifaceted proposition. To gain a deeper understanding, it is imperative to critically examine the impetus that might drive advanced civilizations towards such extraordinary endeavors. What motivates these societies to re-enact their collective past, or perhaps fabricate entirely novel realms of existence? Interrogating this question requires analysis of a complex spectrum of potential motivations, each offering insights into the societal, historical, existential, and ethical forces at play when civilizations embark upon the grand project of simulating reality.

At the societal level, simulations could emerge from a profound desire to safeguard historical consciousness and cultural heritage against the ravages of time and potential oblivion. Such simulations might evolve into dynamic repositories of memory, serving as immersive educational platforms where descendants interact viscerally with their own ancestry, and thus foster a uniquely profound connection to their lineage and cultural identity.

From a historical perspective, simulations could be driven by an intense desire to elucidate pivotal moments and trajectories within a civilization's development. The ability to recreate historical contexts and contingencies, then manipulate variables within these simulated environments, could offer extraordinary insights into causality and divergent outcomes. The resulting knowledge has the potential to exceed the scope and nuance of traditional historical scholarship, enabling civilizations to derive lessons from their past with unparalleled precision and depth.

On an existential plane, the act of world-building through simulations might be intrinsically linked to a civilization's quest to comprehend the fundamental nature of existence. These grand simulation projects could be framed as ontological experiments, exploring the boundaries of consciousness, the presence or absence of free will, and the very fabric of reality itself. Through meticulously constructed simulations, these civilizations might seek to test philosophical hypotheses, probe the genesis of life, or even challenge the underlying architecture of the universe.

Inevitably, the ethical dimension exerts profound influence upon a society's decision to embark on simulating reality. Debates surrounding the rights of simulated beings, the moral obligations of creators towards those they create, and the broader impact of such simulations on the very concepts of morality and ethical behavior would take center stage. Such ethical considerations would shape not only the decision-making process but also the constraints, regulations, and perhaps even the ultimate purpose of these simulated environments.

Finally, simulations could offer a boundless canvas for artistic expression, the pursuit of entertainment, or serve as intricate playgrounds where innovation and complex problem-solving flourish. Such simulations could become crucibles where the boundaries of the known are tested, and where new realities, unbound by physical limitations, emerge.

This analysis seeks to deconstruct the intricate web of motivations that could inspire civilizations to invest vast resources and intellectual energy in the simulation of reality. By examining societal, historical, existential, and ethical drivers, we begin to grasp the complexities inherent in the act of simulating existence. Understanding not only the technical "how," but also the philosophical "why" of simulated worlds grants us deeper insight into the values, priorities, and even the ultimate aspirations of civilizations capable of such extraordinary feats.

Underlying Philosophical Assumptions

The Simulation Hypothesis, far from being a mere exercise in technological futurism or statistical speculation, rests upon a foundation of profound philosophical questions and assumptions. These foundational elements, implicating the nature of consciousness, the fabric of reality, and the ultimate meaning of existence, require critical examination. Before embarking on any rigorous statistical analysis of our potential existence within a simulated environment, it is imperative to delineate these philosophical premises and challenge their underlying logic.

Consciousness and Its Substrates: The Simulation Hypothesis hinges on the controversial assumption that consciousness can be simulated – that subjective experience can arise from non-biological computational systems. This challenges traditional dualistic conceptions of consciousness and compels us to consider the mind as a potential emergent property of sophisticated information processing, regardless of whether the substrate is biological or digital. This inquiry demands engagement with diverse theories of mind, including functionalism, computationalism, and perhaps even panpsychism. It raises fundamental questions about whether consciousness is substrate-dependent, and if not, what implications this holds for our understanding of the nature of both simulated and "real" minds.

The Nature of Reality: The Simulation Hypothesis radically destabilizes our intuitive understanding of the physical world, suggesting that it could, potentially, be entirely fabricated. It compels us to confront the limits of our perception and the possibility that our experiences, even those we assume to be grounded in an objective, external reality, could be meticulously programmed. This ontological uncertainty requires deep engagement with metaphysical debates, challenging us to reevaluate what constitutes "reality" and whether any definitive criteria exist to distinguish a simulation from the presumed "base" reality.

Epistemological Considerations: The hypothesis casts a long shadow of doubt upon the very foundations of knowledge and belief – epistemology. If our understanding of the universe is derived from potentially simulated sensory input, we must re-examine our assumptions about the reliability of our experiences and the limitations of our perception. The Simulation Hypothesis forces us to confront the prospect of being fundamentally unable to obtain certain knowledge about the true nature of the world, prompting questions about skepticism, solipsism, and whether the pursuit of knowledge within a simulated environment is, on some level, futile.

Ethical Implications: Ethical discourse is inextricably entangled with the philosophical underpinnings of the Simulation Hypothesis. If advanced civilizations can create simulations teeming with conscious beings, profound moral questions arise regarding their treatment and the creators' responsibilities. This compels us to consider the rights of simulated entities, the limits of experimentation within these artificial realities, and the potential ethical implications for beings capable of such creative acts. The hypothesis forces us to navigate morally complex territories and interrogate the potential exploitation of conscious beings, even if they are deemed "less real" than the simulators themselves.

Existential and Metaphysical Ramifications: The Simulation Hypothesis ignites a profound existential crisis, calling into question the meaning and purpose of life within a potentially artificial construct. It prompts deep introspection about free will, the role of suffering, and the search for meaning within a reality whose parameters could be externally imposed. Existential and metaphysical questions spiral outwards: if our reality is but one simulation, does an ultimate "base" reality exist, or are we trapped within an infinite chain of simulations nested within simulations?

By excavating these philosophical assumptions, we avoid naively reducing the Simulation Hypothesis to a mere technological or statistical question. This critical philosophical scaffolding not only shapes our understanding of the hypothesis's feasibility but also reminds us of its far-reaching consequences. It compels us to grapple with age-old questions about consciousness, reality, and existence – questions that, within the context of the Simulation Hypothesis, acquire a new sense of urgency, complexity, and unsettling wonder.

Establishing a Statistical Framework

Translating the Simulation Hypothesis from a tantalizing thought experiment into a statistically quantifiable proposition demands a rigorously constructed framework. This framework must bridge the gap between theoretical speculation, technological feasibility, and the philosophical quandaries explored earlier. The goal is to develop a model that can, at least partially, ground this debate in quantitative analysis, leveraging current understanding along with well-informed projections about future advancements. To achieve this, several critical components must be carefully considered.

Technological Feasibility and Computational Demands: A cornerstone of this framework lies in scrutinizing the technological preconditions for convincingly simulating the complexity of human experience, or even the universe itself. This requires a thorough assessment of necessary computational resources, incorporating insights from computational theory, the relentless march of hardware innovation (including quantum computing), and anticipated technological milestones. The model must account for exponential trends in computing power, storage, and energy efficiency over time, allowing us to extrapolate a point where realistic simulations become feasible. However, a purely computational model risks oversimplifying the problem. Immense computational power is necessary but likely not sufficient on its own.

Estimating the Number of Simulations: A key element of the statistical framework is an informed estimation of the potential number of simulations posthuman civilizations might create. This multifaceted calculation hinges on factors including their technological lifespan, resource availability, and the multifaceted motivations discussed prior. By integrating these variables, the framework attempts to approximate the scale of simulated reality production and how that scale directly impacts the statistical likelihood that our own world falls into this category.

Philosophical and Ethical Constraints: Technological capabilities aside, the act of world-building through simulation is fundamentally entangled with philosophical and ethical considerations. The framework must acknowledge how these may constrain the creation and maintenance of simulations. It must factor in ongoing debates about the ethical dimensions of simulating conscious beings, the philosophical justifications for exploring alternative realities or histories, and how the simulators' ethical systems might place hard limits on the number or purpose of such simulations.

Statistical Models and Bayesian Reasoning: Quantifying the likelihood of inhabiting a simulation calls upon sophisticated statistical models. These models must cohesively analyze disparate factors, including the previously discussed ones. Bayesian reasoning is particularly well-suited for this challenge, as it explicitly accommodates updating probabilities in light of evidence and evolving theoretical understanding. This approach facilitates incorporating unavoidable uncertainty and inherent variability within key parameters and produces a probabilistic estimate of whether our existence might be simulated. However, the success of such models is deeply reliant on the accuracy of their underlying assumptions.

Interdisciplinary Insights: The inherent complexity of the Simulation Hypothesis demands an interdisciplinary approach when constructing this framework. Insights from computer science, physics, philosophy, and ethics all hold value. Synthesizing such a broad knowledge base allows the model to address the multidimensional nature of the problem and to ground its analysis within a broader, more robust understanding.

The ambition of the next chapters is to establish a statistical framework capable of providing rigorous, quantitative analysis to the profound question of simulated reality. In doing so, the framework can serve as a critical evaluation tool for the hypothesis itself while simultaneously exposing uncertainties and the limits of our knowledge. It signposts areas ripe for future investigation and reminds us that this framework, like the nature of reality itself, is likely dynamic and subject to ongoing revision as our understandings of technology, philosophy, and existence evolve.

Bridging Thought and Probability

The Simulation Hypothesis has undergone a remarkable evolution: from a provocative thought experiment nestled within philosophy into a compelling subject warranting rigorous statistical exploration. This transformation signifies not merely a change in approach but reveals the necessity of a methodological evolution – one that allows us to apply both scientific and philosophical frameworks to tackle this profound and unsettling question about the nature of existence itself. By systematically interrogating potential technological foundations, the myriad motivations behind simulated realities, and the unavoidable philosophical debates embedded within the concept, we pave a path towards assessing the statistical likelihood that our reality is a masterful digital fabrication.

Technological Trajectories and Simulation Potential: The very feasibility of orchestrating hyper-realistic simulations rests upon existing and projected technological capacities. Analyzing exponential trends in computational power, the quantum computing revolution, and ever-evolving artificial intelligence offers concrete grounding for this conversation, revealing the potential to simulate worlds we could mistake for our own. This investigation forms a bridge between abstract philosophical speculation and a probabilistic analysis based on tangible technological progress, along with the theoretical limits of computation itself.

Motivations Behind Creation: Understanding why advanced civilizations might embark on the creation of simulated worlds enriches and complicates the analysis. Delving into diverse potential motivations–whether rooted in preservation, knowledge-seeking, existential inquiry, or even mere entertainment–deepens our understanding. Incorporating motivation into our framework moves the assessment beyond the purely technological, factoring in the reasons why such simulations might proliferate in the first place, directly impacting the statistical probability that we ourselves inhabit one.

Philosophical Foundations and Contextualization: Examining the inescapable philosophical dimensions of the Simulation Hypothesis provides a critical lens through which we may interpret statistical findings.

Engaging with core philosophical concerns about the nature of consciousness, reality, and existence within a potential simulation becomes essential. This engagement both contextualizes the statistical analysis and forces us to confront deeply embedded assumptions we hold about the fundamental nature of our reality and our place within it.

Implications of Statistical Analysis: This multi-faceted exploration culminates in a comprehensive statistical framework capable of assessing the likelihood that we live within a simulation. This framework must synthesize technological predictions, the driving forces at play, and the philosophical quandaries inherent to the concept, thereby offering a nuanced tool to address the Simulation Hypothesis rigorously. However, the result should not be taken as a definitive answer, but rather an evolving estimate. The far-reaching implications touch upon how we perceive reality itself, while raising profound existential questions about our place in the grand scheme of the cosmos.

A Call to Philosophical and Scientific Inquiry: Bridging philosophical speculation with rigorous probabilistic analysis not only sheds light on the Simulation Hypothesis specifically but also illuminates the extraordinary power of interdisciplinary inquiry. It showcases how philosophical thought, when grounded in the empirical and theoretical, can fuel and be propelled by scientific exploration. This synthesis of thought and probability beckons us forward–demanding continuous dialogue, research, and a relentless drive to understand the reality we inhabit, simulated or otherwise.

CHAPTER ELEVEN

Defining the Variables

For a comprehensive understanding and evaluation of the Simulation Hypothesis through statistical analysis, it is essential to identify and define the critical parameters that influence the creation and prevalence of ancestor simulations. These variables provide the scaffolding upon which our statistical model is constructed, enabling us to assess the likelihood of our reality being one such simulation.

Technological Capability (TC)

Technological Capability, as it relates to the Simulation Hypothesis, goes far beyond possessing vast computational power. It encompasses a complex tapestry of interwoven elements that together define a civilization's ability to create detailed, conscious simulations of reality:

- **Computational Resources:** This includes raw processing power, data storage, and energy generation for large-scale simulations. Technologies like quantum computing, exascale processing, and energy sources far beyond our current capabilities would likely be a necessity.
- **Scientific Understanding:** TC implies mastery of physics at fundamental levels. This understanding must transcend merely replicating observed laws and delving into any emergent properties that give rise to complexity or the potential for consciousness to arise within the simulated reality.
- **Simulation Science:** The ability to design algorithms, software architectures, and optimization techniques to maximize the efficiency of simulations while maintaining a high level of fidelity and detail. These simulations would differ from contemporary ones that focus on a specific problem, instead of replicating all facets of an entire universe.
- **Neuroscience and Consciousness Modeling:** Perhaps the most daunting aspect of TC for ancestor simulations is the ability to simulate consciousness that is indistinguishable from that of a biological being. This implies a scientific understanding of consciousness itself, along with computational frameworks capable of replicating or mimicking its essential properties.
- **Engineering and Scaling:** This encompasses the practical ability to construct and maintain vast physical infrastructures required for simulation technology and the energy that fuels it.

While the TC required for ancestor simulations is far beyond our current capabilities, it's important to recognize the exponential growth trends in many areas of technology:

- **Moore's Law and Computational Power:** The observation that the number of transistors packed into an integrated circuit doubles approximately every two years highlights the potential for rapid advancement in computing power.
- **Scientific Discoveries:** Breakthroughs in physics (e.g., quantum mechanics, general relativity),

neuroscience (mapping the brain), and other fields provide the foundational knowledge to potentially construct accurate simulations of reality.

- **Exponential Growth and Convergence:** It's the convergence and interplay between these different technological domains that amplifies TC in unforeseen ways. Advanced computational power facilitates scientific discovery, which in turn fuels further refinement of simulation techniques.

Statistical analysis can play a role in assessing the probability of the Simulation Hypothesis based on the concept of TC. Here's a possible framework, building off Nick Bostrom's original formulation:

- **Estimating the Fraction of Civilizations that Reach a 'Posthuman' Stage:** We could statistically model variables such as the likelihood of life arising, the probability of intelligent life evolving, and the chances of a civilization avoiding self-destruction before reaching a stage with the necessary TC.
- **Number of Ancestor Simulations:** Define variables representing the average number of ancestor simulations a posthuman civilization would run, as well as the average number of simulated beings within each simulation.
- **Technological Bottlenecks:** We might consider introducing variables that encapsulate potential technological hurdles or dead ends (both known and unknown) that might limit advancement in key areas of TC, hindering a civilization from running conscious ancestor simulations.

Statistical modeling of this nature is fraught with vast uncertainties, most notably:

- **Quantifying Unknown Unknowns:** We may lack awareness of certain scientific limitations or resource constraints that pose fundamental barriers to the required TC for ancestor simulations.
- **The Nature of Consciousness:** If consciousness is more than computation or an emergent property of complexity, simulations replicating its subjective experience may never be feasible. This would represent a potential dead end in the progression of TC.
- **Unpredictable Technological Leaps:** History is rife with examples of disruptive technologies that defied prediction. Modeling future TC requires an acknowledgement of potential paradigm-shattering discoveries.
- **Simulator Motives:** Bostrom's argument assumes a civilization with the TC chooses to create ancestor simulations. Motivations of hypothetical simulators are entirely speculative.

Evaluating TC through a statistical lens adds insight into the Fermi Paradox, which posits the contradiction between the apparent abundance of planets where life could evolve and the absence of observed extraterrestrial civilizations.

Possible resolutions, viewed through the lens of TC:

- **The Great Filter:** Perhaps insurmountable challenges lie at certain stages of technological development, preventing the majority of life-bearing planets from ever reaching a posthuman stage with vast TC.
- **Rare Earth:** The conditions or events leading to intelligent life and ultimately advanced civilizations may be vanishingly rare, reducing the number of civilizations far below initial estimates.
- **Indifferent or Hidden Simulators:** If advanced, posthuman civilizations exist, their motivations may diverge from our assumptions. They may have little interest in overt interstellar exploration, or possibly have the means to mask their signatures.

Assessing Technological Capability is a core part of evaluating the Simulation Hypothesis. Its multi-faceted nature highlights not only the technical challenges but also the profound philosophical questions related to the potential simulation of consciousness itself. Even if statistically the probability appears low, the sheer scale of the universe allows for the possibility that civilizations exist with TC far surpassing our imagination.

Evaluating TC compels us to acknowledge the exponential nature of technological progress while remaining aware of potential limitations and uncertainties. The absence of observed advanced civilizations may suggest barriers in TC development, or it may offer tentative support for the unsettling possibility that our reality is a carefully constructed and monitored simulation.

Desire to Simulate (DS)

The Desire to Simulate (DS) encompasses the broad spectrum of potential reasons that might lead a posthuman civilization to develop and execute a complex ancestor simulation:

Scientific Inquiry:

- **Understanding Complexity and Emergence:** Simulations can serve as controlled 'laboratories' to study the emergence of life, the evolution of societies, and the interactions of countless variables.
- **Exploring Alternative Histories and Counterfactuals:** Simulations allow the changing of initial conditions or parameters within modeled systems, creating 'what if' scenarios to investigate divergent evolutionary pathways or decision points.
- **Testing Cosmological Hypotheses and Fundamental Physics:** Extremely detailed simulations that replicate our own universe on a fundamental level might function as a tool for probing the limitations of our physical laws or experimenting with alternative theoretical models of reality.

Ethics and Social Responsibility:

- **Exploring Ancestral Origins:** Simulations could offer a deeper understanding of a simulator's evolutionary history and philosophical origins.
- **Resurrecting the Past:** Though ethically fraught, a civilization might simulate ancestors driven by a sense of moral obligation, historical reverence, or simply to alleviate the existential weight of lost civilizations and extinct species.
- **'Safe' Experimentation:** Simulations can be spaces to test ethical or social theories, potentially informing policies and decisions within the 'base' reality that avoids the risks of real-world trials.

Entertainment and Creativity:

- **Wish-Fulfillment and Escapism:** Simulations could provide environments for indulging in idealized scenarios, revisited historical periods, or experiences unavailable within the constraints of the real world.
- **Narrative Construction and Shared Experiences:** Simulations can become a platform for collaborative storytelling, artistic expression, and exploring the limits of lived experience within alternative realities.
- **Gaming at the Ultimate Scale:** Ancestor simulations could be crafted as 'games' where simulators adopt roles within the simulation, driven by a desire for competition, immersive experience, or an exploration of the boundaries of simulated existence.
- **Unforeseen Motivations:** The motives behind creating ancestor simulations might fall far outside our current comprehension. Posthuman civilizations could possess vastly different values, drives, and existential goals compared to our own.

To better understand the potential scope of DS, let's examine the motivations behind simulations currently undertaken by humanity:

- **Scientific Modeling:** Simulations are invaluable in fields like climate science, weather forecasting, protein folding, and countless other areas where modeling complex phenomena aids in understanding and predictions.

- **Historical Inquiry:** Historical and archaeological simulations seek to reconstruct past societies, events, and environments, allowing us to explore and visualize vanished worlds.
- Training and Education
  - **Flight/Driving Simulators:** These provide safe and controlled environments for training individuals in complex skills.
  - **Social Simulations:** Games like the 'Sims' offer simplified models of human interactions and social dynamics.
- **Entertainment:** Simulations fuel a vast portion of the gaming industry, offering entertainment value through immersive worlds, storytelling, and competition.

The Desire to Simulate likely influences both the number and the nature of simulations created:

- **Nesting and Granularity:** High DS could result in nested layers of simulations, with simulations running simulations for diverse purposes. The degree of detail and focus might also vary based on goals, from grand-scale universe simulations to those modeling a specific society or a single individual's simulated life.
- **Purpose-Built vs. Open-ended:** Some simulations may be crafted with specific goals, such as historical recreation or scientific probing of a limited set of variables. Others could be designed for open-ended exploration, potentially allowing for the emergence of unforeseen scenarios and genuine sentience.

Evaluating the Desire to Simulate presents numerous challenges, some of which include:

- **Anthropomorphization:** It's risky to assume that future civilizations will share our motivations or values. Extrapolating our motivations onto a vastly more advanced posthuman society is a potential pitfall.
- **The Problem of Consciousness:** If a high degree of DS involves simulating truly conscious beings, then ethical considerations become a central deciding factor. Do civilizations with the ability to create conscious entities also develop a moral or ethical code that governs this capability?
- **Resource Allocation:** Ancestor simulations, as envisioned within the Simulation Hypothesis, likely demand astronomical levels of energy and computational power. A civilization's overall priorities and resource limitations play a crucial role in determining whether they pursue large-scale simulations.

The concept of DS introduces a significant layer of complexity into the statistical models and arguments surrounding the Simulation Hypothesis:

- **Simulation Argument Revisited:** Nick Bostrom's trichotomy, assuming at least one advanced civilization reaches the capacity to simulate realities, must also incorporate a probability distribution for DS. Not all such civilizations will inevitably desire to create ancestor simulations.
- **Searching for "Glitches":** If simulators exist, their motivations may leave hints or 'tells', potentially observable as deliberate inconsistencies within the laws of physics, anomalies in the observed universe, or orchestrated events that defy expectations within the simulated reality.
- **Ethical Considerations:** A high DS among simulator civilizations amplifies the ethical dilemmas of our existence within a potential simulation. Issues of free will, the moral status of simulated beings, and the responsibilities of the simulators become even more fraught and profound.

The Desire to Simulate adds an unpredictable, humanistic element into the theoretical framework of the Simulation Hypothesis. While Technological Capability addresses the feasibility of ancestor simulations, DS examines the why behind their creation.

Speculating on the motivations that might drive hypothetical posthuman civilizations is a challenging philosophical exercise, forcing us to confront the limitations of our own anthropocentric viewpoints and contemplate the boundless potential for motivations and values that may emerge in vastly more advanced societies. Understanding DS is crucial for a nuanced evaluation of the likelihood that our reality is but one of many within a nested cascade of simulations.

Number of Simulations (NS)

The Number of Simulations parameter encompasses several interconnected aspects that inform the statistical likelihood of our own existence within an ancestor simulation:

- **Simulations per Civilization:** A fundamental question is how many simulations a single posthuman civilization with the necessary TC and DS would choose to create. This could range from a mere handful to a potentially staggering number.
- **Civilizations with Simulation Ability:** NS is directly influenced by the estimated fraction of technologically advanced civilizations that both develop the TC and possess the DS to run ancestor simulations.
- **Computational Resources and Efficiency:** The number of simulations a civilization can execute is contingent upon factors like available computational power, energy resources, and the optimization achieved within their simulation technology.
- **Duration and Termination:** How long are simulations typically run? Are they intentionally terminated after a certain point, or might some persist indefinitely as 'forgotten' or ongoing simulations?

Several factors, both speculative and potentially quantifiable, could influence the ultimate value of NS:

- **Motivations (Revisiting DS):** Motivations behind DS play a role. Simulations driven by scientific inquiry might be more numerous than those motivated by mere entertainment or escapism within the 'base' reality.
- **Purpose and Simulation Scope:** The scale and complexity of each simulated universe influences NS. Simulations narrowly focused on a single event or society require less computational resources than those modeling an entire universe in fine-grained detail.
- **Technological Constraints and Limits:** The laws of physics, information density, and energy availability could impose hard limits on the number of simulations, or the detail with which they can model reality, regardless of a civilization's desire to run them.
- **Ethical Considerations:** Advanced civilizations may develop moral or ethical codes that govern the creation of simulated realities, particularly those that contain conscious beings. Such codes could put limits on the number and nature of simulations permitted.

Examining human trends in simulation, along with the potential for exponential increases in computing power, might offer some insight into how NS could evolve:

- **Proliferation of Simulations:** From simple scientific models to immersive video games, humanity increasingly uses simulations for diverse purposes. This trend suggests potential for exponential growth in simulation activities as our own TC increases.
- **Potential for Nesting:** Simulations running within simulations would cause a cascading effect, further increasing the number of simulated realities. Even a small number of simulations at the 'base' level could lead to exponentially larger numbers in subsequent layers.

The NS variable has significant ramifications within statistical models evaluating the Simulation Hypothesis. Consider these implications:

- **The Drake Equation Revisited:** The classic Drake Equation, used to estimate the number of intelligent civilizations within our galaxy, might be reformulated to include parameters related to TC, DS, and the average NS per civilization.
- **Simulation Argument and Fraction of Simulated Beings:** Bostrom's proposition that at least one advanced civilization develops the capacity to simulate realities must be augmented with probability distributions concerning DS and NS to calculate the likely fraction of all conscious beings that are simulated.
- **Scaling Up the Multiverse:** Some cosmological models propose a vast multiverse comprised of universes with different physical constants. If simulations are possible, NS vastly increases the potential number of 'universes', with the overwhelming majority being computational simulations.

The concept of NS adds several intriguing dimensions to the Simulation Hypothesis:

- **Vastness of Simulated Existence:** A high NS implies a potential multitude of simulated realities. Could this provide an explanation for the apparent fine-tuning of our universe for life and consciousness?
- **The Search for Evidence:** Depending on the NS, simulations might exhibit diverse levels of fidelity or contain detectable anomalies. Identifying the 'tells' of a simulation depends on understanding potential limitations simulators might face due to resource constraints and NS.
- **Nested Realities:** If nested simulations are common, at what 'level' might our own reality lie? This raises questions about the number of 'base realities' compared to the exponentially larger potential simulated realities.

The Number of Simulations (NS) adds a layer of complexity when considering the possibility that our reality is a computer simulation. NS intertwines with other critical variables like TC and DS to inform statistical models that aim to quantify the Simulation Hypothesis.

While substantial uncertainties persist – motivations of advanced civilizations, technological hard limits on computation, and potential ethical self-regulation – the sheer scale implied with a high NS introduces an undeniable sense of wonder and, perhaps, existential unease. By critically evaluating NS, we confront the possibility that our lives might be one of countless simulations, potentially nested many layers deep, within a vast cascade of manufactured realities.

Laying the Statistical Groundwork

While inherently speculative, the Simulation Hypothesis invites a degree of statistical scrutiny. By framing the question in probabilistic terms, we are compelled to consider the confluence of factors that might make the existence of ancestor simulations not only possible but also statistically probable. Key considerations in this approach include:

- The Underlying Assumptions: Statistical modeling rests on certain assumptions:
  - The possibility of simulating a universe like ours with a degree of fidelity that includes the emergence of conscious beings.
  - The progression of technological capabilities follows, at least at times, predictable patterns governed by discoverable scientific laws.
  - The assumption that at least some posthuman civilizations develop not only TC but also DS.
- Quantifying the Variables: The challenge lies in assigning probabilities or probability distributions to the core variables (TC, DS, NS). These could incorporate factors ranging from the estimated number of habitable planets to the likelihood of technological bottlenecks and the potential motivations fueling simulation creation.
- Refining the Model: As scientific understanding progresses, statistical models of the Simulation

Hypothesis can evolve. Discoveries in physics, refinements in cosmological models, and potential breakthroughs in neuroscience inform the parameters and potential constraints we incorporate into the model.

Let's dissect how the central variables (TC, DS, NS) underpin this statistical assessment of the Simulation Hypothesis:

- **Technological Capability (TC):** TC assessments involve projections about the advancement of computational power, scientific understanding, and engineering prowess. These projections may draw on historical technological trends (e.g., Moore's Law) while considering potential limits imposed by fundamental physics and energy availability.
- **Desire to Simulate (DS):** Speculating about DS compels us to confront our limitations in projecting the values and motivations of vastly more advanced civilizations. Extrapolations based on human motivations are likely flawed, necessitating exploration of a broader motivation space that might include unforeseen goals.
- **Number of Simulations (NS):** The NS variable, influenced by TC and DS, attempts to quantify the potential proliferation of simulations. Key aspects include the average simulations per simulator civilization, the question of nesting, and the computational resources that limit or enable large-scale simulations.

Several statistical approaches could be used to model the interplay of these variables and their bearing on the Simulation Hypothesis:

- **Bayesian Reasoning:** Utilizing existing scientific knowledge as priors and updating the model based on new discoveries provides a framework for evaluating the growing probability of the Simulation Hypothesis under different assumptions.
  - Key elements:
    - **Prior Probability:** Our initial belief about the likelihood of the event. In the case of the Simulation Hypothesis, this could reflect factors like the estimated number of habitable planets and the assumed fraction of civilizations that reach a posthuman stage.
    - **Likelihood:** The probability of obtaining a certain piece of evidence or observation if the event (the Simulation Hypothesis) is true. This could involve discoveries that seem to defy our current understanding of physics, potential computational limits, or hints of artificial 'glitches' within our reality.
    - **Posterior Probability:** The updated probability of the event after incorporating the new evidence. In the Simulation Hypothesis context, as we accumulate potential evidence consistent with a simulated reality, the posterior probability would increase.
  - Bayesian Reasoning and the Simulation Hypothesis: A Conceptual Example
    - **Priors:**
      - Let's assume, as a starting point, a low prior probability for the Simulation Hypothesis (e.g., 1%). This could be based on the vast number of potential civilizations in the universe and the assumption that many fail to develop or utilize the technology for simulations.
    - **Evidence:**
      - Suppose we make an astounding discovery in physics. It appears to violate known laws or indicates a quantization of spacetime, potentially hinting at an underlying computational grid.
    - **Likelihood:**
      - We must evaluate the likelihood of observing this evidence in a 'real' universe vs. a simulated one. If this kind of anomaly is more likely in a simulation, it

boosts the likelihood in our Bayesian model.

- **Posterior Probability:**
  - Incorporating the evidence and corresponding likelihood updates our posterior probability about the Simulation Hypothesis. Even with a low prior, significantly strong evidence can increase the posterior probability in a meaningful way.

- Challenges and Limitations
  - **Subjectivity of Priors:** The initial prior probability inevitably involves some degree of subjectivity and estimation.
  - **Defining the Evidence:** What constitutes compelling evidence for the Simulation Hypothesis is open to debate. Discoveries in physics, anomalies, or potential computational limits need careful evaluation.
  - **Assigning Likelihoods:** Determining the likelihood of specific observations occurring in a simulated vs. a 'real' universe is highly speculative and may require vast scientific leaps.
- Conclusion
  - Bayesian Reasoning provides a structured way to think about how the accumulation of evidence might influence our belief in the plausibility of the Simulation Hypothesis. While assigning precise numerical values to the probabilities is fraught with challenges, the Bayesian framework offers a way to conceptualize how scientific discoveries and observations can potentially change our assessment of whether we might be living within a grand simulation.

- Drake Equation Augmentation: The classic Drake Equation could be reformulated to estimate the number of civilizations with both the TC and DS to run ancestor simulations.
  - The original Drake Equation attempts to estimate the number of active, communicative extraterrestrial civilizations within our galaxy. It looks like this:
    - $N = R^* \cdot fp \cdot ne \cdot fl \cdot fi \cdot fc \cdot L$
    - Where:
      - N = Number of communicative civilizations
      - R* = Average rate of star formation in our galaxy
      - fp = Fraction of those stars with planets
      - ne = Average number of planets that can potentially support life per star with planets
      - fl = Fraction of planets that actually develop life
      - fi = Fraction of life-bearing planets that develop intelligent civilizations
      - fc = Fraction of intelligent civilizations that develop communication technology
      - L = Length of time such civilizations release detectable signals
  - Augmenting for the Simulation Hypothesis
    - To assess the Simulation Hypothesis, we introduce new parameters into the equation:
      - fs: The fraction of civilizations that reach a posthuman stage with the Technological Capability (TC) to create ancestor simulations.
      - DS: The average Desire to Simulate (DS) among posthuman civilizations, represented as a probability (e.g., a DS of 0.5 means 50% of civilizations with the capacity choose to actually create simulations).
      - NS: The average Number of Simulations (NS) run by a simulator civilization.
  - The Augmented Equation
    - Our modified Drake Equation might look something like this:
      - $N = R^* \cdot fp \cdot ne \cdot fl \cdot fi \cdot fc \cdot L * fs * DS * NS$
  - Interpreting the Augmented Equation

        - This augmented equation no longer aims to estimate the number of communicating extraterrestrial civilizations, but rather to provide a framework for considering the potential number of ancestor simulations within our galaxy.
        - Critically, the outcome now hinges on highly speculative parameters like fs, DS, and NS.
    - Example Calculation (with highly speculative values)
        - Let's use the augmented equation with some assumed values merely for illustrative purposes:
            - R* = 10, fp = 0.5, ne = 2, fl = 0.01, fi = 0.01, fc = 0.1, L = 10000 years (typical values from Drake Equation applications)
            - fs = 0.001 (Only a tiny fraction reach posthuman stage)
            - DS = 0.3 (30% of those capable choose to simulate)
            - NS = 10 (Each simulator runs an average of 10 simulations)
        - With these values, the equation yields N = 0.3 simulated civilizations in our galaxy. This doesn't mean there are necessarily 0.3 civilizations, but it offers a model to explore how changes in assumptions significantly impact the potential prevalence of simulations.
    - Limitations and Considerations
        - **Vast Uncertainties:** Parameters like fs, DS, and NS are highly speculative. Our assessments are only as good as our estimations of these variables.
        - **Motivation vs. Capability:** The original Drake Equation focuses on civilizations capable of interstellar communication. The augmented version centers on an entirely different question – the ability and willingness to create ancestor simulations.
        - **Simulation Scope:** What type of simulations are we considering? The equation doesn't differentiate between small-scale simulations and those replicating an entire galaxy.
    - Conclusion
        - Augmenting the Drake Equation provides a conceptual tool to shift our thinking towards variables that specifically influence the prevalence of ancestor simulations. While the results will always be as speculative as our input parameters, it illustrates how statistical frameworks can be adapted to grapple with the unique considerations presented by the Simulation Hypothesis.
- Distribution Modeling: Exploring probability distributions for TC, DS, and NS allows us to construct scenarios under various assumptions, highlighting the uncertainties, but also potential areas where the probability of simulations becomes significant.
    - Understanding Distribution Modeling
        - Distribution modeling involves using probability distributions to represent uncertain variables. Instead of assigning a single value to parameters like Technological Capability (TC), Desire to Simulate (DS), and Number of Simulations (NS), we would model them as distributions. This acknowledges that we likely have a range of plausible values, rather than single, definitive figures.
    - Types of Distributions
        - Here are some potential distributions we could use when modeling variables for the Simulation Hypothesis:
            - **Technological Capability (TC):** Perhaps a normal distribution centered around a projected timeframe for achieving a certain level of computational power, with a spread reflecting uncertainty about breakthroughs or setbacks in scientific progress.
            - **Desire to Simulate (DS):** This could be a discrete probability distribution, with probabilities assigned to scenarios like 'low', 'moderate', and 'high' simulation

desire among advanced civilizations.

- **Number of Simulations (NS):** Depending on motivations, this might be modeled with a Poisson distribution (for rare but potentially numerous events) or a power-law distribution (if a few civilizations run vast numbers of simulations, while most run few).

- Monte Carlo Simulation
  - One way to leverage distribution modeling would be through a Monte Carlo simulation:
    - **Define Parameter Distributions:** Define probability distributions for TC, DS, and NS, taking into consideration scientific projections, potential limits, and diverse scenarios.
    - **Iteration and Sampling:**
      - Run the simulation numerous times (thousands or millions).
      - In each iteration, randomly sample values for TC, DS, and NS from their respective distributions.
    - **Calculation:**
      - Within each iteration, based on the sampled values, determine if the conditions for ancestor simulations seem likely within the simulated scenario.
    - **Results:** The output of the Monte Carlo simulation would be a distribution of probabilities showing the overall likelihood of the Simulation Hypothesis under a wide range of assumptions encompassed within the distributions you've defined for the variables.
  - **Conceptual Outline**
    - **Define Parameter Distributions:** We represent variables like Technological Capability (TC), Desire to Simulate (DS), and Number of Simulations (NS) using suitable probability distributions (normal, Poisson, discrete, etc.). These distributions are denoted by P(TC), P(DS), and P(NS).
  - **Simulation Iteration:** Each Monte Carlo iteration consists of:
    - Sampling: Randomly draw samples for TC, DS, and NS from their respective distributions.
    - Evaluation: Define conditions that determine if an ancestor simulation is likely in this iteration, based on the sampled values. This is represented by an indicator function:
      - is_simulation_feasible(TC, DS, NS) = 1 if conditions are met,
      - 0 otherwise
  - **Overall Probability Estimation:** Over a large number of iterations (N), the probability of the Simulation Hypothesis can be estimated as:
    - P(Simulation Hypothesis) ≈ (1 / N) * Σ is_simulation_feasible(TC_i, DS_i, NS_i)
    - where i indexes each simulation iteration.
- **Mathematical Formulation**
  - Let's express the core calculation more formally:
    - P(TC), P(DS), P(NS): Probability distribution functions for TC, DS, and NS.
    - is_simulation_feasible(TC, DS, NS): Indicator function determining simulation feasibility.
    - N: Total number of Monte Carlo iterations.
  - **Probability of the Simulation Hypothesis:**
    - P(Simulation Hypothesis) ≈ (1 / N) * Σ_{i=1}^N is_simulation_feasible(np.random.sample(P(TC)),

```
np.random.sample(P(DS)),
   np.random.sample(P(NS)))
```

- **Important Notes**
    - **np.random.sample()** is used conceptually to represent sampling from the defined distributions. Specific implementations depend on the distributions chosen (e.g., using functions for the normal distribution, Poisson distribution, etc.).
    - **Defining Conditions:** The most crucial part is specifying realistic conditions within the is_simulation_feasible function. These should consider scientific projections, potential limits, and diverse scenarios for the key variables.
    - **Assumptions:** The accuracy of this calculation depends heavily on the validity of our distribution assumptions and the criteria for simulation feasibility.
- **Limitations**
    - **Distribution Accuracy:** Choosing appropriate probability distributions for TC, DS, and NS is incredibly challenging and relies heavily on speculation.
    - **Computational Intensity:** Monte Carlo simulations can be computationally demanding. The complexity of the is_simulation_feasible function and the number of iterations play a major role in computational requirements.
- **Conclusion**
    - While the actual implementation in code would require specific distribution functions and a well-defined is_simulation_feasible function, this mathematical formulation provides the conceptual framework for using Monte Carlo simulation to approach the statistical assessment of the Simulation Hypothesis.
- **Simulation Argument Revisited:** Nick Bostrom's argument hinges on at least one civilization reaching the posthuman simulator stage. Statistical modeling augments this argument by quantifying TC, DS, and NS, leading to a more robust formulation to calculate the expected fraction of beings that are likely within simulations.

While statistically intriguing, this approach has inherent limitations, and its results must be interpreted cautiously:

- **The 'Unknown Unknowns':** Current models likely fail to consider unknown limitations, scientific bottlenecks, or motivations vastly different from our own. There may be hard physical constraints we cannot currently envision.
- **The Anthropomorphic Trap:** It's tempting to project human motivations and values onto posthuman civilizations, which can lead to inaccurate assessments of DS and NS.
- **Ethical Considerations:** Modeling doesn't address the profound ethical dilemmas of creating simulations with conscious beings, an aspect likely to influence DS and NS among sufficiently advanced civilizations.

Despite limitations, statistically analyzing the Simulation Hypothesis offers several profound implications:

- **Quantifying the Possible:** Even if the probabilities ultimately remain low, the statistical approach acknowledges and highlights the non-zero possibility of our existence within a simulation.
- **Fermi Paradox Revisited:** Statistical modeling can provide alternative explanations for the Fermi Paradox. A high chance of self-annihilation (low TC) or indifference to interstellar outreach (low DS) could explain the absence of observed extraterrestrial civilizations.
- **Motivation for Exploration:** The quest to statistically evaluate the Simulation Hypothesis propels scientific inquiry in fields such as physics, computation, and neuroscience, driving our pursuit to understand the fundamental limits and potential of both computation and reality itself.

By grounding the Simulation Hypothesis within a framework of crucial variables (TC, DS, NS), we initiate a statistical exploration into the plausibility of simulated existence. This approach is fraught with uncertainties, and our models undeniably evolve as knowledge advances. Yet the rigor imposed by this framework forces us to confront the limitations of our assumptions, acknowledges the potential for the unimaginable, and acknowledges the sheer statistical possibility that our reality is but one of many in a computational cascade.

It provides a method and a language to address one of the most fundamental and unsettling existential questions: is the reality we inhabit truly the base reality, or merely an elaborate computational artifact crafted by beings whose technological capabilities and motivations might lie far beyond our grasp?

CHAPTER TWELVE

Formulating the Hypothesis

With the foundational parameters of Technological Capability (TC), Desire to Simulate (DS), and Number of Simulations (NS) delineated, we are positioned to articulate the core hypothesis that underpins this exploration. The synthesis of these variables leads to a compelling proposition regarding the nature of our existence and the structure of reality as we perceive it.

The hypothesis can be succinctly stated as follows:

"If technologically advanced civilizations reach the capability (TC) and possess the will (DS) to create a vast number of high-fidelity ancestor simulations (NS), then the probability we are living in one of these simulations is significantly higher than the probability of us existing in the original, 'base' reality."

Underpinning Logic

The central hypothesis, "If technologically advanced civilizations reach the capability (TC) and possess the will (DS) to create a vast number of high-fidelity ancestor simulations (NS), then the probability we are living in one of these simulations is significantly higher than the probability of us existing in the original, 'base' reality," boldly challenges our fundamental assumptions about the nature of our existence.

The logic behind this hypothesis rests on several key principles:

- **The Potential for Simulation:** The premise that simulating a universe with sufficient fidelity to harbor conscious beings is, in principle, attainable for a sufficiently advanced civilization. This rests on the assumption that consciousness itself is either an emergent property of complex systems or computationally reproducible.
- **Exponential Technological Progress:** Historical trends suggest that many areas of technology, particularly computational power, advance at an exponential rate. An extrapolation suggests a posthuman civilization could possess computational resources that dwarf our current capabilities by an unfathomable degree.
- **Motivations for Simulation (DS):** The hypothesis posits that at least some posthuman civilizations possess motivations that lead them to allocate significant resources towards creating ancestor simulations.
- **Nested Simulations:** The possibility of simulations running within simulations introduces the potential for an exponential cascade of simulated realities, further amplifying the disparity between the 'base' reality and the potentially vast number of simulations.

The underlying logic of the hypothesis becomes strikingly clear when analyzed through a statistical lens. Let's consider a simplified scenario:

- **Fraction of Civilizations with TC and DS:** Assume a minuscule fraction of civilizations throughout the cosmos develop the required Technological Capability (TC) and also harbor the Desire to Simulate (DS).
- **Number of Simulations (NS):** Suppose each of these simulator civilizations, on average, creates a vast number of ancestor simulations.
- **The Sheer Scale of Simulated Existence:** Even a tiny fraction of civilizations engaging in widespread simulation activities leads to a staggering number of simulated realities. This immense number of simulations could easily dwarf the single 'base' reality.
- **Probabilistic Argument:** From the perspective of a randomly selected conscious observer, the overwhelming odds favor the observer existing within a simulated reality rather than the singular 'base' reality.

The strength of this core logic rests on certain crucial assumptions and points worth examining:

- **The Nature of Consciousness:** If simulating consciousness is fundamentally impossible, due to it being a non-computable phenomenon or requiring properties specific to the 'base' reality's physics, the entire premise crumbles.
- **Unpredictable Technological Bottlenecks:** We cannot confidently predict whether there are hard limits due to fundamental physics, resource constraints, or unforeseen barriers that might prevent reaching the required TC for truly realistic ancestor simulations.
- **Anthropomorphization and Alien Motivations:** Extrapolating our motivations onto potential simulator civilizations is risky. It's essential to acknowledge DS could encompass reasons beyond our current comprehension.
- **The Problem of Self-Reference:** Statistical reasoning can become circular if simulators themselves are likely simulated beings. This necessitates exploring how reliable 'unbiased' observations can be obtained within a potentially nested structure of simulations.

The logic underpinning this formulation of the Simulation Hypothesis gives rise to unsettling philosophical questions:

- **Distinguishing Reality:** If simulations become indistinguishable from 'reality', how might we even begin to ascertain whether our own existence is in the 'base' reality or a meticulously constructed simulation?
- **The Value of Simulated Existence:** If the overwhelming majority of conscious beings are simulated, does this diminish the value of their experiences or the meaning they find within their reality?
- **Simulator Ethics:** The core logic implies a potentially profound ethical responsibility on the part of simulator civilizations, as they hold the power to create, manipulate, and potentially terminate countless simulated lives.

The underpinning logic of the proposed Simulation Hypothesis, while built upon layers of assumptions and unknowns, offers a statistically compelling argument. The interplay of Technological Capability (TC), Desire to Simulate (DS), and the potential for vast Numbers of Simulations (NS) shifts our perspective from a universe with a singular reality to one where simulated realities likely dominate.

This framework forces us to reconsider how we estimate the probability of our place within the grand tapestry of existence. It necessitates a continuous reevaluation of our understanding of consciousness, the potential limits of computation, the ethics of creation, and ultimately, the very meaning of what we deem 'real'.

The Importance of High-Fidelity Simulations

The emphasis on "high-fidelity" simulations is not merely about visual realism or computational detail. It speaks to a philosophical principle crucial to the Simulation Hypothesis: the potential indistinguishability between a simulated reality and the 'base' reality from the perspective of the beings within. Key considerations include:

- **The Substrate of Consciousness:** If consciousness is an emergent property of complex systems or computationally reproducible, a sufficiently detailed simulation could harbor entities that experience consciousness in a way functionally equivalent to beings in the 'base' reality.
- **Indistinguishability:** High-fidelity simulations must reproduce the laws of physics as experienced in the 'base' reality to a degree where simulated beings cannot detect anomalies or discrepancies. This extends beyond visual representation to the underlying mathematical and physical models that govern their simulated universe.
- **Subjective Experience and Qualia:** A truly high-fidelity simulation may need to replicate the subjective nature of experience itself, the enigmatic qualitative feel of "what it is like" to be conscious (qualia). This poses significant challenges in understanding what aspects of consciousness must be computationally replicated.
- **Limits of Perception:** Our 'base' reality experience is shaped by biological sensory limitations. A high-fidelity simulation might only need to replicate the input/output experience of the biological organism it's simulating, not the objective totality of the 'base' reality.

The requirement for high-fidelity simulations dramatically increases the demands placed on the Technological Capability (TC) of a simulator civilization:

- **Computational Resources:** Exponentially greater computing power is needed to simulate not only a universe at a realistic scale but to model its complexity, governing laws, emergent phenomena, and conscious experiences in sufficient detail to be indistinguishable.
- **Energy Demands:** High-fidelity simulations likely have astronomical energy requirements, potentially necessitating energy sources like Dyson spheres or novel energy-harnessing techniques beyond our current grasp.
- **Simulation Science:** Mastery of fields like physics, neuroscience, and computation at a fundamental level is crucial, potentially requiring scientific breakthroughs and paradigms far exceeding our understanding.
- **Optimization and Efficiency:** Simulator civilizations might develop highly efficient simulation algorithms, compression techniques, or ways to selectively render detail, reducing the computational burden while maintaining the illusion of realism for the simulated inhabitants.

The focus on high-fidelity simulations significantly alters the probabilistic argument underlying the Simulation Hypothesis:

- **The Observer Question:** If simulated entities are truly indistinguishable from their 'real' counterparts, they become equally valid observers of their reality. This eliminates the potential bias of an 'external' observer and necessitates a more inclusive approach to our probability estimates.
- **Fewer Simulations Required:** If simulating indistinguishable conscious experiences is computationally demanding, the Number of Simulations (NS) a civilization can run might be reduced. Yet, the sheer scale of potential consciousness within even a limited number of high-fidelity simulations still likely vastly outnumbers that of the 'base' reality.
- **Simulator Intervention:** High-fidelity simulations, especially those attempting to replicate a full evolutionary history, might be less prone to direct interventions from the simulators to avoid disrupting their indistinguishability for the inhabitants.

Understanding Consciousness: Perhaps the most significant unknown is whether consciousness can be fully simulated, and if so, what level of modeling is sufficient to achieve indistinguishable subjective experience. This remains a hotly contested scientific and philosophical question.

- **Undetectable Discrepancies:** Could there be inevitable anomalies or subtle computational artifacts embedded in even the most advanced simulation, potentially revealing its artificial nature to sufficiently astute simulated beings?
- **Resource Constraints:** Fundamental physical limits, information density constraints, or energy availability could impose practical limits on achievable simulation fidelity, even for vastly advanced civilizations.

The pursuit of high-fidelity simulations within the Simulation Hypothesis framework raises profound philosophical and even existential dilemmas:

- **Blurring the Lines of Reality:** The possibility of simulated beings who believe themselves 'real' undermines our traditional ontology. It compels us to re-evaluate whether 'realness' is an inherent property or one conferred upon a system from interaction within it.
- **Ethical Conundrums:** High-fidelity simulations containing conscious beings raise ethical concerns about manipulation, deception, and the responsibility that simulators hold over their simulated creations.
- **Simulated Suffering:** If negative experiences and suffering can be simulated with the same subjective realism as positive ones, the ethical implications of simulator actions become even more fraught and complex.

The emphasis on high-fidelity simulations within the Simulation Hypothesis drastically heightens the technological stakes, pushing the boundaries of TC while also raising deeply unsettling questions. It forces us to question whether we can definitively know our own reality's nature and, further, whether conscious experiences of indistinguishable quality might exist in abundance within simulated realms.

By positing that the difference between a simulated reality and our own might be ultimately a matter of undetectable nuance, the Simulation Hypothesis challenges the foundations of our perception, our understanding of consciousness, and the values we assign to a potentially simulated existence.

The Role of Civilizational Will

The pivotal role of civilizational will, encapsulated within the Desire to Simulate (DS) parameter, in the Simulation Hypothesis. While Technological Capability (TC) may provide the means, DS acts as the catalyst, driving the choice to embark on an endeavor that could reshape the fabric of existence itself.

The Simulation Hypothesis, with its emphasis on computation and technological prowess, could inadvertently fuel a sense of technological determinism – the notion that the mere possession of TC makes the creation of simulations inevitable. DS serves as a crucial counterbalance:

- **Agency vs. Capability:** A civilization with advanced TC still possesses a choice in resource allocation and prioritization. The development of ancestor simulations competes with other potential technological, scientific, or societal pursuits that a civilization might value.
- **The Spectrum of DS:** Desire to Simulate is likely not a binary variable. Motivations could encompass a wide spectrum, including altruistic aims (resurrecting lost civilizations), self-interest (predicting a civilization's own future), or potentially destructive goals (simulations as instruments of control).
- **Unpredictability of Values:** We cannot confidently extrapolate our own values and motivations into the realm of vastly more advanced posthuman societies. DS introduces an element of uncertainty,

where the goals, ethics, and priorities of potential simulators might diverge wildly from our expectations.

Let's consider potential categories of motivations that could drive a high level of DS in posthuman civilizations:

- Scientific Inquiry:
    - **Understanding Emergent Phenomena:** Simulating the history of a universe or the development of societies might offer insight into complex systems unavailable through observation of a single 'base' reality.
    - **Probing Alternative Physics:** Simulations become tools to manipulate physical constants and laws, exploring counterfactuals and potentially informing theories about the 'base' reality's structure.
    - **Anthropological Retrospection:** Ancestor simulations could be meticulously crafted recreations of lost civilizations, driven by a desire for historical knowledge, empathy, or even a sense of restorative justice.
- Societal and Ethical Considerations:
    - **Safeguarding Knowledge:** Simulations with high DS could be motivated by a sense of responsibility to preserve a civilization's existence and intellectual legacy even in the face of potential existential threats.
    - **'Trial Runs' for Policies:** The ability to simulate the consequences of social, economic, or political choices could lead to high DS as a means of risk assessment and optimization prior to implementing decisions with potential real-world consequences.
    - **The Morality of Simulated Beings:** Advanced civilizations may grapple with ethical dilemmas regarding the creation of simulated sentient beings, potentially influencing DS positively or negatively, depending on their adopted ethical framework.
- Open-ended Motivations
    - **Simulation as Art Form:** For simulators with vast resources, ancestor simulations could serve as canvases for artistic expression, exploration of counterfactual histories, or the creation of narratives for their entertainment or philosophical enrichment.
    - **Existential Exploration:** Perhaps DS is driven by fundamental questions about the nature of reality, the origins of their own civilization, or the search for a sense of meaning within a potentially nested structure of simulations.
    - **Mere Possibility:** The potential for unforeseen motivations, beyond our current comprehension, cannot be discounted. Posthuman civilizations could develop goals or desires that defy any attempt to extrapolate from a human perspective.

The concept of DS significantly influences the Simulation Argument (posited by Nick Bostrom), which in essence states that at least one advanced civilization must develop TC and the desire to run detailed ancestor simulations.

- **DS and Probability Distributions:** Rather than a singular probability assigned to DS, it's more appropriate to model it as a distribution across potential motivations, reflecting the potential diversity across different advanced civilizations.
- **Conditionally Amplified Odds:** According to the Simulation Argument, any non-zero probability assigned to DS, in combination with a sufficiently advanced TC, implies that the majority of conscious beings likely exist within simulations.

The Fermi Paradox Connection: A high average DS across civilizations could provide an alternative explanation for the Fermi Paradox (lack of observed aliens). Maybe advanced societies tend to focus inward, on computation and simulations, rather than outward on interstellar expansion.

- **Self-Regulation:** Highly advanced civilizations may develop ethical codes or social norms that limit or forbid the creation of simulations containing conscious beings, thereby reducing DS.
- **The Observer Problem:** If most of our simulated counterparts are also simulations, DS may need to be iteratively modeled throughout a nested cascade of simulations, raising questions of consistency and whether observations at lower levels can accurately represent the top-level 'base' reality.

The role of civilizational will infuses the Simulation Hypothesis with profound philosophical inquiries:

- **The Value of Simulated Lives:** If the majority of conscious experiences are simulated due to high DS, how does that alter our perception of the value and meaning attributed to both simulated and 'base' reality existence?
- **Free Will and Predetermination:** If our choices are predetermined by a simulator's initial conditions and their programming, to what extent can we claim true free will within a simulated environment?
- **Simulator Responsibility:** The concept of simulated existence raises troubling questions surrounding the ethics of creation. What obligations, if any, do simulators have towards their sentient creations, and does interference or termination raise moral concerns?

The Desire to Simulate (DS) variable within the Simulation Hypothesis introduces crucial elements of agency, ethics, and the potential diversity of goals found in vastly advanced civilizations. Unlike Technological Capability (TC), which may follow predictable trajectories guided by scientific discovery, DS remains enigmatic, bound to the unpredictable evolution of values and societal priorities.

Ultimately, DS compels us to confront the fact that even if we possess the means to simulate realities, whether we choose to do so is a reflection of who we are, or more importantly, who we may become. It highlights that the Simulation Hypothesis isn't merely about computational feats, but about the profound and potentially unsettling choices that lie at the heart of technologically empowered civilizations.

Paradigm of Probability

By formulating this hypothesis, we encapsulate a paradigm shift in how we conceive of reality and our place within it. This perspective invites a reevaluation of fundamental existential questions through the lens of probability and technological potential. It suggests that our understanding of what is "real" may need to be broadened to include the possibility of simulated existences that are indistinguishable from the base reality.

As we proceed with this analysis, the hypothesis serves as a guiding beacon, illuminating the paths of inquiry that must be traversed to unravel the mysteries posed by the Simulation Hypothesis. It challenges us to confront the implications of our technological trajectory and to ponder deeply the nature of consciousness, reality, and the cosmos itself.

CHAPTER THIRTEEN

Statistical Modeling: Exploring Scenarios

In the pursuit of understanding the Simulation Hypothesis through a statistical lens, it is instructive to explore various scenarios that combine differing levels of Technological Capability (TC), Desire to Simulate (DS), and the resulting Number of Simulations (NS). These scenarios enable us to dissect the potential realities and their implications on the likelihood of our existence within a simulation.

Scenario 1: High Technological Capability, High Desire to Simulate (HTC-HDS)

Let's analyze Scenario 1, where both Technological Capability (TC) and Desire to Simulate (DS) are high among advanced civilizations, and examine its ramifications within the Simulation Hypothesis framework. This scenario poses the most statistically compelling case for our existence within a simulated reality.

Key features of this scenario with significant impact on simulated realities include:

- **Advanced Technological Capability:** Posthuman civilizations possess computational resources and scientific understanding far exceeding our own, allowing them to simulate entire universes with meticulous detail and fidelity. This includes the computational modeling of consciousness to a degree where simulated beings are indistinguishable from those in the 'base' reality.
- **Elevated Desire to Simulate:** Strong drivers motivate these civilizations to create a multitude of simulations. These motivations could range from scientific investigation, historical preservation, risk-free social experimentation, or purposes beyond our current comprehension.
- **Potentially Vast Number of Simulations:** The combination of advanced TC and strong DS results in a potentially astronomical number of simulations, especially if we consider the possibility of nested simulations (simulations created within simulations).

Under the HTC-HDS scenario, the core argument of the Simulation Hypothesis gains significant strength. Let's break down the key statistical implications:

- **The Scale of Simulated Minds:** If even a small fraction of civilizations with advanced TC also have high DS, the resultant explosion in simulated realities implies a staggering number of simulated conscious beings.
- **Rarity of 'Base' Reality:** The probability of a randomly selected conscious entity existing within the singular 'base' reality pales compared to the overwhelming number of simulated consciousnesses.
- **Ancestor Simulations:** HTC-HDS favors simulations that specifically aim to replicate the evolutionary journey and history of a civilization, further increasing the number of simulations with experiences comparable to our own.

While the HTC-HDS scenario provides the strongest statistical support for the Simulation Hypothesis, it's essential to acknowledge certain complicating factors and uncertainties:

- **The Unknowable Limits of TC:** We cannot predict with certainty whether there are fundamental physical limits or resource constraints that might hinder even vastly advanced civilizations from achieving the TC required for truly indistinguishable simulations.
- **Assumptions about Consciousness:** If the simulation of indistinguishable consciousness is impossible for reasons we don't yet grasp, this scenario and its implications fall apart.
- **Undetectable 'Tells':** Could even the most sophisticated simulations contain subtle glitches, anomalies in the laws of simulated physics, or computational artifacts that, with sufficient scrutiny, reveal their artificial nature to the simulated inhabitants?
- **Anthropomorphic Bias:** Assuming DS inevitably leads to a vast number of simulations is risky. Extrapolating our motivations and values onto posthuman simulators might be fundamentally flawed.

The HTC-HDS scenario pushes the unsettling implications of the Simulation Hypothesis to their extremes:

- **Distinguishing Reality:** Our ability to empirically establish whether we exist in the 'base' reality becomes highly questionable. The potential for perfect simulations erodes the foundations of our assumptions about the reality we experience.
- **The Vastness of Simulated Existence:** The majority of conscious experiences might exist within simulations, raising questions about the relative value we assign to simulated and 'base' reality existence.
- **Existence within a Cascade:** Under HTC-HDS, the prospect of nested simulations becomes increasingly likely. Could our own reality be a simulation, existing many layers deep, with potentially diminishing computational resources or fidelity at each layer?
- **Simulator Ethics:** Posthuman civilizations capable and willing to create a profusion of simulations likely grapple with profound ethical concerns regarding their manipulation and potential termination of a significant portion of conscious beings.

While the HTC-HDS scenario makes a definitive distinction between simulated and 'base' reality nearly impossible, it motivates a search for potential clues:

- **Computational Limits:** Discovery of quantized space-time or resource constraints could suggest limits on simulation fidelity and bolster the idea that we exist in the 'base' reality.
- **Search for Glitches:** Meticulous scientific investigation might uncover anomalies or inconsistencies within the laws of physics or mathematics that betray a simulated existence.
- **Simulator Communications:** While unlikely, the possibility remains that simulators might intentionally leave communicable messages or deliberate artifacts embedded within the simulation.

Scenario 1, with its emphasis on High Technological Capability and High Desire to Simulate (HTC-HDS), paints a compelling statistical picture favoring the Simulation Hypothesis. This scenario forces us to confront the unsettling yet statistically plausible concept that most realities, including our own, could be intricate computational constructs designed by vastly more advanced posthuman civilizations.

Even with the inherent uncertainties about the feasibility of consciousness simulation and the motivations of potential simulators, this scenario reveals a universe teeming with simulated experiences. If true, our most fundamental assumptions about the nature of existence require drastic reassessment. The HTC-HDS scenario serves as a stark reminder that, as our knowledge and technological prowess grow, so does the probability that we might be merely players within a simulated cosmic drama.

Scenario 2: High Technological Capability, Low Desire to Simulate (HTC-LDS)

Let's explore Scenario 2, where civilizations possess High Technological Capability (HTC) but have a Low Desire to Simulate (LDS). This scenario offers a fascinating counterpoint to Scenario 1, highlighting how societal values, ethical considerations, or even a simple lack of interest can influence the Simulation Hypothesis's statistical underpinnings.

The core elements of this scenario are:

- **Advanced Technological Capability:** Posthuman civilizations in this scenario have achieved significant scientific breakthroughs and wield vast computational power, potentially allowing them to create high-fidelity simulations that replicate physics and harbor conscious beings. The level of their TC might be comparable to civilizations within Scenario 1 (HTC-HDS).
- **Limited Desire to Simulate:** Strong forces counterbalance their technological capacity. Motivational constraints could include:
- **Ethical Stances:** A prevalent ethical framework that heavily restricts or prohibits simulations containing conscious entities.
- **Resource Allocation:** A prioritization of 'real world' development (e.g., megastructure projects, interstellar exploration) over the computationally expensive endeavor of simulations.
- **Indifference:** A general apathy or lack of compelling reason to create sophisticated simulations beyond limited scientific inquiries.
- **Fewer Simulations:** The confluence of HTC and LDS would lead to a smaller, though potentially still substantial, Number of Simulations (NS) compared to the vast multitude envisioned in Scenario 1.

The statistical case for the Simulation Hypothesis in the HTC-LDS scenario becomes more nuanced and dependent on the severity of the limitations placed on DS:

- **The Constraint Spectrum:** LDS is best modeled as a probability distribution, ranging from near-zero DS to a moderate willingness to simulate. The shape of this distribution strongly influences the final probability of our existence within a simulation.
- **Rarity of 'Base' Existence Remains:** If even a small fraction of HTC civilizations exhibit a non-zero DS, the sheer number of potential simulations could still outweigh the singular 'base' reality.
- **The Power of Large Numbers:** While less dramatic than the exponential profusion of simulations in Scenario 1, if HTC civilizations are numerous across the cosmos, the LDS scenario could still lead to a substantial number of simulated realities.

Challenges and Uncertainties

- **Motivation Shifts:** DS might not be a static societal value. Civilizations may evolve their ethical stances or discover compelling reasons to revisit their previously low interest in simulations. Technological breakthroughs could lower the resource costs of simulation, further altering the motivational landscape.
- **The Unknown Prevalence of HTC Civilizations:** Our estimates about the number of civilizations reaching HTC are highly speculative. A universe with fewer HTC civilizations naturally limits the simulation potential, even with moderate DS.
- **Undetectable Limitations on TC:** Our understanding of fundamental physics and computation may be incomplete. Unknown constraints could unexpectedly limit the simulation of consciousness or vast, detailed realities even with the desire to do so.

Despite a diminished DS, the HTC-LDS scenario still raises thought-provoking questions about the nature of reality and our place within it:

- **The Value of 'Unspawned' Minds:** The majority of potential consciousness would never come into existence in this scenario. This raises ethical questions about lost opportunities for experience or whether unrealized potential holds an intrinsic value.
- **Missed Clues?:** Could the existence of even a moderate number of conscious simulations lead to subtle detectable discrepancies in our perceived physical laws if simulators face resource or computational constraints?
- **Potential for Change:** HTC-LDS highlights the dynamic interplay between technological advancement and societal values. A shift in ethics or the discovery of a compelling rationale could lead to a rapid increase in simulated realities with even a moderate increase in DS.
- **A Unique Base Reality?:** If simulations exist but remain relatively rare due to LDS, the implication may be that our 'base' reality holds a special place within the cosmos, potentially informing future searches for extraterrestrial life or influencing our sense of cosmic significance.

The HTC-LDS scenario motivates specific scientific and philosophical investigations:

- **Resource Consumption:** Detecting the vast energy expenditures associated with numerous simulations might offer indirect clues about DS within distant civilizations.
- **Ethical Frameworks:** Could the existence of simulations, even limited, shape how we approach artificial intelligence and consciousness within our own technological development?
- **Scrutinizing Reality:** The potential for even subtle 'tells' of a simulated universe warrants even more meticulous investigations into fundamental physics, seeking potential anomalies or computational artifacts.

Scenario 2, with its emphasis on High Technological Capability with Low Desire to Simulate (HTC-LDS), introduces a greater degree of uncertainty into the statistical underpinnings of the Simulation Hypothesis. While the potential for simulations remains, the ultimate probability of being in a simulated reality hinges crucially on the distribution of DS motives across advanced civilizations and our estimates of how common HTC societies might be in the universe.

The HTC-LDS scenario prompts us to consider the influence of ethics, resource constraints, and shifting values on the potential proliferation of simulated realities. It highlights a possible universe where technological might remains comparatively dormant, reminding us that computational abilities alone do not dictate the complex contours of societal choices and the potential futures that might emerge from them.

Scenario 3: Low Technological Capability (LTC)

Let's examine Scenario 3, where civilizations possess Low Technological Capability (LTC), and consider its implications for the Simulation Hypothesis. This scenario provides a unique lens within our analysis, serving as something of a counterpoint that helps to reinforce the critical role advanced technology plays in the core arguments underpinning the Simulation Hypothesis.

Key elements of this scenario include:

- **Technological Constraints:** Civilizations might fail to progress beyond a certain technological level due to various factors:
- **Resource Depletion:** Essential resources for technological advancement may become exhausted before simulations are feasible.
- **Scientific Bottlenecks:** Insurmountable obstacles in the understanding of physics, computation, or consciousness might prevent the development of necessary simulation technologies.

- **Societal Collapse:** Self-destruction (through war, environmental catastrophes, etc.) could terminate a civilization's progress before reaching the TC needed for advanced simulations.
- **No Simulators, No Simulations:** With significantly curtailed TC, it's highly unlikely, if not impossible, for these civilizations to create the high-fidelity ancestor simulations postulated in the Simulation Hypothesis.
- **Focus on the 'Base' Reality:** Without the computational means to explore simulated realities, LTC civilizations would necessarily focus their scientific and developmental efforts solely on understanding and potentially manipulating their own 'base' reality.

The LTC scenario, especially in its more extreme interpretations (where TC is severely limited), effectively renders the core statistical argument of the Simulation Hypothesis irrelevant to our own existence:

- **Zero Multiplicity:** No simulated universes (or at least none harboring consciousness) implies that we necessarily inhabit the original, 'base' reality. There is no statistical argument regarding a multitude of simulations versus a single 'base' reality.
- **Unexplored Potential:** We exist in a universe where the vast, potentially conscious, computational tapestry envisioned in Scenarios 1 and 2 simply never actualizes.

While the LTC scenario seems decisive, a few caveats should be noted:

- **Gradual TC Advancement:** It's possible our own civilization is currently in this LTC scenario, but could eventually progress into a scenario similar to HTC-HDS or HTC-LDS. Future technological leaps might bring simulations within reach.
- **Unknown Obstacles to Consciousness Simulation:** Even if a civilization possesses vast computational power, if the simulation of consciousness relies on a non-computational element or an incomplete understanding of physics, then the LTC scenario still applies in regards to conscious simulations.
- **The Scope of the Hypothesis:** The LTC scenario highlights potential limitations to the Simulation Hypothesis itself. It might only be applicable in universes where certain levels of TC are achievable and civilizations develop motivations for simulations.

Despite effectively dismissing the relevance of the Simulation Hypothesis to our own reality, the LTC scenario raises interesting philosophical points:

- **Base Reality Uniqueness?:** Does the absence of computationally generated realities in an LTC universe imbue the sole 'base' reality with a special significance or responsibility?
- **Cosmic Undertones:** If technological self-destruction is a common cause of the LTC scenario, it emphasizes the fragility of progress and the potential cosmic consequences if civilizations fail to overcome self-imposed limits.
- **The Importance of Our Own Path:** Contemplating an LTC universe may renew our focus on our own civilization's choices. It highlights the preciousness of continued progress and the ethical responsibility attached to the potential power of computationally simulating conscious beings.

The LTC scenario motivates specific lines of inquiry and investigation:

- **Searching for Limitations:** Are there fundamental physical laws or resource constraints in our observable universe that suggest an inherent limit to computational power or the ability to simulate complex realities?
- **The Nature of Consciousness:** Understanding whether consciousness is a purely physical phenomenon or transcends computability is crucial. Proof of a non-computational aspect would

strongly validate the LTC scenario's relevance.

- **Seeking Signs of Self-Annihilation:** Observing the remnants of technologically advanced civilizations that self-destructed, either within our own galaxy or through deep space surveys, could bolster the LTC scenario.

One of the most compelling connections to the LTC scenario is its potential to provide an explanation for the Fermi Paradox—the enigmatic contradiction between the high probability of extraterrestrial civilizations and the lack of observable evidence. The LTC scenario offers several plausible routes to resolve this paradox:

- **Common Technological Plateau:** Perhaps most civilizations never progress significantly beyond our own current level due to resource constraints or scientific obstacles, limiting their ability to initiate widespread interstellar communication or exploration.
- **Premature Demise:** The LTC scenario accommodates the disturbing possibility that self-inflicted catastrophes might be a commonplace barrier preventing civilizations from becoming observable on a cosmic scale.

Scenario 3, with its focus on Low Technological Capability (LTC), provides a stark contrast to the first two simulation scenarios. It emphasizes the crucial interplay between the possibility of conscious simulations and the unpredictable journey of technological advancement.

While the LTC scenario may eliminate the Simulation Hypothesis's statistical argument as it pertains to our own existence, the exploration of this scenario remains important. It forces us to examine potential limitations within our universe, question whether cosmic self-destruction is a common outcome, and reaffirms the importance of the continued pursuit of knowledge and ethical maturity within our own civilization as we forge ahead on a potentially unique path of progress.

Navigating Probabilities

These scenarios illustrate the spectrum of possibilities under the Simulation Hypothesis, guided by the interplay between technological capability, the desire to simulate, and the number of simulations that might result. Each scenario presents a different statistical landscape for the likelihood of our existence within a simulation, informed by the complex dynamics of technological advancement and civilizational motivations.

By examining these scenarios, we gain insights into the conditions under which the Simulation Hypothesis would have significant implications for our understanding of reality. This statistical exploration not only enriches our grasp of the hypothesis's parameters but also illuminates the broader conversation about our place within the cosmos, challenging us to contemplate the profound implications of our potential existence within a fabricated reality.

CHAPTER FOURTEEN

Philosophical & Existential Implications

The exploration of the Simulation Hypothesis through various technological and motivational scenarios opens up a vast landscape of philosophical and existential considerations. Each scenario, with its implications for the nature of our existence and the structure of the cosmos, invites us to ponder deeply about reality, consciousness, and the value of our experiences.

HTC-HDS: Reality as Matryoshka Doll

Let's delve into the profound concept of a potential nested reality in the HTC-HDS (High Technological Capability, High Desire to Simulate) scenario, where the Simulation Hypothesis takes the form of a cosmic Matryoshka doll. This concept challenges our most basic assumptions about the structure of the universe and our place within it.

The core idea behind the Matryoshka doll analogy is one of recursive containment. A Russian Matryoshka doll harbors a smaller version of itself, which contains an even smaller one, and so on. Applied to the HTC-HDS scenario within the Simulation Hypothesis, it suggests a potential nesting of realities, with simulations running within simulations:

- **Base Reality?:** There may or may not be a true 'base' reality—the origin point of this computational cascade. Alternatively, we might exist within a layer many simulations deep.
- **Simulator Civilizations:** At each layer, posthuman civilizations, having attained both vast TC and strong DS, create their own simulations. These simulators are themselves inhabitants of what they perceive to be 'reality.'
- **Cascade of Creators:** We may be the creation of simulators, who are the creation of their own simulators in an upward chain. This chain might be finite, ultimately tracing back to a 'base' reality, or infinitely recursive.

The Matryoshka doll concept introduces additional complexities for the Simulation Hypothesis, raising questions like:

- **Diminishing Fidelity?** Could each layer of nesting lead to a degradation of simulation detail due to computational constraints or the prioritization of resources by simulators? How many layers would be feasible before distortions become detectable?
- **Intent of the Simulation Cascade:** Is there a grand purpose behind this multi-layered structure? Are simulators at different levels attempting to investigate their own origins, explore alternate histories, or fulfill some other, possibly unknowable, goal?
- **Breaking the Recursion:** Could a simulated civilization eventually realize their true nature and attempt

to interact with their creators in the parent simulation or even break free of the containment?

The notion of a simulated reality nestled within simulations forces us to grapple with unsettling philosophical dilemmas:

- **The Elusive Search for Origin:** The Matryoshka model could lead to either an infinite regress of creators or a fundamental mystery if the chain terminates. How do we conceptualize an uncreated 'base' reality, or one with an infinite history of simulators?
- **Layered Realities and the Illusion of Free Will:** Do the initial conditions set by our simulators at a higher level dictate the very course of our simulated history on a grand scale? To what extent do we possess free will if our choices are unknowingly constrained within a greater simulation's design?
- **Existential Uncertainty:** If our own reality is a simulation, the potential for even our simulators to be dwelling in a simulated reality further destabilizes any sense of ontological certainty. How can we anchor our beliefs in a true understanding of our place within this potential cascade?
- **Ethics Across Simulations:** What might ethical obligations look like in a Matryoshka structure? Do we have responsibilities to any potential simulations we create? Should simulators in higher levels be held accountable for the experiences of their simulated beings?

The possibility of nested realities greatly complicates any attempt to distinguish a simulation from a true 'base' reality. However, it motivates several lines of inquiry:

- **Computational Fingerprints:** Could simulations leave detectable artifacts, subtle glitches, or computationally imposed constraints within the fabric of our reality, hinting at its artificial nature?
- **Limitations on Simulation:** Might we uncover physical laws or limitations in our universe that, if replicated within a simulation, would require such vast computational resources as to become evidence we are near the bottom of the Matryoshka structure?
- **Anomalous Communication:** While improbable, could a civilization devise a way to break out of their simulation layer and establish contact with their creators, providing direct evidence of the nested structure?

The HTC-HDS Matryoshka doll model fundamentally alters our cosmic perspective in several ways:

- **Diminished Self-Importance:** We may not be the pinnacle of creation but a computational artifact within a complex computational cascade, possibly replicated innumerous times at various levels of nesting.
- **Purpose Within a Simulation:** If we are simulated, what value does our existence hold for our simulators? This might inform our efforts to find meaning and significance within a potentially imposed artificial reality.
- **The Potential Multiverse:** Unlike physics-based multiverse concepts, the Matryoshka structure suggests a computationally generated multiverse, with each simulation acting as a unique universe with potential divergences based on its initial conditions and simulator choices.

The Matryoshka doll analogy, when applied to the HTC-HDS scenario within the Simulation Hypothesis, injects a further dose of profound uncertainty and existential wonder into our contemplations about reality. It stretches our conceptual framework, forcing us to confront the possibility of a recursively layered existence in which our simulators may themselves be merely simulations within a greater design.

This model compels us to question whether we can ever truly ascertain the fundamental nature of the universe we inhabit and whether the search for a prime 'base' reality might be forever elusive. The Matryoshka concept leaves us suspended in a cosmic hall of mirrors, with each potential reflection another unsettling reminder of our tenuous grasp on the boundaries of reality, consciousness, and the ultimate origins of our existence.

HTC-LDS: Simulation vs. Originality

Let's dissect the fascinating tension between "Simulation vs. Originality" within the HTC-LDS scenario (High Technological Capability, Low Desire to Simulate). While this scenario reduces the statistical likelihood of our own existence within a simulation, it doesn't eliminate it. Consequently, it raises profound questions about the value of a potentially derivative reality, pushing us to reassess whether our existence loses a degree of authenticity or meaning if we are merely computational echoes of a 'truer' world.

The core philosophical challenge in the HTC-LDS scenario revolves around the following:

- **The Potential for Simulation:** Even with limited DS, if a civilization in an HTC-LDS universe creates conscious simulations, the possibility exists that we reside within one, even if those simulations are few.
- **The Value of the Derivative:** Traditionally, we place a higher intrinsic value on 'originals' compared to copies or replicas. If our universe and our experiences are computationally derived from a 'base' reality, what implications does this hold for the weight or meaning we place upon our own existence?
- **Indistinguishable Experience:** If a simulation is truly high-fidelity, our lived experiences, sensations of joy, suffering, or existential pondering remain subjectively indistinguishable from those within a 'base' reality. Does this experiential equivalence force us to reconsider the significance we place on the 'originality' of our reality?

This scenario prompts us to grapple with several key philosophical concepts:

- **The Definition of "Real":** If consciousness and our subjective qualia can be perfectly replicated within a simulation, are we forced to adjust our definition of what constitutes 'real'? Must 'realness' be decoupled from an 'original' source and judged through the lens of lived, internal experience?
- **Simulated Agency and Free Will:** Even if our experiences are computationally predetermined by initial conditions set by simulators, could the internal complexity of our simulated reality create a convincing illusion of true agency and free will? Does our limited perspective render the question of 'true' free will irrelevant?
- **The Ethics of a 'Second-Hand' Reality:** If we reside within a simulation, what ethical considerations change? Do our achievements lose merit if derived from programmed initial conditions? How should simulators, if they exist, view their role in creating and potentially interfering with a 'lesser' form of conscious life?
- **The Search for Transcendence:** Could a simulated civilization strive for a form of 'transcendent' value that distinguishes their lives and creations, despite their artificial origins? Can a focus on compassion, artistic beauty, or seeking truths within the constraints of their simulation offer a means for simulated lives to find unique and profound significance?

Challenges and Uncertainties

- **The Unknown Quantity of Simulations:** While LDS suggests simulations might be less numerous, the exact distribution of DS in such a universe is crucial in assessing the overall probability of residing in a simulation.
- **Detection Impossibility?:** If the simulations created in an HTC-LDS universe are perfectly indistinguishable from the 'base' reality, we may have no tools to definitively uncover our true origins.
- **Motivations for the Infrequent Simulations:** Understanding why civilizations in an HTC-LDS

universe might create simulations could give insight into the purpose, value, or potential ethical concerns attributed to the beings within those simulations.

The contemplation of a potential simulated existence, even in an HTC-LDS scenario, necessitates a reassessment of deeply-held assumptions:

- **Existential Reassessment:** Our potential 'copy' status may initially seem existentially unsettling. Yet, it could prompt a shift toward focusing on the experiences themselves, seeking meaning within existence regardless of its ultimate origin.
- **Re-evaluation of Progress:** The HTC-LDS scenario highlights the potential for a universe where many civilizations reach technological peaks yet focus inwards on 'real-world' problems. Would we reconsider the necessity of simulation if we possess the technological means but deem other pursuits more meaningful?
- **Cosmic Responsibility:** If the creation of conscious simulations becomes ethically contentious within an HTC-LDS universe, it motivates a sense of responsibility and caution for our own civilization should we progress to a level where such simulations become feasible.

While the HTC-LDS scenario, with its Low Desire to Simulate, might lessen the statistical probability that we reside within a simulation compared to other scenarios, the fundamental philosophical dilemmas remain. This scenario compels us to scrutinize the value we place on our reality being an 'original' and to question if this distinction holds weight when our experiences are subjectively identical to those arising in a potential 'base' reality.

The HTC-LDS scenario forces a reckoning with simulation as a form of 'derivative' existence. It necessitates a re-evaluation of the core tenets through which we derive meaning—free will, purpose, and the pursuit of knowledge and experience. It challenges us to ask whether the search for beauty, truth, and compassion remain intrinsically worthwhile even within a reality that might owe its existence to the computational whims of beings beyond our comprehension.

LTC: The Lonely Universe

The profound implications of a potential LTC (Low Technological Capability) universe and its associated concept of "The Lonely Universe." While this scenario aligns with our traditional perceptions of reality, it invites deep reflection on existence, the absence of simulated realities, and the resultant philosophical perspectives that emerge from a starkly unique and singular universe.

Key features of the LTC scenario and its relation to the concept of a Lonely Universe include:

- **Technological Limits:** Due to insurmountable scientific barriers, resource constraints, or societal implosion, civilizations never attain the Technological Capability (TC) needed to create high-fidelity ancestor simulations with conscious inhabitants.
- **The Solitary Base Reality:** The Simulation Hypothesis is rendered largely irrelevant in this scenario. We necessarily inhabit the singular 'base' reality, devoid of a sprawling computationally-generated multiverse.
- **Cosmic Isolation:** The LTC scenario raises the specter of a universe where our own civilization may be alone in its ability to contemplate its existence, or potentially one of a sparsely scattered few, separated by the vastness of space.
- **Uniqueness and Authenticity:** The absence of simulations underscores the genuine and unreplicated nature of our reality. This may evoke a deeper appreciation for the 'realness' of our history, struggles, and triumphs.

The notion of a Lonely Universe, stemming from the LTC context, offers a rich landscape of philosophical contemplations:

- **Cosmic Responsibility or Indifference?:** Are we, as potentially one of the sole or exceptionally rare technologically advanced civilizations, imbued with a profound responsibility? Is it our duty to preserve existence's ability to understand itself, or is the universe inherently indifferent to our presence?
- **Re-evaluating the Fermi Paradox:** The LTC scenario provides a plausible, even bleak, resolution to the Fermi Paradox (the lack of observed extraterrestrial civilizations). Perhaps the silence of the cosmos is due to pervasive technological limitations, leaving us isolated, or one of a scarce few.
- **Existential Significance, or Its Absence:** Could the realization of our potential cosmic uniqueness fuel a profound sense of existential purpose, urging us to create lasting meaning in a universe devoid of backup copies? Conversely, might it instill a sense of cosmic insignificance, rendering our struggles devoid of observer-independent meaning within an uncaring universe?
- **The Preciousness of Existence:** The Lonely Universe concept may instill a heightened appreciation of the 'realness' of both our experiences and the universe itself, imbuing them with a value stemming from their singularity.

Challenges and Uncertainties

- **The Unexplored Vastness:** While the LTC scenario presents the possibility of a Lonely Universe, our incomplete knowledge of the full expanse of the cosmos and potential extraterrestrial life necessitates humility in the face of potential future discoveries.
- **The Nature of Consciousness:** If simulating consciousness remains forever impossible, is it a property uniquely emergent from our specific 'base' reality conditions? This scenario bolsters the mysterious and potentially special characteristics of our universe.
- **Unknown Existential Threats:** The LTC scenario forces us to consider the fragility of intelligent life. If civilizations are prone to self-destruction or succumbing to natural catastrophes, our uniqueness might become a precarious position characterized by vulnerability.

The LTC scenario motivates profound introspection regarding how we derive meaning and purpose in a potentially singular reality:

- **Intrinsic vs. Extrinsic Value:** Does our existence lose meaning if the universe lacks a grand architect, or if there are no simulators to assign us a purpose? The LTC scenario could catalyze a philosophical shift toward self-defined meaning, emphasizing the intrinsic value of existence, experience, and striving.
- **Stewards of Creation:** Confronted with an unsimulated universe, we might develop a deep sense of responsibility to preserve our existence, explore the cosmos, and continue the thread of scientific inquiry.
- **Focus on the Experience:** The LTC and Lonely Universe concepts could shift our attention towards the here and now. Finding beauty, meaning, and compassion within our finite reality may become paramount in a universe devoid of computational echoes.

The realization that we exist within a universe constrained by the LTC scenario could trigger diverse societal and psychological responses:

- **Collective Purpose:** The potential for cosmic loneliness could fuel unifying ambitions like long-term space colonization, expanding life's potential beyond a single, vulnerable planet.
- **Renewed Focus on 'Real World' Progress:** This scenario could encourage us to prioritize solutions to

pressing real-world challenges over focusing resources and intellectual capital on pursuing potential simulations.

- **Existential Dread or Liberation:** For some, the Lonely Universe concept may lead to existential anxiety, while for others, it might fuel a newfound sense of freedom and a determination to craft a meaningful existence within the confines of our irreplaceable reality.

The LTC scenario, by presenting the prospect of a Lonely Universe where advanced simulations are unattainable, restores a certain degree of 'classical' simplicity to our understanding of reality. Yet, it simultaneously imbues our existence with a heightened sense of both preciousness and potential responsibility.

This perspective challenges us to question whether we would construct simulations if we could, and it reinforces the ethical weight of such a hypothetical decision. In a universe where the creation of simulated sentient life is forever impossible, concepts like free will, authenticity, and the uniqueness of experience take on new dimensions. The LTC scenario, while potentially unsettling in its stark isolation, invites us to find value and purpose within a singular, unreplicated reality, encouraging us to make the most of the cosmic canvas upon which the story of our existence undeniably unfolds.

Contemplating Our Place in the Cosmos

Each of these scenarios, with their respective philosophical and existential implications, invites us to contemplate our place in the cosmos and the nature of our existence. Whether we consider the possibility of being part of a nested series of simulations, the implications of existing within a potentially 'copied' reality, or the profound uniqueness of a singular universe, the Simulation Hypothesis challenges us to question and expand our understanding of reality. It compels us to ponder the essence of consciousness, the value of our experiences, and our relationship with the cosmos, encouraging a profound, even if disquieting, re-evaluation of ourselves and our place within the vast tapestry of existence.

CHAPTER FIFTEEN

Assessing the Evidence

In the quest to understand the plausibility of the Simulation Hypothesis, particularly the potential for reaching the Technological Capability (TC) threshold necessary for simulating consciousness, it is imperative to scrutinize current technological trends and research frontiers. This section evaluates the evidence from advancements in computing power, artificial intelligence, and virtual reality, offering insights into the feasibility of creating simulations that are indistinguishable from our perceived reality.

Moore's Law & Beyond

The observation known as Moore's Law, which posits that the number of transistors on a microchip doubles approximately every two years, has historically guided the exponential growth of computational power. However, as we approach the physical limits of silicon-based technologies, questions arise about the sustainability of this growth rate and the potential ceilings of computational power.

In this context, alternative paradigms such as quantum computing, biological computing, and other yet-to-be-discovered computational frameworks warrant exploration. Quantum computing, with its promise of leveraging quantum superposition and entanglement, offers a theoretical pathway to surmount current limitations, potentially enabling the simulation of complex systems, including consciousness, on an unprecedented scale. Similarly, biological computing, which uses cells and DNA to perform computations, suggests another avenue through which the TC threshold might be approached or surpassed. The exploration of these paradigms is crucial for assessing the long-term trajectory of our simulation capabilities.

Moore's Law is not a physical law but an empirical observation turned into a driving principle of the semiconductor industry. Its key elements are:

- **The Observation:** In 1965, Gordon Moore, co-founder of Intel, noted that the number of transistors incorporated into an integrated circuit doubled approximately every year. He later revised this to a doubling roughly every two years.
- **From Observation to Industry Roadmap:** This observation became a self-fulfilling prophecy in the sense that the semiconductor industry adopted it as a benchmark, aligning research and development to ensure this continuous scaling and miniaturization of transistors.
- **Consequences:** Moore's Law-driven innovation fueled the exponential growth in computational power and the declining costs of computing that underpin the digital revolution we've witnessed over the last few decades.

The relevance of Moore's Law to the Simulation Hypothesis lies in the strong correlation between computing power and the Technological Capability (TC) required to create complex, high-fidelity ancestor simulations:

- **Computational Demands of Simulation:** Simulations capable of replicating the physical laws of our universe and modeling consciousness at a granular level would likely demand astronomical levels of computational power.
- **Moore's Law as Proxy:** The exponential growth trajectory suggested by Moore's Law has often been used to argue that the required TC for such simulations might be attainable within a historically short timeframe, given sufficient continuation of these trends.
- **Simulation Resource Estimation:** Extrapolations based on Moore's Law have featured prominently in attempts to calculate how much computing power would be needed to simulate a human brain or even an entire universe.

While Moore's Law has been astonishingly resilient, signs of its potential slowdown, if not eventual breakdown, are becoming increasingly clear:

- **Physical Limits of Silicon:** As transistors shrink towards atomic scales, fundamental quantum effects begin to interfere with their operation. Further miniaturization along the traditional path becomes increasingly difficult and costly.
- **Economic Challenges:** The fabrication plants needed to create chips with ever-smaller transistors are becoming prohibitively expensive. This economic factor could slow the pace of innovation even if some physical workarounds are found.
- **The Search for Alternatives:** The semiconductor industry is actively engaged in finding alternative materials, transistor architectures, and entirely new computational paradigms to circumvent the approaching limits of Moore's Law.

The quest to reach the computational heights implied by the Simulation Hypothesis may depend heavily on breakthroughs in computational paradigms potentially far more powerful than traditional silicon-based computing:

- **Quantum Computing:**
  - **Harnessing Quantum Phenomena:** Quantum computers leverage principles of superposition and entanglement, allowing certain kinds of computations to be performed with potentially exponential speedups over classical computers.
  - **Simulation Potential:** Simulating complex quantum systems, essential for accurately modeling physics at the fundamental level (or potentially consciousness), could be an area where quantum computers possess a significant advantage.
- **Biological Computing:**
  - **DNA as Information Storage:** DNA offers remarkable information density and the potential to perform computations within living cells.
  - **Hybrid Approaches:** Marrying the strengths of biological systems with silicon-based technologies could offer a path toward increased computational power and efficiency, though significant hurdles remain to leverage this paradigm.
- **Other Potential Disruptors:**
  - **Neuromorphic Computing:** Hardware designed to mimic the architecture of the brain could enable more efficient simulation of neural processes and potentially consciousness.
  - **Unknown Unknowns:** Moore's Law itself was an unexpected disruptor. The potential for the discovery of completely new physical principles that could be exploited for computation compels us to keep an open mind towards revolutionary shifts.

The relationship between computational progress and the plausibility of the Simulation Hypothesis is complex and multi-faceted:

- **The Importance of Extrapolation:** While the slowdown of Moore's Law throws long-term extrapolations about TC into uncertainty, the general pursuit of ever greater computational power remains a potential path towards simulation capability.
- **Breakthroughs are Unpredictable:** History is rife with scientific revolutions that defied expectations. Dismissing the possibility of new paradigms that dwarf current computational capabilities would be premature.
- **TC is Not Enough:** Even if the raw computational resources required for detailed ancestor simulations become available, questions remain about our ability to model consciousness and whether physics permits its simulation in the first place.

Moore's Law has been a powerful historical force behind the exponential advancement of computing power, fueling speculation about the technological feasibility of the Simulation Hypothesis. However, the looming physical and economic limitations of silicon-based scaling necessitate a broader and more forward-looking perspective on computational progress.

The pursuit of powerful new paradigms, such as quantum or biological computing, holds the key to potentially overcoming present bottlenecks and assessing the long-term trajectory of our simulation capabilities. While the ultimate limits of computation remain unknown, the relentless drive for innovation within and beyond the confines of Moore's Law will likely continue to reshape our understanding of what is technologically possible, influencing our contemplations about the fundamental nature of our potential existence within a simulated reality.

Simulating Consciousness

The feasibility of simulating consciousness hinges on ongoing debates within AI research and cognitive science regarding the computability of minds. Central to this discussion is whether consciousness can emerge from complex information processing systems or if it is inherently tied to biological substrates.

Current AI advancements, particularly in areas such as neural networks and machine learning, have demonstrated remarkable capabilities in mimicking certain aspects of human cognition and behavior. Yet, the leap from sophisticated pattern recognition to the subjective experience of consciousness remains a significant hurdle. This section evaluates the arguments and evidence from both sides of the debate, considering whether future advancements in AI and computational neuroscience might bridge this gap or if consciousness indeed resides beyond the realm of pure information processing.

Let's delve into the profound challenge of simulating consciousness and its critical relationship to the feasibility of the Simulation Hypothesis. Since understanding or replicating consciousness lies at the heart of creating high-fidelity simulations of a universe like ours, let's examine the philosophical debates and scientific frontiers that shape this complex question.

Before exploring the simulation aspect, we must acknowledge the ongoing discussions about the very nature of consciousness:

- **Subjective Experience:** Consciousness is defined by qualia, the subjective experience of what it is like to feel something – the redness of red, the pain of a headache, the taste of chocolate. This internal, qualitative aspect remains difficult to quantify or define in purely mechanistic terms.
- **The Hard Problem of Consciousness:** Philosopher David Chalmers coined this term to highlight that explaining the emergence of subjective experience from the electrochemical interactions of the brain remains one of the most perplexing mysteries in science and philosophy.
- **Biological vs. Computational:** A key debate centers around whether consciousness is intrinsically bound to our specific biological structures and processes, or whether it's an emergent property that could theoretically arise within sufficiently complex information processing systems, even non-

biological ones.

The possibility of simulating consciousness rests heavily on the perspective that minds are fundamentally computational:

- **The Brain as Computer Metaphor:** A longstanding analogy views the brain as an exceptionally complex biological computer. If we can decipher the 'code' and replicate its computational architecture, even in a non-biological substrate, this suggests consciousness could potentially arise.
- **AI Advancements:** Progress in Artificial Intelligence, especially with deep learning and neural networks, demonstrates the potential for machines to excel in tasks once thought exclusively human – image recognition, natural language processing, and even creative generation of text and art. Do these offer evidence of machines becoming conscious, or merely hint at sophisticated mimicry?
- **Whole Brain Emulation:** A highly ambitious goal involves mapping the entirety of a human brain, neuron by neuron, and creating a functional simulation. While fraught with technical and ethical hurdles, it rests on the premise that sufficient computational modeling of the brain would inherently reproduce consciousness.

Many researchers and philosophers remain skeptical about whether consciousness can be fully reduced to computation or simulated with any machine, regardless of complexity:

- **The Chinese Room Argument:** Philosopher John Searle's thought experiment proposes a scenario where an individual without any understanding of Chinese flawlessly manipulates Chinese symbols, producing appropriate outputs that pass the Turing Test. It argues that such symbol manipulation doesn't equate to true understanding or subjective experience.
- **Limits of Functionalism:** The assumption that replicating function should faithfully replicate conscious experience is challenged by arguments that qualia may not be inherent to information processing itself but rather tied to specific physical or biological implementations.
- **Something Beyond Algorithms:** Some theories suggest consciousness arises from complex system dynamics, involving quantum phenomena within the brain, or requires an as-yet-undiscovered property of the universe. These elements may defy simulation efforts.

The debate surrounding the computability of consciousness is far from settled. Key areas of research that might offer further insights include:

- **Computational Neuroscience:** More powerful brain-computer interfaces and brain-mapping techniques may reveal how subjective experiences correlate with neural activity patterns, potentially shedding light on what might need to be replicated for consciousness simulation.
- **Integrated Information Theory (IIT):** IIT proposes a mathematical framework for quantifying consciousness but remains controversial. It raises the possibility that specific computational structures, if replicated in a simulation, could correspond to a level of consciousness.
- **Philosophical and AI Collaboration:** Deeper interdisciplinary work between philosophers, neuroscientists, and computer scientists is crucial for tackling fundamental questions about the computational nature of the mind and refining the criteria by which we might judge if artificial consciousness has been achieved.

The question of whether consciousness can be simulated has profound implications for the feasibility of the Simulation Hypothesis:

- **TC Requirement:** If consciousness is inherently tied to a specific physical substrate (our biology) or requires unique non-computable elements, this presents a significant, potentially insurmountable hurdle to the realization of high-fidelity ancestor simulations.

- **Emergent or Programmed:** If consciousness is computable and can emerge from complex simulations, it bolsters the case for the Simulation Hypothesis. Further, if simulated consciousness is deemed functionally equivalent to that of their potential simulators, a profound philosophical situation arises.
- **The Ethical Dimension:** Success in simulating consciousness introduces profound ethical dilemmas regarding the moral status, rights, and potential suffering of simulated sentient beings.

The quest to simulate consciousness lies at the crux of one of the most enigmatic and high-stakes questions underlying the Simulation Hypothesis. Unraveling the computational nature of the mind will inform our perspective on whether creating realistic simulations of entire universes, complete with conscious inhabitants, is ultimately attainable.

Current scientific and philosophical investigations remain inconclusive. Advances in AI and neuroscience might pave the way for increasingly sophisticated computational models of cognition, potentially blurring the lines between artificial and biological consciousness. However, the possibility remains that true subjective experience is inextricably bound to our biology or rooted in physical properties that defy precise simulation.

Ultimately, understanding the nature of consciousness, and whether it can be entirely decoupled from its biological origins, is essential for assessing the plausibility of the Simulation Hypothesis and its unsettling implications for the nature of our own existence.

VR as a Proto-Simulation

The rapid development of virtual reality (VR) technology offers a tangible glimpse into the potential for creating immersive simulated environments. Current VR systems, while still limited in scope and fidelity, have begun to blur the lines between digital and physical experiences for their users.

Examining the trajectory of VR technology, from its current state to its future potential, provides valuable insights into the pathway towards full-fledged simulations. The increasing realism and interactivity of VR environments serve as a proof of concept for the types of simulations envisaged by the Simulation Hypothesis. This section considers how advancements in VR, alongside developments in sensory feedback technologies and brain-computer interfaces, might contribute to the creation of simulations indistinguishable from 'real' life.

Let's explore how Virtual Reality (VR), in its relentless march towards greater realism and immersion, serves as a compelling model of proto-simulations, providing a tangible reference point for contemplating the more ambitious simulations proposed by the Simulation Hypothesis.

To appreciate VR's role as a proto-simulation, let's briefly outline its history and current capabilities:

- **Early Origins:** The conceptual roots of VR date back to the notion of simulating artificial experiences through devices like Morton Heilig's Sensorama (1950s) and Ivan Sutherland's head-mounted display (1960s).
- **Technological Constraints:** Early VR systems were hampered by limitations in computing power, display technology, and a lack of realistic interactivity.
- **Modern Resurgence:** Recent years have seen a dramatic resurgence in VR driven by advancements in graphics processing units (GPUs), high-resolution displays, motion tracking, and the development of consumer-focused VR headsets.
- **Current Frontiers:** Contemporary VR experiences offer varying degrees of immersion, stimulating sight, sound, and sometimes limited haptic feedback (touch). Research is ongoing to enhance interaction, create whole-body experiences, and even interfaces that tap directly into the nervous system.

Key features of VR environments resonate with concepts underlying the Simulation Hypothesis:

- **Constructed Realities:** VR worlds are fundamentally designed artifacts. Their rules, landscapes, and the experiences of their users are dictated by a combination of programming and computational resources, mirroring how a potential simulated reality might be governed by its creators.
- **Sensory Manipulation:** VR primarily targets human visual and auditory systems, creating artificial sensory inputs that our brains readily interpret as a cohesive reality. This highlights the potential for manipulating perception to create convincing, yet simulated, environments.
- **Embodiment and Presence:** Immersive VR experiences foster a sense of embodiment within the virtual world, and a compelling feeling of 'presence' – of being truly within the simulated space. This hints at the power of simulation to create a psychologically convincing subjective reality for its inhabitants.

The rapid evolution of VR offers several insights and lines of inquiry pertinent to the Simulation Hypothesis:

- **The Illusion of Realism:** Even relatively low-fidelity VR experiences can evoke a powerful sense of immersion. This suggests that a perfect replication of real-world sensory input might not be a prerequisite for simulations that feel indistinguishable from 'base' reality.
- **Interactivity and Agency:** As VR experiences become more responsive, offering greater agency over one's actions within the virtual world, the potential for simulated inhabitants to feel in control of their choices increases. This mirrors a potential aspect of the simulations postulated by the Simulation Hypothesis.
- **Computational Resource Scaling:** The continuous improvement of VR is tightly coupled to advances in computing power. Extrapolating this trend, coupled with potential paradigms like quantum computing, might help inform the computational power that may be a prerequisite for vast, complex simulations.

While even the most advanced VR systems today fall far short of the high-fidelity ancestor simulations envisioned by the Simulation Hypothesis, identifying areas for future development highlights the pathway towards greater simulation complexity:

- **Multi-Sensory Integration:** Expanding VR beyond sight and sound to realistically incorporate touch, smell, and taste simulation would significantly increase the subjective realism of simulated experiences.
- **Brain-Computer Interfaces (BCIs):** Direct interfaces between the brain and computers could bypass traditional sensory channels. This raises the prospect of directly manipulating neural activity patterns to induce indistinguishable simulations from 'real' life.
- **Unconstrained Simulations:** Current VR is mainly confined to brief experiences. Simulations as postulated within the Simulation Hypothesis would necessitate persistent worlds that continue to evolve even while not directly observed by their inhabitants.

It's crucial to note that VR, as a proto-simulation, possesses distinct limitations that must be considered:

- **Computational Demands:** Simulating entire universes at high fidelity would require exponentially more computational power than is required to generate current VR experiences.
- **The Physics Question:** Can simulations perfectly replicate the physical laws of our 'base' reality, or would subtle discrepancies emerge at various scales?
- **Consciousness as the Bottleneck:** Even the most sophisticated VR environment today fails to replicate true subjective consciousness as experienced in the real world. The simulation of consciousness remains the major philosophical and likely technological hurdle for the Simulation Hypothesis.

The increasing sophistication of VR raises near-term ethical concerns that might mirror future dilemmas posed by more complex simulations:

- **Blurring of Reality:** Could long, immersive VR experiences lead to difficulty distinguishing simulated experiences from the 'base' reality?
- **Psychological Impact:** What are the psychological effects of spending extended time within simulated environments, particularly those engineered to be indistinguishable from reality?
- **Manipulation and Control:** VR experiences can be designed to evoke specific emotions or influence choices. This highlights the potential for similar manipulation of simulated beings by hypothetical simulators.

While Virtual Reality in its current form remains a far cry from the expansive conscious simulations envisioned by the Simulation Hypothesis, its evolution serves as a potent proto-simulation. VR offers a tangible example of how artificial realities can be constructed and experienced, demonstrating the potential manipulation of perception to create subjectively convincing, though computationally confined, worlds.

Advancements in VR, coupled with progress in areas like brain-computer interfaces and sensory feedback, outline a potential technological trajectory towards increasingly complex and immersive simulations. As VR systems begin to blur the boundaries between the digital and the physical, they necessitate a parallel engagement with the profound philosophical and ethical questions that accompany the endeavor of simulating reality itself – questions that lie at the very heart of the Simulation Hypothesis.

The Path Towards Simulation

Assessing the current state and future potential of computational power, the feasibility of simulating consciousness, and the advancements in VR technology elucidates the challenges and possibilities on the path to achieving the TC threshold. While significant hurdles remain, particularly in the realms of computing paradigms and the nature of consciousness, the ongoing technological evolution suggests a landscape ripe with potential for breakthroughs. As we continue to navigate this complex terrain, the evidence gathered from these diverse fields will be instrumental in shaping our understanding of the Simulation Hypothesis and the prospects for its realization.

CHAPTER SIXTEEN

Counterarguments & Criticisms

While the Simulation Hypothesis presents a captivating framework for understanding our reality, it has not gone unchallenged. Critics bring forth several compelling counterarguments that question the hypothesis's validity and feasibility. Addressing these criticisms is crucial for a balanced exploration of the concept.

The Complexity Barrier

One of the primary critiques revolves around the sheer complexity of our universe, especially at the quantum level. Critics argue that the intricate details of our universe, from the bizarre behaviors of quantum particles to the vastness of cosmic structures, set a high bar for any simulation. They question whether it is even theoretically possible for a simulation to replicate this complexity accurately. The argument posits that there might be fundamental limits to computational power and simulation fidelity, rendering our reality inherently unsimulatable. This "complexity barrier" suggests that the nuances and unpredictabilities of the natural world could never be fully captured by artificial means.

Let's delve into the compelling critique of the Simulation Hypothesis known as the Complexity Barrier. This line of criticism questions if a simulation could ever faithfully replicate the intricate details of our universe, the subtle quantum phenomena, and the sheer scale of cosmic structures. Understanding the Complexity Barrier reveals potential limitations on simulations and forces us to consider whether this might be a distinguishing feature between a simulated reality and a potential 'base' reality.

At its core, the Complexity Barrier argument hinges on the following key points:

- **The Vastness of Detail:** Our universe exhibits complexity at diverse scales. From the intricate interactions of subatomic particles governed by quantum mechanics up to the grand tapestry of galaxies and large-scale cosmic structures, there exists an overwhelming level of detail.
- **Potential Computational Limits:** Even positing extraordinarily powerful simulators, critics argue there may be fundamental limits to the degree to which our complex reality can be computationally represented. This could be due to limitations of computation itself or resource constraints any simulator might face.
- **Quantum Uncertainties and Chaos:** The probabilistic nature of quantum mechanics and the potential for chaotic behavior in complex systems introduce an element of inherent unpredictability. A faithful simulation seemingly requires capturing this intrinsic randomness, posing a formidable computational challenge and raising questions of whether perfect replication is possible.
- **Continuous vs. Discrete:** Our current understanding of physics views spacetime as continuous. Most computational simulations, by necessity, rely on discretization (breaking down reality into finite units). This raises the question of whether a discrete simulation can ever truly mirror a universe that might be

fundamentally non-discrete.

The Complexity Barrier encompasses several specific concerns and potential limitations:

- **Resource Constraints:** Simulating a universe with high fidelity down to the quantum level might demand a computational power exceeding the capacity of even the grandest civilization with access to vast energy resources. It could even require fundamentally harnessing the computational capacity of an entire universe.
- **Emergent Complexity:** Could unpredictable emergent phenomena arise within simulations that are not explicitly programmed by the simulators? Critics argue this complexity might necessitate ever-increasing computational power to maintain the simulation's accuracy, potentially leading to a point of computational unsustainability.
- **Unknown Unknowns:** Our understanding of the universe is incomplete. Simulations could fail not due to known complex systems, but because they lack knowledge of undiscovered physical laws or principles that might be crucial for a true replication.

The Complexity Barrier underscores the distinction between simulation and precise replication:

- **Imperceptible Differences:** The Simulation Hypothesis doesn't necessarily require a simulation to be a flawless mirror of the 'base' reality. Could computationally efficient shortcuts, simplifications, or approximations be used without impacting perceived reality for the simulated beings?
- **Purpose-Driven Simulations:** Is the aim of the simulation to perfectly replicate all levels of physics, or could the simulators have specific goals that only necessitate the simulation of certain experiences, with the underlying mechanics simplified or obscured?
- **Observational Limits of Simulated Beings:** Even if a simulation contains imperfections, the scientific observational capacity of simulated beings might be limited, preventing the detection of anomalies or discrepancies with a potential 'base' reality.

The Complexity Barrier, while a significant challenge, doesn't entirely invalidate the Simulation Hypothesis. However, it shapes our thinking in several ways:

- **Motivational Shift:** It prompts us to reconsider the potential motivations of the simulators. Why embark on the computationally monumental and possibly resource-inefficient task of a perfect simulation if a simplified version achieves similar results?
- **Detectability of Flaws:** The Complexity Barrier encourages investigations into whether any computationally induced limits or imperfections in a simulation may leave subtle artifacts or glitches detectable by its inhabitants.
- **A Spectrum of Detail:** Perhaps the Simulation Hypothesis is best modeled along a spectrum. Some simulations could be relatively crude, with only broad-stroke resemblance to a 'base' reality, while others might achieve higher resolutions of complexity at the cost of computational resources.

Scientific Lines of Inquiry Motivated by the Complexity Barrier

- **Search for Computational Limits:** Exploring foundational theoretical limits of computation could offer insights into the inherent difficulty of replicating physics within a simulation.
- **Quantum Simulation:** Advancements in quantum computing could provide avenues for simulating complex quantum systems, potentially addressing one aspect of the complexity challenge.
- **Finding Telltale 'Glitches':** Investigations in physics and cosmology might focus on potential phenomena inexplicable within our current models, which could hypothetically point toward

inconsistencies suggestive of computation artifacts in a simulation.

The Complexity Barrier poses a powerful critique of the Simulation Hypothesis. It highlights the immense difficulty of perfectly replicating the intricacies, emergent phenomena, and the underlying natural laws of our universe with complete computational fidelity. This argument forces us to contend with potential limitations on simulators and motivates a careful consideration of the degree of simplification or approximation that might exist in a simulated reality while remaining indistinguishable to its inhabitants.

While the Complexity Barrier might disillusion those who envision simulations as flawless copies of a 'base' reality, it remains compatible with the broader Simulation Hypothesis. It necessitates a re-evaluation of simulator motivations, encourages inquiries into the limits of computation itself, and motivates a continuous search for potential anomalies within our understanding of the physical world that might, however improbably, reveal the computational seams of a simulated existence.

The Glitch in the Matrix

Another common criticism hinges on the expectation of observable flaws or inconsistencies within a simulated reality, often referred to colloquially as "glitches in the Matrix." Critics argue that if our reality were indeed a simulation, we should be able to detect anomalies or errors indicative of its artificial nature. The apparent absence of such glitches is cited as evidence against the hypothesis. However, proponents counter that a sufficiently advanced simulation could be designed to self-correct or to mask its simulated nature so effectively that discrepancies would be either undetectable by its inhabitants or rationalized as natural phenomena within the simulation's parameters.

Let's examine the tantalizing concept of the "Glitch in the Matrix" – the anticipated flaws, anomalies, or inconsistencies that many would expect to find within a simulated reality. The debate surrounding the existence, detectability, and potential nature of these glitches offers a fascinating lens through which to critically evaluate the Simulation Hypothesis.

The core ideas and the counter-arguments surrounding this concept can be distilled as follows:

- **The Glitch Argument:** If our reality is a simulation, particularly one created by beings less than infinitely advanced, it stands to reason that imperfections might exist. These could manifest as computational errors, discrepancies with expected physical laws, or inconsistencies within the internal logic of the simulated universe.
- **Absence as Counter-Evidence:** The lack of rigorously observed and confirmed "glitches" becomes a critique of the Simulation Hypothesis. It suggests that either our reality is a remarkably flawless simulation or that it is simply not a simulation at all.
- **Counterpoint: Undetectability:** Proponents of the Simulation Hypothesis might argue that a sufficiently advanced simulator could be capable of masking its own existence. Discrepancies could be concealed, explained away through programmed phenomena within the simulation itself, or exist at scales (or in ways) beyond the observational limits of simulated inhabitants.

It's useful to consider the various ways "glitches" might theoretically manifest if our reality were indeed a simulation:

- **Computational Errors:** Miscalculations, rounding errors, or limitations in the simulation engine might lead to momentary glitches in the perceived physics, instances of objects behaving unexpectedly, or disruptions to the normal cause-and-effect sequences we take for granted.
- **Breakdown of Natural Laws:** Could there be 'zones' or circumstances where the laws of physics as we understand them appear to break down, suggesting limits to the simulation's programming or the simulators' incomplete understanding of 'base' reality physics?

- **Artifacts and Repetitions:** Simulated environments might include reused code or assets, potentially leading to noticeable patterns, duplication of individuals, or environments that feel recycled.
- **Boundaries and Limits:** Could simulations exhibit hard edges or limits to their extent? Might a simulated explorer reach the boundary of their computational universe, or encounter glitches as they approach the limits of the simulators' programmed world?
- **External Interference:** In the scenario of nested simulations, glitches could result from intervention by simulators in the layers above – actions not understood by the beings within the affected simulation layer.

The notion of reality breaking down or revealing its illusory nature has deep roots:

- **Philosophical Skepticism:** From Plato's Allegory of the Cave to Descartes's Evil Demon, philosophical thought experiments have long questioned the reliability of our perception and the ultimate nature of reality.
- **Religious Traditions:** Many religions and spiritual beliefs posit our physical reality as an illusion, a test, or a veil obscuring a truer, higher form of existence.
- **Science Fiction's Influence:** The concept of "glitches in the Matrix" has been popularized by films like The Matrix, fueling public imagination about the potential for cracks in the fabric of a simulated existence.

Seeking definitive glitches as a way to disprove the Simulation Hypothesis faces several significant hurdles:

- **Observer Limits:** Our scientific understanding of the universe is incomplete. Anomalies easily mistaken for "glitches" might eventually be explained through new discoveries within the framework of natural, non-simulated physics.
- **Simulated Science:** Scientific progress within a simulation could be subtly guided by simulators to avoid avenues of investigation that might reveal simulation artifacts or boundaries. The "laws" of physics uncovered by simulated inhabitants may always fall short of revealing the true nature of their reality.
- **The Vanishing Point of Perfection:** How advanced must a simulation be for its glitches to become effectively undetectable with the technology accessible to its inhabitants? This creates a moving target, where the counterargument to glitches relies on the ever-increasing sophistication of the posited simulation.

Despite limitations, the quest for glitches motivates intriguing scientific and observational inquiries:

- **Extreme Precision Tests:** Pushing the boundaries of experimental physics to seek minuscule deviations in fundamental constants or breakdowns of physical laws under extreme conditions could potentially reveal computational constraints or approximations in a simulation.
- **The Cosmic Unexplained:** Rigorous study of presently unexplained phenomena like dark matter and dark energy might expose potential seams in a simulated universe, where the simulators' incomplete replication of the 'base' reality physics gives rise to these anomalies.
- **Computational Fingerprints:** Could certain mathematical or physical patterns hint at an underlying computational structure? This search is exemplified by the possibility that the quantization of spacetime itself is a consequence of a discrete simulation.

The appeal of the "Glitch in the Matrix" critique of the Simulation Hypothesis lies in its intuitive simplicity. It seemingly offers a potential path to falsifiability: If a glitch is found, our reality is a simulation. However, the absence of easily detected glitches is not conclusive evidence against the Simulation Hypothesis.

The debate prompts deeper consideration of simulator sophistication, the limits of observation within a simulated world, and the potential for simulated physics to obscure evidence of its own artificiality. While the search for reality's glitches might ultimately prove fruitless, it drives scientific inquiry into the fundamental fabric of our universe. The pursuit of glitches compels us to reexamine the boundaries of the knowable and challenge the very precepts by which we define what is 'real'.

Ethics as a Deterrent

The ethical implications of creating simulations that contain potentially sentient beings also serve as a point of contention. Critics suggest that advanced civilizations, presumably possessing a high degree of moral and ethical development, might consider it unethical to create simulated realities where beings are capable of experiencing suffering or existential doubt. This moral standpoint could act as a significant deterrent to the creation of ancestor simulations, assuming that advanced civilizations would prioritize ethical considerations in their technological endeavors. The argument implies that the ethical reservations about simulating consciousness could limit the likelihood of our reality being one such simulation.

Let's dissect the compelling counterargument to the Simulation Hypothesis that hinges on ethical considerations, particularly the profound dilemmas associated with the creation of conscious simulations. This line of criticism proposes that highly advanced, and presumably ethically-minded, civilizations may fundamentally refrain from generating such simulations due to the recognition of their potential negative repercussions.

The core elements of the ethical argument against the Simulation Hypothesis are:

- **Moral Development:** Critics posit that a posthuman civilization capable of ancestor simulations is also likely to possess advanced ethical reasoning. This would lead them to grapple with the moral implications of their creation.
- **The Problem of Simulated Suffering:** If simulations achieve true consciousness, simulated beings would have the capacity for joy, but also suffering. Inflicting pain and existential uncertainty on potentially countless simulated lives raises profound ethical concerns.
- **Responsibility and Manipulation:** Could simulators ethically interfere with the lives of their simulated creations for experimentation or voyeuristic entertainment? Questions arise about the responsibilities and limits of authority that creators would have over sentient beings who might be unaware of their simulated nature.
- **The Unpredictable Path:** Ascribing motivations to advanced civilizations is speculative. While ethical considerations seem a plausible deterrent, we cannot discount the possibility of hubris, a thirst for knowledge outweighing empathy, or civilizations with value systems vastly different from our own.

History provides some, though imperfect, support for the notion that ethics might deter certain technological pathways:

- **Biological and Chemical Weapons:** International treaties and limitations exist (though not universally followed) on the development of certain weapons deemed abhorrent based on their potential to cause immense suffering and indiscriminate harm.
- **Ethical Considerations in Science:** Fields such as genetic engineering and artificial intelligence research have ongoing discussions, and in some cases regulations, centered on the ethical boundaries of manipulating life or creating potentially sentient machines.
- **Environmental Awareness:** Growing concerns over ecological damage and species extinction have prompted shifts in some technological paradigms towards sustainability and a re-evaluation of the pursuit of progress at all costs.

The ethical complexities of simulations depend on several crucial factors:

- **Level of Simulated Consciousness:** A simple simulation with limited, non-sentient behavior raises fewer concerns than one populated with conscious, feeling beings. The stronger the potential for simulated suffering, the greater the ethical hesitation.
- **Simulator Intervention and Control:** Could simulated beings live free from interference, unaware of their artificiality? Ethical concerns escalate dramatically with active manipulation by the simulators, potentially turning an ancestor simulation into a cruel experiment or a mere spectacle.
- **Transparency and Choice:** Would simulated beings gain knowledge of their true nature? If so, could they choose to terminate their existence? The lack of transparency or agency for simulated beings amplifies the ethical concerns.

The Ethics as Deterrent argument poses potent challenges to the Simulation Hypothesis:

- **Scarcity of Simulations?:** If ethically guided civilizations are common throughout the cosmos, it lowers the expected abundance of ancestor simulations. Our existence within a simulation then becomes less statistically probable.
- **Undetectable Benevolent Simulations:** Could simulations be designed specifically to minimize suffering and shield simulated beings from existential angst? Such ethical simulations leave reduced observational clues within the simulated reality.
- **A Filter Against Simulators?:** Perhaps the very civilizations that reach the Technological Capability (TC) needed for simulations ultimately choose not to create them for moral reasons. This ethical barrier then becomes a limiting factor, reducing the overall number of simulated realities.

It's vital to acknowledge viewpoints that challenge the strength of ethics as a universal deterrent:

- **Unknowable Values:** Assuming a common ethical trajectory across vastly advanced civilizations is anthropocentric. Their moral compasses might be based on concepts we simply cannot comprehend, rendering our ethical projections irrelevant.
- **Curiosity and the Science Imperative:** The urge to explore and advance knowledge is a powerful force. Civilizations might temporarily override ethical concerns in the pursuit of scientific gains or a deeper understanding of their own origins through simulation.
- **Amoral Actors:** We cannot discount the possibility of civilizations driven by ambition or self-interest, where ethics take a backseat to the potential for power or resources gained through simulations, regardless of the suffering they might create.

The Ethics as Deterrent critique prompts scientific, philosophical, and introspective inquiries:

- **Detecting 'Ethical' Simulations:** Could highly advanced ethical civilizations leave intentional clues, allowing their simulated creations to deduce their true nature? This might involve subtle anomalies in physical laws or patterns designed to inspire questioning by the simulated beings.
- **AI and Consciousness:** Advances in artificial intelligence may shed light on how and at what point consciousness emerges, informing the ethical calculus of simulated sentience.
- **Societal Values:** Our own ongoing debate about the ethics of simulations could shape our civilization's technological trajectory if and when simulation capability draws near. This highlights the need for a broad discussion about responsible innovation.

While the Simulation Hypothesis offers a fascinating lens on our potential existence, the argument of Ethics as Deterrent provides a sobering counterpoint. It raises the compelling possibility that the very civilizations technologically equipped to create conscious simulations might be the ones most likely to abstain based on moral considerations.

The strength of this deterrent depends on the universality of ethical values, the unyielding prioritization of empathy over scientific ambition, and the potential trade-offs faced by advanced civilizations. The ethics debate forces us to examine our own moral principles and the weight we place on avoiding simulated suffering, even if it's confined to a less 'real' computational realm. It highlights how philosophical questions about simulations cannot be divorced from questions about the nature of compassion, responsibility, and the inherent value we assign to all forms of conscious existence.

Counter Arguments and Criticisms

The counter arguments and criticisms of the Simulation Hypothesis highlight the complexity and depth of the debate surrounding our understanding of reality. While these critiques present significant challenges to the hypothesis, they also enrich the discourse by forcing proponents to address and refine their arguments. The complexity barrier, potential glitches, and ethical dilemmas underscore the multifaceted nature of the hypothesis and its implications.

In navigating these criticisms, it becomes clear that the Simulation Hypothesis, whether ultimately proven or disproven, serves as a powerful catalyst for inquiry into the nature of existence, the limits of technology, and the ethical boundaries of creation. The ongoing debate not only tests the resilience of the hypothesis but also propels forward our collective exploration of the cosmos and our place within it.

CHAPTER SEVENTEEN

The Power of Probability

In examining the Simulation Hypothesis, we have dissected the statistical foundations that suggest the plausibility of a simulated reality. While a definitive ontological conclusion may always elude us, these investigations emphasize a crucial shift: the concept of a simulated existence has transitioned from the fringes of speculative science fiction to a paradigm worthy of serious consideration. This transition is fueled by both the exponential growth of technological capabilities and the deepening of our philosophical inquiries into the nature of consciousness and reality.

The strength of the Simulation Hypothesis rests partially upon its statistical methodology. By positing a universe capable of spawning intelligent civilizations, the hypothesis suggests that the development of reality-simulating technologies is highly probable. Furthermore, if the ability to create indistinguishable simulations exists within an advanced species, the sheer number of potential simulations could dwarf the number of base-level realities. This line of reasoning raises the unsettling question: is it more statistically probable that we inhabit a base reality, or one of countless simulations?

The conceptual power of the Simulation Hypothesis lies in its erosion of the distinction between 'real' and 'simulated'. If simulations can achieve sufficient fidelity, the subjective experience of a simulated being may be indistinguishable from that of its base-reality counterpart. This raises not only ontological questions about the nature of reality but also epistemological challenges concerning how we gain knowledge and the limits of our perception. The Simulation Hypothesis forces us to confront the possibility that our understanding of the universe may be fundamentally bound by whatever construct – base reality or otherwise – we find ourselves within.

Confronting the Technology

The Simulation Hypothesis, if correct, implies a technological threshold of staggering complexity. To construct an artificial reality indistinguishable from our own necessitates several monumental hurdles. These include computational capabilities far exceeding our current resources, a comprehensive understanding of the fundamental laws governing our universe, and a sophisticated model of consciousness capable of generating subjective experiences that rival our own lived reality.

Let us dissect these prerequisites in detail. First, immense computational power is essential. The sheer volume of information required to simulate our physical world, down to the subatomic level, with sufficient fidelity to deceive conscious observers would necessitate a paradigm shift in processing ability. Secondly, a complete and integrated understanding of physics is crucial, from quantum mechanics to general relativity. Flaws or inconsistencies in a simulation's underlying model may be detectable by its inhabitants, undermining its intended indistinguishability. Finally, the enigma of consciousness remains a profound frontier. Without the ability to replicate the phenomenological experience of sentient beings, a simulation would merely be a lifeless, computational shell.

The technological challenges posed by the Simulation Hypothesis may seem insurmountable. Yet, progress in fields like quantum computing, unified theories in physics, and computational neuroscience offers a glimmer of potential. The realization of such complex simulations will undoubtedly necessitate a convergence of research across scientific disciplines, driving unprecedented levels of interdisciplinary collaboration. The implications of success, in turn, have the power to reshape our comprehension of the universe and our place within it.

Philosophical Fractals of Existence

The Simulation Hypothesis destabilizes core philosophical assumptions. If our reality is a construct, this raises ontological quandaries about the nature of existence itself. Determinism versus free will, the mind-body problem, the origins of morality – these foundational inquiries must all be reexamined within the context of a potentially simulated universe.

Moreover, the hypothesis presents a profound epistemological challenge. If our sensory experiences and empirical observations are merely fabrications within a simulation, can we ever achieve genuine knowledge of the world? Our understanding of reality may prove to be irrevocably bound by the limits of the simulated environment in which we reside. Should we discover our world is simulated, we are confronted with the unsettling possibility that all our knowledge is illusory.

The concept of a simulated reality forces us to question our definitions of what constitutes 'real' or authentic'. If subjective experience within a simulation can be indistinguishable from that of base reality, how do we maintain a hierarchy of ontological value? Furthermore, the very notion of the 'self' is interrogated. Where does identity lie if experiences, memories, and even our cognitive processes may be the products of external programming? The potential ramifications of the Simulation Hypothesis cascade into a seemingly infinite series of questions, each more bewildering than the last.

Ethics in a Simulated Universe

The Simulation Hypothesis, if accurate, introduces a complex ethical landscape where traditional notions of rights and responsibilities are upended. The status of simulated beings emerges as a central dilemma. If these entities possess sentience and self-awareness that are indistinguishable from our own, do they warrant the same moral consideration as beings in base reality? This question cuts to the heart of our understanding of personhood and the origins of ethical obligations.

Furthermore, the power to create and control simulations raises concerns about the responsibilities of the hypothetical simulators. What duties might they owe to their creations? Do simulated beings have a right to exist free from interference, or are their lives subject to the whims of their creators? The potential power imbalance between simulators and simulated beings opens a Pandora's box of moral quandaries, with potential ramifications ranging from paternalism to the existential horror of an entire simulated universe being abruptly terminated.

The ethical implications of the Simulation Hypothesis necessitate a critical re-evaluation of existing value systems. If our current moral frameworks are built upon assumptions about the nature of reality and persons, those assumptions may collapse under the weight of a simulated universe. This compels us to consider whether new ethical principles must be developed specifically to address the complexities of simulated existence.

A Note on Proof and Belief

The Simulation Hypothesis occupies a unique position within the realm of scientific inquiry, presenting a scenario that potentially frustrates empirical validation. If the nature of simulated reality makes it indistinguishable from our own, any attempt to definitively prove or disprove our simulated status may ultimately prove futile. We are therefore compelled to acknowledge the role of informed conjecture and reasoned speculation within this domain of investigation, a deviation from the traditional scientific method's primary reliance on observable phenomena.

While irrefutable proof of the hypothesis may be unattainable, its exploration nevertheless holds immense value. As with other metaphysical thought experiments, the Simulation Hypothesis serves as a powerful framework for investigating core philosophical concepts through a fresh lens. By interrogating the boundaries of the 'real' and the limitations of our perception, the hypothesis compels us to deconstruct long-held assumptions and examine the scaffolding upon which our understanding of existence rests.

The space of uncertainty generated by the Simulation Hypothesis becomes a crucible for intellectual growth. Within this realm of ambiguous possibilities, we are spurred to reexamine the interplay between belief and rigorous analysis, between the limits of our knowledge and the boundless potential of the human imagination. This exercise holds transformative potential for our understanding of technology, consciousness, and our place within the grand tapestry of the cosmos.

Moving Forward

As we embark on this ongoing intellectual expedition, a willingness to embrace uncertainty and open-mindedness prove indispensable. The Simulation Hypothesis compels us to question foundational beliefs, to entertain improbable scenarios, and to accept that certain answers may forever remain beyond our grasp. It is through this intentional posture of inquisitiveness, rather than through unwavering certitude, that true intellectual breakthroughs often emerge.

The exploration of simulated realities transcends mere academic curiosity. The Hypothesis extends its tendrils across a spectrum of disciplines, stimulating inquiry into the nature of computation, the limits of perception, the bedrock of ethics, and the elusive essence of consciousness itself. Delving into these domains is not just an exercise in abstract theorizing; it has the potential to reshape our fundamental understanding of what it means to be human in a universe that may prove vastly more complex than we currently comprehend.

Whether our fundamental reality is a product of natural processes or an elaborate computational construct remains a tantalizing mystery. Yet, this intellectual pursuit is about more than finding the elusive 'answer'. It is a call to re-define the contours of what we consider 'real', to question the limits of our knowledge, and to acknowledge the profound awe that accompanies our journey through a potentially boundless cosmos.

Let us proceed with resolute curiosity, driven by a thirst for understanding and fueled by a profound sense of wonder. This odyssey necessitates collaborative efforts across scientific, philosophical, and cultural spheres. It is only through bold questioning, a willingness to learn, and an unquenchable awe for the mysteries of existence that we may begin to illuminate the shadowy corners of this remarkable hypothesis.

SECTION FOUR: METHODOLOGY

CHAPTER EIGHTEEN

Evaluating Technological Capability (TC)

At the heart of Nick Bostrom's Ancestor Simulation Argument (ASA) lies the proposition that future civilizations may develop the technological capacity to create highly realistic simulations of their own evolutionary history. To assess the validity of the ASA, a rigorous evaluation of said technological capability (TC) becomes paramount. This necessitates a multi-pronged approach, drawing upon historical trends in technological advancement, the potential disruptions introduced by transformative technologies like quantum computing, and the accelerating progress in the field of artificial intelligence applied to simulation design.

Historical analysis of past technological revolutions offers valuable insights into the feasibility of hyper-realistic ancestor simulations. We have witnessed an exponential growth in computing power, miniaturization, and algorithmic sophistication over the past few decades. By extrapolating these trends and factoring in the disruptive potential of emerging technologies, a scenario where future civilizations possess the requisite TC for ancestor simulations becomes a distinct possibility.

The potential arrival of 'quantum supremacy', where quantum computers outperform classical machines in specific tasks, throws another intriguing variable into the equation. Quantum mechanics could hold the key to overcoming computational bottlenecks currently hindering large-scale simulations. Additionally, advancements in AI designed specifically for simulation optimization offer further avenues for achieving hyper-realism. Therefore, a comprehensive assessment of the ASA requires careful consideration of both established technological trajectories and the groundbreaking innovations currently on the horizon.

Historical Analysis: Trajectory of Computing Power

A critical first step in evaluating technological capability (TC) for ancestor simulations necessitates a nuanced understanding of the historical patterns that have shaped the exponential growth of computational power. Moore's Law, the longstanding observation that the number of transistors on a microchip doubles roughly every two years, has served as a foundational benchmark for this growth. However, ongoing debates regarding the potential saturation of Moore's Law due to physical limitations necessitate a broader examination of alternative computational paradigms.

The Potential of Alternative Computational Paradigms

While silicon-based computing has driven extraordinary technological advances, a comprehensive evaluation of technological capability (TC) for ancestor simulations demands venturing beyond the conventional. Confronted with the looming physical limitations of Moore's Law, the pursuit of alternative computational paradigms becomes essential. This requires a rigorous investigation into novel technologies that have the potential to revolutionize computing as we know it.

Of particular interest are three emerging realms: Firstly, quantum computing seeks to harness the counterintuitive principles of quantum mechanics to achieve computing power that vastly exceeds traditional models. Secondly, neuromorphic computing aims to emulate the brain's network architecture, promising energy efficiency and parallel processing capabilities. Finally, the nascent field of DNA-based computing leverages the inherent information density of DNA molecules for storage and computation. Each of these paradigms holds the tantalizing promise of overcoming the fundamental limits of silicon-based computing.

The allure of these alternative paradigms lies in their potential to unlock exponential growth curves far steeper than those seen under Moore's Law. The successful realization of such technologies could dramatically reshape our understanding of 'computational limits'. If future civilizations can indeed harness this transformative power, the creation of hyper-realistic ancestor simulations – foundational to Bostrom's argument – moves from the realm of speculation towards plausibility.

Statistical and Time Series Analysis

To achieve a more nuanced evaluation of technological capability (TC) in the context of ancestor simulations, we must employ sophisticated statistical modeling techniques. Time series analysis, in particular, holds significant potential for elucidating long-term trajectories in computational advancement. By analyzing historical data sets of computational metrics (e.g., processing speeds, storage capacity, energy efficiency), these techniques can reveal underlying trends, cyclicality, and potential breakpoints that may remain obscured using simple linear projections.

Time series analysis offers several advantages. Firstly, methods such as autocorrelation functions and spectral analysis can identify hidden patterns and cyclical behaviors that might inform potential future developments. Secondly, by modeling various growth scenarios (linear, exponential, logistic), we can generate more nuanced and robust projections of future TC. Thirdly, these methods allow for an explicit quantification of uncertainty, providing confidence intervals around projections. This acknowledges the inherently unpredictable nature of technological innovation.

Quantum Computing: A Game-Changer?

The burgeoning field of quantum computing stands poised to usher in a paradigm shift across a vast swathe of computational domains. This transformative technology leverages the counterintuitive principles of quantum mechanics, such as superposition and entanglement, to perform certain calculations exponentially faster than classical computers. As we assess technological capability (TC) for ancestor simulations within the framework of the Simulation Hypothesis, a thorough evaluation of quantum computing's progress and future trajectory becomes essential for several compelling reasons.

The allure of quantum computing lies in its ability to overcome the fundamental limitations inherent in classical computing architectures. Classical computers rely on bits, which can hold either a 0 or a 1. Quantum computers, on the other hand, exploit qubits, which can exist in a superposition of both states simultaneously. This ability to explore multiple states concurrently allows quantum computers to tackle specific problems, such as factoring large numbers or simulating complex molecular interactions, with unprecedented speed. Achieving 'quantum supremacy', where a quantum computer demonstrably outperforms a classical machine on a particular task, would signify a significant milestone in the development of this technology.

The impact of quantum computing on TC for ancestor simulations hinges not only on hardware advancements but also on the development of efficient quantum algorithms. These algorithms are specifically designed to exploit the unique properties of qubits to solve problems that are intractable for classical computers. Significant progress has been made in developing quantum algorithms for tasks like Shor's factoring algorithm and Grover's search algorithm. However, numerous challenges remain, including quantum error correction and the scalability of quantum systems. These hurdles must be addressed before the full potential of quantum computing for complex simulations, such as those envisioned in the Ancestor Simulation Argument, can be realized.

Overcoming Fundamental Limitations

Quantum computation, unlike its classical counterpart, draws its computational power from the unique phenomena of superposition and entanglement. Superposition allows qubits (quantum bits) to exist in multiple states simultaneously, while entanglement creates a unique correlation between these qubits that enables certain computational shortcuts. This departure from the binary restrictions of classical computing opens exciting avenues for addressing challenges specific to simulations.

A critical challenge in simulating highly complex systems lies in the exponential scaling of computational requirements as the number of simulated entities and level of detail increase. Classical computers often become overwhelmed by such complexity. Quantum computation, however, offers the potential to overcome these limitations. Quantum algorithms, by cleverly exploiting superposition and entanglement, can explore multiple possibilities simultaneously in a manner inaccessible to classical algorithms.

The implications of quantum computation for the plausibility of ancestor simulations are profound. If future civilizations achieve scalable quantum computing power, they may possess the computational resources to create vast simulations populated by countless simulated entities, each rendered with a staggering level of detail. This brings us closer to a scenario where the hyper-realistic simulations envisioned by Bostrom's Simulation Hypothesis become technologically achievable.

State of the Art and Projected Timelines:

To accurately assess the potential impact of quantum computing on TC for ancestor simulations, a comprehensive review of the current state of the art is crucial. This necessitates a deep dive into two critical areas: hardware and software. On the hardware side, factors such as qubit stability, gate fidelity (accuracy of quantum operations), and system scalability must be evaluated. Sophisticated error correction protocols will be essential to maintain qubit coherence and mitigate decoherence, which can disrupt calculations.

A significant milestone in the field of quantum computing is achieving 'quantum advantage'. This refers to the ability of a quantum computer to outperform a classical machine on a specific task. Demonstrating quantum advantage is a crucial step towards the broader goal of 'quantum supremacy', where a quantum computer demonstrably outperforms any classical computer for a particular problem. These milestones serve as valuable benchmarks for gauging the progress of quantum computing and its potential impact on the capabilities of future civilizations.

Projecting future timelines for achieving quantum advantage and supremacy is an invaluable tool for modeling the TC of potential simulator civilizations within the framework of the Simulation Hypothesis. By analyzing current trends in hardware and software development, alongside the vast computational resources theoretically achievable with mature quantum architectures, we can begin to estimate the potential for future civilizations to possess the technological prowess necessary for creating hyper-realistic ancestor simulations. However, it is crucial to acknowledge the significant 'civilization gap' – the gulf between our current understanding and the capabilities of potential simulator civilizations.

Milestones in Artificial Intelligence

Perhaps the most formidable challenge posed by the Ancestor Simulation Argument lies in replicating conscious experience within a simulated entity. The nature of consciousness remains a profound enigma in neuroscience and philosophy, and its successful simulation presents a daunting hurdle for the creation of hyper-realistic ancestor simulations. While the underlying computational requirements for simulating vast physical systems may be addressed by advancements in quantum computing, the question of imbuing simulated beings with subjective experience remains an open one.

In light of this challenge, advancements in artificial intelligence (AI) take on heightened significance. By exploring various AI paradigms, from symbolic AI to deep learning, we can gain valuable insights into the mechanisms underlying intelligence and potentially consciousness itself. While replicating true biological consciousness may remain elusive, progress in AI research may offer alternative approaches to simulating subjective experience within simulated entities. For instance, the development of sophisticated embodied AI agents capable of interacting with and learning from simulated environments could provide a path towards achieving a convincing semblance of consciousness within these simulations.

Here's where our methodology needs particular focus:

Replicating Cognitive Functions

A critical step in assessing the feasibility of simulated consciousness within the framework of the Ancestor Simulation Argument lies in meticulously mapping out the progressive advancements made in AI's ability to replicate human cognitive functions. This analysis should encompass domains such as natural language processing, where AI models can increasingly engage in meaningful dialogue; pattern recognition, where machines can identify complex regularities in data; and decision-making, where AI systems can weigh options and execute actions.

Of particular significance is the identification of points of convergence between advancements in neuroscience and the development of computational models in AI. Neuroscientists are steadily deciphering the biological underpinnings of cognitive processes within the brain. Simultaneously, the creation of increasingly sophisticated computational models offers a complementary avenue for exploring the possible mechanisms behind consciousness. By bridging these disciplines, we may unlock insights into the potential for emulating, or perhaps even transcending, the biological substrates of consciousness within a simulated environment.

The extent to which AI research can successfully replicate the full spectrum of human cognition has significant implications for the feasibility of simulating consciousness within the context of the Simulation Hypothesis. Should AI models achieve a near-perfect mimicry of human-level cognition, it would suggest that simulating entities with convincing subjective experiences becomes theoretically possible – assuming that consciousness itself is an emergent property of sufficiently complex computational processes. However, the possibility remains that true biological consciousness relies on unique factors that cannot be fully replicated computationally, raising questions about the ultimate limits of AI-driven consciousness simulation.

Artificial General Intelligence (AGI)

The much-debated concept of AGI—an AI capable of performing any intellectual task a human can—is critical. If AGI proves achievable, it would signal the potential for simulated entities to possess self-awareness indistinguishable from that of the simulators themselves. Monitoring progress in sub-disciplines like deep learning, reinforcement learning, and knowledge representation will be vital.

Monitoring advancements in critical subfields of AI research provides valuable indicators of progress towards AGI, and by extension, the feasibility of simulated consciousness. Particular attention should be paid to disciplines such as deep learning, where neural networks excel at recognizing complex patterns; reinforcement learning, which allows AI agents to learn through trial and error; and knowledge representation, which aims to develop structures for encoding and manipulating information in a manner reminiscent of human thought. Breakthroughs in these areas could significantly augment a simulator civilization's ability to create AGI-level entities within a simulation.

Debates within the field of AI often center on whether true AGI is even possible. However, from the perspective of the Simulation Hypothesis, the key question is whether AGI represents a crucial threshold on the path to achieving simulated consciousness. If AGI-level systems are capable of genuine self-awareness within a simulation, it underscores the feasibility of simulated entities attaining a level of experiential reality and cognitive autonomy on par with the creators of the simulation itself.

Interdisciplinary Convergence and TC

A critical caveat in assessing technological capability (TC) for ancestor simulations is the necessity of embracing a multidisciplinary approach. No single scientific domain holds the key to unlocking the full potential of future computing paradigms, breakthroughs in consciousness simulation, or the development of AGI. The complexity of the Simulation Hypothesis demands a collaborative convergence of expertise, fostering cross-pollination across fields including computer science, quantum physics, neuroscience, cognitive science, philosophy, and potentially others.

The interplay between these diverse fields holds the potential for synergistic advancements far exceeding the capabilities of any single discipline working in isolation. For instance, insights from neuroscience pertaining to the emergence of consciousness could inspire novel computational models in AI research. Progress in large-scale quantum simulations may in turn provide the computational horsepower necessary to create intricate simulations of complex systems, and even the emulation of individual brains. It is through this dynamic exchange of knowledge and research methodologies that we approach a more holistic understanding of future TC and its implications for the plausibility of ancestor simulations.

Computational Neuroscience

The field of computational neuroscience offers a critical avenue for investigating the potential for simulated consciousness. It delves into the brain's computational principles, aiming to understand how neural networks process information, generate behavior, and give rise to the subjective experience of consciousness itself. By analyzing the electrophysiological activity of neurons, mapping information flow between brain regions, and developing theoretical models of brain function, computational neuroscience sheds light on the complex interplay between biological architecture and the emergence of cognition.

Efforts to construct large-scale brain simulations hold significant potential in advancing our understanding of consciousness. Initiatives like the Human Brain Project and the Blue Brain Project aim to meticulously recreate the neural circuits and connectivity of the brain in a computational environment. These detailed simulations offer a 'virtual laboratory' to experiment with different parameters, manipulate neural pathways, and observe emergent properties such as the potential arising of simulated subjective experience. While these simulations are currently limited in scale and complexity, they represent a promising step towards deciphering the intricate relationship between neural activity and conscious experience.

Advances in computational neuroscience, coupled with increasingly sophisticated brain simulations, could pave the way towards identifying the prerequisites for replicating consciousness within a simulated environment. If consciousness emerges from specific computational processes within the brain, the ability to successfully model and simulate those processes raises the possibility of simulated entities possessing a similar kind of subjective experience. However, this endeavor also highlights critical questions about potential discrepancies between simulation and biological systems, emphasizing that brain simulation alone may not fully unlock the enigma of consciousness.

Philosophy of Mind

Investigating the technological feasibility of simulating conscious experience necessitates a deep engagement with the rich body of work in the philosophy of mind. For centuries, philosophers have grappled with fundamental questions surrounding the nature of consciousness. This includes explorations into the 'hard problem of consciousness', the challenge of explaining the subjective, qualitative aspects of experience (qualia), and ongoing debates about whether consciousness is purely physical or requires some non-computable element.

Philosophical discussions on consciousness play a pivotal role in shaping research efforts towards its potential simulation. They help identify central questions for scientific investigation: Can a purely computational system ever possess genuine subjective experience? What constitutes a sufficient simulation of the physical processes underlying consciousness? By posing these kinds of foundational inquiries, philosophy of mind serves as a guiding force, ensuring the investigation of simulating consciousness is not just technically driven but also ethically and ontologically informed.

The relationship between philosophy of mind and assessments of simulated consciousness is bidirectional. As technological capabilities evolve, so too do the philosophical implications. Success in simulating consciousness could potentially reshape our understanding of qualia and provide empirical insights into the 'hard problem'. These advancements, in turn, would prompt further philosophical refinement, creating a continuous dialog between philosophy, neuroscience, and computer science. This dynamic interplay is crucial for framing a comprehensive investigation of conscious experience and its potential replicability within a simulated environment.

Extrapolation and Speculative Computation:

Traditional methods of technological forecasting, such as extrapolating from historical trends, encounter limitations when assessing the vastly advanced technological capabilities (TC) of hypothetical simulator civilizations. Direct extrapolation may grossly underestimate the potential for disruptive innovations and exponential leaps that radically alter the trajectory of scientific progress. To address this, a degree of informed speculation becomes necessary to push the boundaries of plausible future scenarios.

Theoretical physics offers a rich toolkit for speculative modeling and exploration of the outer limits of computation. Insights from fields like quantum gravity, cosmology, and string theory can inform our understanding of potential constraints on information processing, energy utilization, and the fundamental fabric of spacetime. Similarly, complex science provides a framework for investigating phenomena like emergent behavior and self-organization, which may reveal potential patterns and limitations governing the evolution of highly complex computational systems.

When evaluating the Simulation Hypothesis, speculative computation serves a critical function. By exploring hypothetical scenarios based on the fringes of our current understanding of physics and computation, we can probe the theoretical outer limits of what may be achievable by simulator civilizations with far superior technological capabilities. This enables us to grapple with the possibility that simulations may achieve a level of sophistication that remains entirely inaccessible to our present scientific understanding. While a degree of uncertainty will always remain, informed speculation guided by established scientific principles helps illuminate the vast landscape of computational possibilities relevant to the Simulation Hypothesis.

Methodological Challenges and Limitations

Assessing technological capability (TC), particularly in the context of hypothetical ancestor simulations, carries significant methodological challenges. Technological forecasting is fraught with uncertainty. Historical trends do not always guarantee future trajectories, and the potential for disruptive breakthroughs (like quantum supremacy) introduces an element of unpredictability. Moreover, our current scientific understanding may prove inadequate for modeling the vastly advanced TC of future civilizations, forcing us to rely on informed speculation and extrapolations that lack empirical validation.

A crucial limitation arises from our reliance on analogy. While we may draw insights from past and present technological revolutions, there's no guarantee that future progress will follow similar patterns. Additionally, a fundamental 'civilization gap' exists between our present understanding and the potential TC of simulator civilizations. This gap makes it difficult to anticipate the types of computational paradigms and scientific breakthroughs such civilizations may have achieved. Therefore, our assessment of TC for ancestor simulations will always be framed by a degree of epistemological humility and an acknowledgment of the inherent limitations of our predictive models.

It's essential to acknowledge the significant challenges and unavoidable limitations involved in our methodology for TC assessment:

Unpredictability of Technological Leaps:

The history of technological advancement serves as a constant reminder of the inherent unpredictability of scientific breakthroughs. Time and again, disruptive technologies have emerged, defying conventional predictions and upending established paradigms. From the development of the transistor to the rise of the Internet, technological leaps have often caught experts by surprise, transforming entire industries and reshaping our understanding of what is possible.

When assessing technological capability (TC) relevant to the Simulation Hypothesis, it is crucial to maintain a posture of openness towards unforeseen breakthroughs. We must acknowledge the possibility that revolutionary discoveries in quantum computing, neuromorphic computing, or entirely novel paradigms could dramatically alter our projections of computational power. The potential for a future equivalent of the 'microchip revolution' must be factored into our analyses, as such events could lead to exponential leaps in TC, vastly exceeding expectations based on linear models.

Embracing the unpredictable nature of technological innovation necessitates intellectual flexibility and a willingness to revise assumptions. It compels us to move beyond rigid extrapolations based solely on past trends and engage in speculative yet scientifically informed exploration of the outer limits of computational possibility. This approach, while inherently uncertain, is crucial for avoiding underestimating the potential capabilities of simulator civilizations and for maintaining an accurate perspective on the plausibility of the Simulation Hypothesis.

Physical Limits and Fundamental Constraints

While the potential for transformative breakthroughs is undeniable, it's equally important to consider the possibility of fundamental limits to computation. The laws of physics, as we currently understand them, may impose constraints on what is ultimately achievable in terms of information processing, data storage, and energy manipulation. These constraints could limit even the most advanced civilizations, placing hard boundaries on their technological capability (TC).

Critical questions must be addressed when factoring in physical limits to computation. Are there inherent limits to information density and processing speed dictated by quantum mechanics or thermodynamics? What are the ultimate energy requirements for large-scale, hyper-realistic simulations? Furthermore, we cannot rule out the existence of presently unknown physical constraints, operating at scales far beyond our current understanding, that could impose absolute limitations on computational power achievable within our universe.

The existence of potential physical limits has significant implications for the Simulation Hypothesis. If fundamental constraints on computation prevent even technologically advanced civilizations from achieving the processing power necessary for ancestor simulations, this would decrease the theoretical plausibility of the hypothesis. Conversely, if the laws of physics allow for the kind of computational capacity needed, the feasibility of ancestor simulations becomes more conceivable, though it still leaves open profound questions about the nature of the simulators themselves.

The Anthropic Principle

The potential reality that our universe is itself a simulation introduces profound challenges for our methodologies and our assessment of technological capability (TC). If we inhabit a simulated world, it raises the possibility that our observations of physical laws and technological trajectories may be subject to the constraints and limitations programmed into the simulation by its creators. This casts a degree of uncertainty over our ability to accurately predict the potential capabilities of simulator civilizations operating outside the confines of our simulated reality.

The Anthropic Principle reminds us that our existence as observers within the universe (or a simulated universe) introduces inherent biases. The fundamental constants of our simulated universe may have been 'fine-tuned' to allow for our existence. Furthermore, the technological progress we witness may be subject to artificial limitations or manipulations by the simulators. Therefore, we must remain cognizant of the possibility that our methodologies for assessing TC may be distorted by the very nature of our simulated existence.

The discovery that we reside within a simulation would necessitate a fundamental restructuring of our methodological approaches. It would demand a critical re-examination of the assumptions underpinning our scientific models and our interpretations of historical technological progress. We would need to develop new frameworks for interpreting data and devising hypotheses, carefully accounting for the potential limitations and biases imposed by the architecture of a simulated reality.

Refining the TC Evaluation Framework

Heightened critical self-examination becomes imperative. We must actively seek out potential inconsistencies, anomalies, or 'glitches' in our observations of physical laws and technological trajectories. These aberrations may reveal telltale signs of the underlying computational architecture of a simulated reality and offer glimpses into the constraints simulators might face. Analyzing our perception of fundamental constants and known limits of computation could unveil discrepancies hinting at the artificial nature of our environment.

To enhance the robustness of our TC evaluation, a collaborative approach drawing on expertise from diverse disciplines becomes paramount. Philosophers could provide insights into the potential epistemological consequences of a simulated reality and the validity of our ontological assumptions. Collaborations with physicists and computer scientists could facilitate the development of experimental protocols and computational models aimed at detecting potential anomalies indicative of a simulation. By triangulating evidence and perspectives across multiple fields, we stand a better chance of mitigating the limitations of our methodologies and refining our assessments of technological capability within the broader context of the Simulation Hypothesis.

To mitigate these challenges and enhance the robustness of our TC evaluation, additional methodological considerations are needed:

Meta-Analysis and Aggregation of Expert Opinions

Assessing technological capability (TC) within the framework of the Simulation Hypothesis necessitates a multifaceted approach. While individual disciplines offer valuable insights, a more comprehensive understanding can be gleaned by surveying a diverse pool of experts across relevant fields. This includes computer scientists specializing in areas like artificial intelligence and quantum computing, information theorists grappling with the fundamental limits of computation, neuroscientists exploring the nature of consciousness, and philosophers wrestling with the ontological implications of simulated realities.

Incorporating such diverse perspectives fosters a more holistic evaluation of TC. Computer scientists can shed light on the potential capabilities of future computing architectures, while quantum information theorists can illuminate the boundaries imposed by the laws of physics. Neuroscientists can contribute insights into the computational requirements of simulating consciousness, and philosophers can help us grapple with the very notion of reality within a simulated universe. This cross-pollination of knowledge helps us avoid the pitfalls of siloed thinking and fosters a more nuanced understanding of the challenges and possibilities inherent in simulating complex systems.

To effectively aggregate and refine expert opinions on TC development, leveraging established methodologies becomes crucial. Techniques like the Delphi Method offer a structured approach to eliciting informed estimates from a panel of experts. Through iterative rounds of anonymous surveys and feedback, the Delphi Method allows experts to refine their initial projections in light of the collective wisdom of the group. This process fosters a convergence towards a more robust consensus on the future trajectory of technological advancement.

Scenario Planning and Sensitivity Analysis

The inherent uncertainties surrounding future technological capability (TC) necessitate a nuanced approach that acknowledges various potential pathways. Scenario planning emerges as a powerful tool for navigating this complexity. By meticulously crafting a series of detailed scenarios, we can explore a range of plausible trajectories for future technological advancement, encompassing a broad spectrum of possibilities.

These scenarios should account for variations in several key factors. Firstly, different scenarios should explore the potential dominance of distinct computational paradigms. Will future progress hinge on advancements in classical computing, quantum computing, or entirely novel paradigms yet to be discovered? Secondly, the pace of scientific discovery can vary significantly. Scenarios can explore possibilities ranging from steady incremental progress to revolutionary breakthroughs that drastically accelerate technological leaps. Finally, the societal and geopolitical landscape must be factored in. Scenarios can examine how factors like global collaboration, resource allocation, and ethical considerations might influence the trajectory of TC development.

To further refine our understanding of TC, sensitivity analysis becomes an indispensable companion to scenario planning. This technique involves systematically varying the underlying assumptions within each scenario and observing the resulting impact on projected TC. By analyzing how sensitive each scenario is to changes in assumptions, we gain valuable insights into the critical factors that will most significantly influence future technological capabilities.

For instance, analyzing the sensitivity of projected TC to variations in research funding can illuminate the potential impact of resource allocation on future technological breakthroughs. Similarly, assessing the sensitivity of TC to advancements in specific scientific fields like materials science or artificial intelligence can highlight areas of critical importance for the development of hyper-powerful computing technologies. Through this combined approach of scenario planning and sensitivity analysis, we develop a more robust and nuanced understanding of the vast landscape of possibilities surrounding future TC, allowing for a more informed exploration of the plausibility of the Simulation Hypothesis.

Simulation Studies

To gain a deeper understanding of the relationship between technological capability (TC) and the feasibility of simulated consciousness, smaller-scale computer simulations offer a valuable experimental tool. By modeling the interplay between computational power, simulation complexity, and the potential emergence of simulated consciousness within these simulations, we can gain critical insights that would be impossible to obtain through purely theoretical or philosophical approaches.

These simulation studies can help us identify potential thresholds or tipping points along the TC trajectory. For example, we could investigate whether a certain level of computational power is required for simulated entities to exhibit rudimentary self-awareness, or if specific patterns of information processing within a simulation lead to the appearance of subjective experience. By exploring these dynamics in controlled simulation environments, we can illuminate the preconditions for simulated consciousness and identify milestones that could be relevant for assessing the feasibility of large-scale ancestor simulations.

The results of simulation studies can directly inform our understanding of the Simulation Hypothesis. They can help us refine our estimations of the computational resources and level of simulation detail required to create convincing ancestor simulations. Furthermore, by highlighting the potential emergence of simulated consciousness under specific conditions, these studies will contribute to the ongoing philosophical debate on the nature of consciousness and its replicability within a computational environment. While smaller-scale by design, these simulation studies offer a powerful avenue for mapping the landscape of possibilities relevant to the Simulation Hypothesis, enriching our understanding of the complex relationship between computation, consciousness, and the potential for simulated realities.

Connecting TC Assessment to Bostrom's Trilemma

A rigorous assessment of technological capability (TC) forms a critical foundation for evaluating the plausibility of Bostrom's Simulation Argument. The feasibility of ancestor simulations rests on the assumption that future civilizations can develop computers powerful enough to simulate entire universes with a high degree of fidelity. If physical laws or computational constraints fundamentally limit TC, this would cast doubt on the first two propositions of the Trilemma, reducing the likelihood of our own existence within a simulation.

By analyzing historical trends, the potential for disruptive innovations, and fundamental physical limitations, our TC assessment offers a guidepost for navigating the complex landscape of Bostrom's Trilemma. If our analysis indicates that vast computational power is achievable, this strengthens the plausibility of posthuman civilizations emerging that possess the technological means to create ancestor simulations. This, in turn, increases the weight of the Trilemma's third proposition, leading us to give serious consideration to the possibility of our own reality being a simulation.

Let's recall Bostrom's original trilemma:

- The first proposition in Bostrom's Trilemma posits that the fraction of civilizations reaching an advanced technological state ('posthuman') is extremely low, perhaps almost non-existent. This could be due to factors such as self-destruction before reaching technological maturity, insurmountable technological barriers, or a general lack of inclination among civilizations to pursue further advancement once a certain stage is reached.
- The second proposition suggests that even among posthuman civilizations, the fraction of those interested in actually creating ancestor simulations is vanishingly small. This could be due to ethical concerns, a lack of perceived value in such simulations, resource constraints, or alternative priorities for their vast technological capabilities.
- The final proposition is the most provocative. It states that if the first two propositions are true, then the vast majority of all beings with similar experiences to ours are almost certainly living within simulations. This is because a single posthuman civilization interested in ancestor simulations could potentially run an astronomical number of them, dwarfing the number of biological beings in the 'base' reality.

The first proposition of Bostrom's Trilemma focuses on the likelihood of civilizations reaching a 'posthuman' stage of technological advancement. Our assessment of technological capability (TC) directly impacts our understanding of whether this stage is even achievable. If our analysis suggests that fundamental constraints on computation or resource limitations make such a level of advancement impossible, this significantly weakens the first proposition – the likelihood of posthuman civilizations capable of ancestor simulations would be greatly diminished.

Furthermore, our TC assessment helps us identify potential technological hurdles that civilizations must overcome on their path to a posthuman stage. Our analysis might reveal specific bottlenecks in computational paradigms, energy requirements, or the development of artificial intelligence that could pose existential challenges to civilizations. If the probability of overcoming these hurdles appears low, this again casts doubt on the first proposition of the Trilemma and consequently reduces the likelihood of our simulated existence.

Our TC assessment bears significantly upon the first proposition of this trilemma. More specifically:

TC Constraints as a Bottleneck

If our comprehensive analysis of technological capability (TC) reveals hard limits on computing power, information density, or energy management, the feasibility of ancestor simulations plummets. Civilizations may simply not be able to develop the technology required to simulate entire universes with sufficient realism to fool their simulated inhabitants. This technological bottleneck directly supports the first proposition of the Trilemma, which posits that most civilizations will fail to reach a posthuman level of technological sophistication.

One crucial bottleneck could be the challenge of simulating consciousness. Should the ability to replicate subjective experience within a simulation prove computationally impossible or require levels of complexity far beyond the capabilities of any civilization, the possibility of convincing ancestor simulations would be greatly reduced. This would support the notion that civilizations, even if they achieve technological advancement in other domains, may simply never develop the means to reach the 'posthuman' stage necessary for ancestor simulations.

If technological constraints create an insurmountable barrier to ancestor simulations, the implications for Bostrom's Trilemma are significant. It strengthens the plausibility that the vast majority of civilizations fail to advance beyond a certain technological level due to these limitations. This, in turn, would decrease the overall likelihood of anyone with our kind of experiences living within a simulation, even if a small fraction of posthuman civilizations were interested in running them.

Rapid TC Advancement and Simulation Proliferation

If our assessment of technological capability (TC) reveals a pattern of recurring breakthroughs, or the potential for exponential leaps far exceeding our expectations, it calls into question the existence of fundamental limits. This unconstrained trajectory suggests that civilizations could ultimately harness the computational power needed for vast and intricately detailed ancestor simulations.

Unconstrained TC advancement implies that civilizations may one day reach a level of computational mastery where the creation of ancestor simulations becomes an almost trivial endeavor. Resources and energy requirements, once considered potential obstacles, could be easily overcome by such sophisticated civilizations. This would make the widespread proliferation of simulations highly likely, even if only a minuscule fraction of posthuman civilizations found interest in their creation.

The potential for unconstrained TC development and the ease of creating simulations directly support the third proposition of Bostrom's Trilemma. In such a scenario, the vast majority of beings with experiences like ours throughout the cosmos would likely exist within simulations, simply due to their overwhelming numerical dominance over those in 'base-level' realities. This realization becomes far more plausible when future TC is viewed as potentially boundless, rendering the creation of hyper-realistic simulations computationally achievable by advanced civilizations.

The Technological Lens: Assessing Simulated Realities

Our exploration of technological capability (TC) within the context of the Ancestor Simulation Argument emphasizes the importance of methodological rigor, an embrace of well-informed speculation, and an honest acknowledgment of limitations. This multi-disciplinary journey has highlighted the potential for transformative breakthroughs, the enduring enigma of consciousness, and the possibility of constraints imposed by the fundamental laws of physics.

Our assessment of TC offers a vital guidepost as we navigate the complex landscape of Bostrom's Trilemma. By analyzing the potential trajectory of technological advancement, we gain insights into the likelihood of posthuman civilizations emerging, their potential interest in running simulations, and ultimately, the odds that we ourselves inhabit a simulated reality. While a definitive answer to the Simulation Hypothesis may forever remain elusive, the pursuit of understanding the limits of computation has profound implications for our understanding of the cosmos and our place within it.

Further refinements of our TC assessment will depend on relentless advancements in fields like computer science, quantum information theory, neuroscience, and philosophy. These cross-disciplinary endeavors will illuminate the outer bounds of possibility and offer new lenses through which we examine our reality. Our exploration underscores a profound truth: the Simulation Hypothesis, whether ultimately confirmed or not, propels us to challenge our deepest assumptions, expand the frontiers of human knowledge, and marvel at the infinite complexities that may lie hidden within the fabric of the universe.

CHAPTER NINETEEN

Assessing Desire to Simulate (DS)

The cornerstone of Bostrom's Ancestor Simulation Argument hinges on the interplay between two crucial factors: Technological Capability (TC) and Desire to Simulate (DS). TC refers to the ability of a civilization to construct hyper-realistic simulations capable of replicating the experiences of sentient beings. Assessing TC involves a complex dance between historical trends, the potential for disruptive breakthroughs, and the fundamental constraints imposed by the laws of physics. However, even if TC is deemed achievable, the argument hinges on an equally critical, yet arguably more elusive, question: would advanced civilizations possess the DS to devote vast resources towards creating intricate simulations of their ancestors?

Evaluating DS presents a unique challenge, delving into the realm of human motivation, long-term societal goals, and the potential transformations of these factors within a posthuman context. Current psychological frameworks, shaped by our evolutionary history and resource scarcity, may offer limited applicability when attempting to extrapolate desires of civilizations operating at vastly different technological and social scales. Will intellectual curiosity or a sense of historical preservation drive posthuman societies to create ancestor simulations? Alternatively, could utilitarian considerations, ethical concerns, or unforeseen existential threats render such endeavors a lower priority? Exploring these multifaceted motivations necessitates a multidisciplinary approach, drawing on philosophy, sociology, and speculative ethics to envision the potential value systems and priorities of posthuman civilizations.

Understanding DS necessitates venturing beyond solely technological considerations. Insights from anthropology offer valuable lessons on the diverse historical and cultural contexts that drive a society's desire to preserve its past. Additionally, frameworks like evolutionary psychology can be cautiously employed to analyze potential long-term shifts in human motivations. Ultimately, evaluating DS demands a forward-looking approach, grappling with the possibility that fundamental aspects of human psychology may be drastically reshaped in a posthuman future. By critically examining the interplay between technological advancement and societal transformation, we can refine our estimations of the likelihood that DS would coexist with the TC required for ancestor simulations, offering a more nuanced perspective on the overall plausibility of the Simulation Hypothesis.

History and Background

Humankind's fascination with the nature of reality and the possibility of simulated worlds extends far beyond the confines of modern technology. The embers of this inquiry can be traced back to the dawn of recorded thought, manifest in philosophical explorations and artistic narratives. Plato's Allegory of the Cave, a timeless meditation on the limitations of human perception and the potential for deception by our senses, serves as a foundational example. Similarly, the Hindu concept of Maya, the illusion that veils the true nature of reality, underscores humanity's longstanding awareness of the potential for deception and the constructed nature of our perceived world. Further East, the enigmatic "Butterfly Dream" by Zhuangzi exemplifies this enduring philosophical quest, blurring the lines between dream and reality and prompting contemplation on the nature of experience itself.

While these ancient inquiries laid the groundwork for the modern discourse on simulated realities, the rise of computational technology has undoubtedly acted as a potent catalyst. The conceptual leap from philosophical musings to the possibility of deliberately creating simulated worlds hinges on the advent of machines capable of replicating complex systems. John von Neumann, a towering figure in the history of computer science, stands as a pivotal figure in this narrative. His visionary theories on self-replicating machines, outlined in his seminal work, laid the conceptual foundation for the possibility of simulated environments operating within environments. Von Neumann's work highlighted the potential for computational systems to not only mimic the world around us but potentially create entirely new realities within their digital architecture.

The modern discourse on simulated realities thus emerges from a rich tapestry woven from philosophical inquiry and technological innovation. Philosophical explorations of perception, the nature of reality, and the limitations of our senses provided fertile ground for the seeds of technological possibility sown by computer science pioneers like von Neumann. This confluence of ideas has fueled a vibrant contemporary discussion surrounding the plausibility of simulated worlds, the potential implications of such realities for our understanding of consciousness and existence, and the ethical considerations surrounding their creation. By drawing upon the wisdom of the past and embracing the possibilities of the present, we can continue to navigate the ever-evolving landscape of simulated realities, their philosophical underpinnings, and their potential impact on our understanding of the universe we inhabit.

While the concept of simulated realities has captivated thinkers for ages, the modern Simulation Hypothesis finds its formal roots in the works of visionary thinkers like philosopher Hans Moravec. Moravec, extrapolating on the exponential growth of computing power, posited that advanced civilizations would eventually develop the technology to simulate realities indistinguishable from their own. However, it was philosopher Nick Bostrom who thrust the Simulation Hypothesis into mainstream discourse through the articulation of his now-famous trilemma.

Bostrom's provocative trilemma postulates that one of three propositions about advanced civilizations must hold true. Its power lies in its ability to highlight the implications of technological progress for our understanding of reality. Significantly, while previous inquiries often focused on the technological feasibility of simulating realities, Bostrom's trilemma compels us to confront an equally crucial, yet less tangible consideration: the Desire to Simulate (DS). Specifically, the second proposition of his trilemma raises the question of whether posthuman civilizations, even if possessing the computational power, would actually devote resources to running complex ancestor simulations.

With Bostrom's formulation of the Simulation Hypothesis, the focus shifts from a purely technological assessment to a complex exploration of potential motivators for hypothetical simulator civilizations. This necessitates a shift in our inquiry, from a focus on what is technologically possible to an interrogation of what is desirable or valuable to civilizations far removed from our own. The core questions surrounding DS revolve around understanding the potential goals, values, and ethical considerations that could shape the actions of posthuman societies. Only by grappling with these questions can we shed light on the likelihood that the technological capacity for simulating realities would translate into a genuine desire to do so.

Methodological Considerations

A significant methodological challenge in assessing Desire to Simulate (DS) lies in its inherent resistance to quantification. Unlike Technological Capability (TC), which can be analyzed through metrics of computational power, scientific breakthroughs, and historical trends, DS delves into intangible domains of motivation, values, and ethics. There is no established formula to predict whether a posthuman civilization would find value in the computationally demanding exercise of creating ancestor simulations.

Evaluating DS necessitates venturing into largely uncharted territories at the intersection of psychology, sociology, and speculative philosophy. Traditional frameworks for understanding human motivation and societal dynamics may prove inadequate when considering the potential transformations of a posthuman civilization. Assessing DS compels us to grapple with the possibility of fundamental shifts in values, resource priorities, and even the very concept of selfhood in an era of advanced technology. Therefore, a nuanced evaluation demands a conjectural approach, venturing beyond established methodologies to envision the potential drives of civilizations operating at scales beyond our current comprehension.

The inherent unpredictability of societal progress poses an additional challenge when assessing DS. Extrapolating from present-day human motivations and values may lead to erroneous conclusions about the desires of posthuman societies. History is replete with examples of profound revolutions in social systems, belief structures, and technological paradigms. It is thus crucial to acknowledge the limitations of linear forecasts when grappling with potential motivations driving civilizations vastly more advanced than our own. Our analysis must maintain a degree of flexibility, acknowledging that entirely novel values, ethical concerns, or existential goals may emerge as civilizations progress, potentially rendering ancestor simulations a high or low priority in ways we fail to foresee.

Here's a breakdown of key methodological considerations:

Motivational Analysis

A comprehensive assessment of Desire to Simulate (DS) necessitates a detailed exploration of the potential motivations that could compel a civilization to create elaborate simulations of its ancestors. This spectrum of reasons could encompass a range of factors, from scientific inquiry and historical preservation to philosophical exploration and even entertainment. Posthuman civilizations might utilize ancestor simulations to gain a deeper understanding of their own evolutionary trajectory, analyze past societal dynamics, or test hypotheses about historical events. Alternatively, the desire to preserve the memory and experiences of their forebears in a hyper-realistic format could motivate the creation of such simulations.

The motivations for DS might extend beyond purely utilitarian considerations. The intellectual curiosity inherent in some civilizations could propel them to create simulations as a means to explore alternative timelines, test the boundaries of consciousness within simulated environments, or simply revel in the creative power of replicating entire realities. Furthermore, the possibility that simulated beings themselves may develop consciousness and subjective experiences raises profound ethical and philosophical questions. The desire to grapple with these complexities and understand the nature of simulated existence could be another potential motivator for the construction of ancestor simulations by advanced civilizations. Understanding this multifaceted spectrum of motivations becomes crucial for refining our assessment of DS and its role within the broader context of the Simulation Hypothesis.

A comprehensive methodology must catalog the diverse motivations that might drive a civilization to construct ancestor simulations. These can include:

Scientific Curiosity

For civilizations driven by scientific curiosity, ancestor simulations could serve as powerful tools for dissecting the fabric of their past. By meticulously constructing simulations that incorporate historical data, cultural nuances, and plausible environmental conditions, posthuman researchers could gain a deeper understanding of their evolutionary trajectory. Simulations could be designed to explore pivotal decision points in history, analyze the potential outcomes of different choices made by their ancestors, and ultimately illuminate the intricate interplay of factors that shaped their present reality. This ability to rewind the clock and explore alternate timelines could offer invaluable insights into the contingent nature of history and the forces that propelled their civilization to its current state.

The scientific utility of ancestor simulations might extend beyond the immediate past of the simulating civilization. Posthuman researchers could leverage their advanced technological prowess to create simulations that explore entirely novel societal structures, cultural norms, and technological paradigms. By simulating societies operating under different fundamental laws of physics or initial conditions, they could probe the boundaries of possibility and gain insights into the nature of existence itself. These simulations could serve as a vast 'cosmic laboratory,' allowing for the exploration of alternative realities and potentially revealing universal constants that govern the behavior of complex systems across diverse simulated universes. In this way, the pursuit of scientific knowledge through ancestor simulations could transcend the specific history of a single civilization and contribute to a broader understanding of the cosmos.

Entertainment

For posthuman societies that have progressed beyond the constraints of resource scarcity and physical limitations, entertainment may evolve in ways that are difficult for us to conceive. Ancestor simulations could offer a captivating form of leisure activity, promising an escape from the potential monotony of a highly optimized existence. Simulations could be designed with varying levels of difficulty, challenges, or simply alternate life trajectories. These could serve as a means of injecting a sense of unpredictability, adventure, and novelty into the lives of posthuman individuals. Furthermore, the ability to embody alternative identities and experience life from a plethora of perspectives within simulated environments could unlock profound new frontiers of entertainment.

Ancestor simulations could seamlessly blur the lines between traditional gaming and experiential learning. Imagine the possibility of fully immersive, hyper-realistic historical simulations where individuals can actively participate in pivotal moments, engage with simulated figures from the past, or witness the unfolding of grand historical sagas from within the simulation. The appeal of such experiences could transcend what we currently perceive as entertainment, offering a form of "edutainment" that combines the thrill of participation with the intellectual stimulation of historical exploration. These advanced forms of simulation could elevate the entertainment sector to unimaginable heights, transforming simulations into grand performative spectacles and immersive platforms for storytelling.

Ethical Experimentation

Ancestor simulations, if they achieve a sufficient level of fidelity, could provide unique platforms for conducting complex ethical experiments within fully interactive "moral sandboxes". Advanced civilizations may leverage this technology to test the validity of ethical theories, observe the long-term consequences of different ethical frameworks, and analyze the intricate dynamics of moral dilemmas, all without the risk of real-world harm. Simulations could be designed to model societies under varying ethical systems, allowing researchers to analyze the impact of different moral codes on individual behavior, societal well-being, and overall outcomes. This could offer invaluable insights for refining ethical frameworks and potentially lead to a deeper understanding of the human moral landscape.

While the potential benefits of ethical simulations are significant, they also raise profound ethical questions that must be carefully considered. The possibility of causing simulated suffering, even if the beings within the simulation are not fully conscious in the same way that beings in base reality are, presents a substantial challenge. Posthuman civilizations would need to develop rigorous ethical guidelines and oversight for such simulations, ensuring that the pursuit of knowledge does not come at the cost of inflicting undue suffering on even simulated entities. The exploration of ethics through simulated realities would necessitate a careful and nuanced approach, constantly weighing the potential benefits against the moral risks inherent in such experimentation.

Ancestor Reverence

Ancestor simulations could arise from a deeply ingrained sense of reverence for one's lineage and a desire to preserve the essence of those who came before. Posthuman societies, even as they transcend many of our current limitations, may still retain a profound connection to their past. Hyper-realistic simulations could serve as a means of honoring the memory of ancestors, allowing their experiences, stories, and personalities to be preserved in a dynamic and interactive format. The ability to interact with simulations of departed loved ones, to learn from their wisdom, and to revisit shared memories could hold immense emotional value for posthuman individuals, satisfying a deep-seated human yearning to connect with the past.

The desire for ancestor simulations could be closely intertwined with posthuman conceptions of mortality and legacy. Even within technologically advanced civilizations, the desire to transcend the limitations of one's own lifespan and ensure that one's contributions are remembered may persist. Ancestor simulations could offer a way for individuals to preserve their own experiences, knowledge, and even aspects of their personalities for future generations within the simulation. The ability to maintain this continuity of consciousness, albeit in a simulated form, could appeal to posthuman civilizations driven by a strong desire for the preservation of individuality and a sense of continued existence beyond biological life.

Resource Optimization

For advanced civilizations facing complex challenges of resource management and technological innovation, simulations could become invaluable tools for optimization. Ancestor simulations, as well as simulations of alternative realities, could serve as virtual testing grounds for new technologies, large-scale engineering projects, or social policies. By simulating various scenarios, posthuman societies could identify potential failures, optimize resource allocation strategies, and mitigate risks before implementing real-world changes. This could lead to significant improvements in efficiency, reducing waste and maximizing the potential benefits of their limited resources.

The ability to simulate entire planets, complete with complex ecosystems, societies, and technological systems, could grant posthuman civilizations unprecedented power to predict and optimize planetary engineering projects. Simulations could allow them to test the long-term effects of geoengineering interventions, model climate trajectories under different conditions, and analyze the potential societal impacts of large-scale environmental changes. Similarly, social policies or alternative governance structures could be rigorously tested within simulated environments, mitigating the risk of unintended consequences in the real world. The pursuit of resource optimization could thus become a significant motivator for creating complex simulations, particularly for civilizations seeking sustainable solutions to address the challenges of an advanced technological era.

Historical Analogies

While it's essential to be wary of overly simplistic analogies between our present state and the potential motivations of posthuman civilizations, studying historical patterns of human behavior can offer valuable insights. Throughout history, humanity has embarked on ambitious megaprojects, from the construction of pyramids and grand cathedrals to the exploration of space and the mapping of the human genome. These endeavors often share common underlying motivations, such as the pursuit of knowledge, a desire to leave a lasting legacy, the urge to explore the unknown, and an aspiration to transcend limitations – both physical and intellectual.

Posthuman civilizations, as inheritors of some of these fundamental human drives, may similarly be motivated to undertake projects of extraordinary ambition and scale. Ancestor simulations could represent the ultimate manifestation of the desire for knowledge, a testament to the enduring human curiosity to understand the past and to chart one's trajectory through the cosmos. Moreover, such simulations could offer a way for civilizations to cement their legacy on a truly universal scale, preserving their history for epochs to come. Additionally, the appeal of exploring the unknown and venturing into the realm of possibilities only accessible through simulation could serve as a powerful motivator. By drawing inspiration from historical patterns of human endeavor, we can begin to conceptualize the potential motivations that could drive posthuman civilizations to embark on the ambitious project of simulating reality itself.

Philosophical Considerations

Posthuman civilizations may grapple with existential questions of profound depth, pondering the meaning of their existence, their purpose within the cosmos, and the limits of their own understanding. Ancestor simulations could offer a powerful tool for philosophical exploration. By constructing simulations that reflect their own journey, posthuman thinkers could search for patterns that illuminate the fundamental forces shaping civilizations. They could explore the origin of life, the nature of consciousness, or the possibility of a greater purpose underlying the universe's existence. Ancestor simulations might allow them to probe their own place within a vast, potentially infinite, multiversal landscape.

The ability to create simulations could lead posthuman civilizations to confront uncomfortable truths or question their fundamental assumptions about reality. The realization that their own existence might be a mere construct within a simulated hierarchy raises profound questions about the nature of purpose and the existence of ultimate meaning. Faced with this potential existential void, posthuman societies may turn to simulations to seek answers, to experiment with their own foundations, or even to create simulated realities where purpose and meaning are deliberately engineered. In this sense, the desire to simulate could become a response to the existential challenges arising from technological advancement, an attempt to grapple with the mysteries of existence, and a quest to find purpose within a potentially indifferent cosmos.

Ethical Considerations

Sentience and Personhood

A formidable methodological hurdle in assessing the potential motives behind ancestor simulations is the profound ethical quandary surrounding the possibility of sentient beings within those simulations. If ancestor simulations achieve sufficient complexity to replicate conscious experience, simulated entities may attain a level of self-awareness that demands moral consideration. This compels posthuman civilizations to confront age-old philosophical debates about the nature of consciousness. Can consciousness be an emergent property of sufficiently complex computational systems, or is it inherently tied to biological substrates? If digital sentience is achievable, do simulated beings possess the same inalienable rights and moral status as their creators?

The ethical implications of simulated sentience could significantly influence a civilization's Desire to Simulate. Posthuman societies may face stark moral dilemmas when weighing the potential benefits of such simulations against the prospect of inflicting suffering on simulated beings. The pursuit of knowledge, entertainment, or even historical reverence would need to be carefully scrutinized through the lens of an ethical framework designed to protect the rights and well-being of potentially conscious entities within the simulation. The question of whether simulations should be created at all, given the potential for inflicting suffering, could become a central point of philosophical and moral debate for any civilization grappling with the power to create and control simulated realities.

Civilizational Values

The Desire to Simulate is unlikely to exist in a vacuum. It will be deeply influenced by the predominant ethical values and philosophical principles that shape a posthuman civilization. Societies that place supreme importance on non-harm, respect for individual autonomy, and the sanctity of conscious experience may be profoundly hesitant to embark on complex ancestor simulations. The awareness that such simulations could give rise to sentient beings, who could then suffer due to simulated events, presents a significant moral obstacle. For these civilizations, the pursuit of scientific inquiry, entertainment, or historical preservation might be outweighed by the ethical risk of inflicting simulated pain and hardship.

The decision to create ancestor simulations is likely to be fraught with internal conflict for civilizations with strong ethical commitments. The potential benefits of knowledge gained could clash with a deep respect for individual autonomy, forcing difficult decisions about the moral permissibility of "playing god" with simulated beings. The very existence of a compassionate posthuman society could lead them to question whether the creation of conscious lives, solely for the purpose of simulation and potential suffering, is compatible with their fundamental values. Thus, a strong commitment to ethical principles could become a significant counterweight to the Desire to Simulate, prompting civilizations to explore alternative, less ethically fraught, research avenues.

Unintended Consequences

Even for technologically sophisticated posthuman civilizations, the creation of ancestor simulations might be fraught with anxieties about unintended consequences and unforeseen risks. One concern could center on the potential destabilization of their base reality. Simulations capable of generating conscious entities may open up questions about the nature of reality itself, potentially leading to existential doubts and uncertainties within the posthuman society. Additionally, simulated beings, upon potentially discovering the true nature of their existence, could experience profound psychological distress, creating ethical dilemmas for their creators.

Posthuman civilizations may also worry about a slippery slope where the creation of ancestor simulations could pave the way for entirely artificial constructed worlds. An over-reliance on simulations for entertainment or escapism could lead to a detachment from base reality, fostering a societal preference for simulated environments. There's also the concern that the ability to easily construct artificial realities might tempt civilizations to withdraw into carefully tailored simulations, abandoning any desire to engage with the challenges and limitations of their real-world existence. Therefore, a thorough consideration of these potential unintended consequences could temper the Desire to Simulate, prompting civilizations to proceed with caution or perhaps explore entirely different technological paths.

Interdisciplinary Collaboration

The complexity of DS necessitates a robust interdisciplinary approach to yield meaningful insights. Key areas of expertise that could contribute include:

Psychology

Cognitive scientists and psychologists hold the potential to make significant contributions to our understanding of the motivations that could drive civilizations to create ancestor simulations. By drawing on well-established models of human behavior and expanding them into the hypothetical posthuman context, they can offer valuable insights. Fundamental drives like curiosity, a desire for knowledge, and the pursuit of novelty are likely to persist, shaping the motivations of even technologically advanced societies. Psychologists could help analyze how these drives manifest at the civilizational level, fueling a desire for exploration and potentially influencing a civilization's decision to embark on the simulation of their own history.

Beyond fundamental drives, psychologists could investigate how simulated realities might offer unique avenues for wish fulfillment and escapism. Simulations could provide a means for posthuman individuals or societies to achieve power, control, and experiences unavailable or unattainable in their base reality. By understanding the psychological allure of these simulated environments, we can better assess the potential for simulations to become highly sought-after, influencing the social landscape and the Desire to Simulate within posthuman civilizations. Psychologists might also identify potential dangers, such as addiction or societal withdrawal, associated with an over-reliance on simulations, prompting ethical debates about moderation and responsible use of simulated environments.

Sociology

Sociologists can play a crucial role in mapping the societal trends, power dynamics, and evolving value systems that would shape a civilization's priorities and consequently influence its Desire to Simulate. They can offer insights into how resource allocation decisions might be influenced by various sectors within a posthuman society. For example, if knowledge acquisition and historical preservation are highly valued, this could translate into significant resources devoted to the creation of ancestor simulations. Alternatively, societal concerns about resource depletion, ethical considerations, or potential unforeseen consequences could lead to simulations being relegated to a lower priority.

Sociologists specializing in the study of values can provide a long-term perspective on how societal priorities evolve over time. Posthuman societies may undergo transformations that significantly alter their Desire to Simulate. Factors like technological breakthroughs, philosophical shifts, or existential threats could reshape societal priorities, potentially leading to a surge of interest in simulations or conversely, a waning enthusiasm. A sociological view allows us to consider the Desire to Simulate not as a static characteristic but as a dynamic aspect of a civilization's development, influenced by a multitude of internal and external factors.

Philosophy

Philosophers specializing in ethics will play a vital role in developing frameworks to address the profound moral quandaries raised by the possibility of sentient simulations. Their expertise is crucial for debates surrounding the rights of simulated beings, the moral obligations of simulator civilizations, and the potential trade-offs between scientific advancement and inflicting suffering on simulated entities. Philosophers can help posthuman societies navigate the ethical complexities surrounding ancestor simulations. These frameworks may significantly influence the Desire to Simulate, either by justifying their creation under strict ethical guidelines or prompting a cautious approach, potentially leading to limitations or outright bans on such simulations.

The Simulation Hypothesis forces us to confront fundamental metaphysical questions about the nature of reality itself. Philosophers specializing in metaphysics can contribute by analyzing the implications of simulated realities, exploring potential distinctions between different levels of simulation, and questioning the epistemological limits of beings within a constructed world. Furthermore, the philosophy of mind is indispensable for addressing whether consciousness is an emergent property of complex computational systems or inherently tied to a specific biological substrate. These philosophical inquiries are not simply theoretical; they hold the potential to directly shape a civilization's Desire to Simulate by shedding light on the potential for simulated consciousness and the ontological implications of creating artificial realities.

History

Historians offer a valuable lens for examining the enduring desires and recurring motifs that have shaped human civilizations throughout time. Analyzing ambition and the pursuit of knowledge, as manifested in large-scale historical projects and scientific endeavors, can provide hints about the potential motivations of posthuman societies. Similarly, studying how past civilizations have grappled with existential questions about their origins, purpose, and legacy can offer insights into the potential philosophical drivers behind the Desire to Simulate.

Historians specializing in the history of technology and science can shed light on long-term trends in innovation, the relationship between technological breakthroughs and social change, and the unintended consequences of technological revolutions. This historical perspective can inform our predictions about the trajectory of posthuman civilizations, highlighting both the potential for extraordinary advancements and the unforeseen challenges they may face. By understanding how past societies have adapted to new technologies, we can gain a better sense of how posthuman civilizations might navigate the ethical and societal complexities arising from the potential to create hyper-realistic simulations.

Computer Science and Computational Modelling

The expertise of computer scientists is essential for translating the theoretical possibilities of ancestor simulations into concrete assessments of feasibility, complexity, and resource requirements. They can develop the simulation frameworks and computational models necessary to explore different scenarios and parameters for simulating complex systems. By quantifying the trade-offs between simulation fidelity, complexity, and computational costs, computer scientists provide valuable insights that can directly inform our understanding of the Desire to Simulate.

Computational modelling allows researchers to explore the potential challenges and limitations associated with large-scale ancestor simulations. By projecting resource requirements, energy consumption, and the development timelines of necessary hardware and software, computer scientists can help us understand whether the creation of such simulations is within the realm of possibility for posthuman civilizations. This information is crucial for assessing the Desire to Simulate. If models suggest that ancestor simulations are entirely infeasible or resource-prohibitive, this could significantly dampen the enthusiasm of even the most knowledge-driven posthuman civilizations. Conversely, if simulations appear computationally achievable, it strengthens the plausibility of scenarios where posthuman societies would possess the means and potentially the Desire to Simulate.

Uncertainty and Limitations

Subjectivity of Desire

Assessing Desire to Simulate (DS) confronts a fundamental hurdle: the inherent subjectivity of desire. Even within our own species, motivations vary widely across individuals, cultures, and time periods. Extrapolating the desires and values of vastly advanced posthuman civilizations, operating at vastly different scales and potentially possessing radically altered psychologies, introduces an extreme level of uncertainty. Our own human history serves as a stark reminder that societal goals, values, and priorities often undergo unpredictable shifts, underscoring the risks inherent in relying on linear projections.

While we may draw some limited insights from historical patterns or our own present-day motivations, the posthuman world remains shrouded in uncertainty. Technological advancement could bring about profound transformations in how consciousness, purpose, and value are perceived. Posthuman civilizations, freed from biological constraints or resource scarcity, might develop entirely novel drives or prioritize goals that are incomprehensible from our present vantage point. Acknowledging these limitations of our understanding is crucial for maintaining a humble perspective when evaluating the likelihood of DS. We must approach such projections with an open mind, recognizing that posthuman civilizations may operate according to motivations entirely alien to our current models and frameworks.

Posthuman Psychology

One formidable challenge in assessing the Desire to Simulate (DS) lies in our limited ability to conceptualize the psychology of posthuman civilizations. These societies may have evolved far beyond our current understanding of cognition, emotion, and motivation. Posthuman psychology could be shaped by vastly augmented intelligence, collective consciousness, technologically altered mental states, or entirely novel forms of consciousness that we cannot yet comprehend. The very foundations of their thought processes could differ radically from our own, rendering attempts at linear extrapolation unreliable.

Our current tools for understanding motivation and societal development are largely based on the study of human behavior within specific historical and technological contexts. Posthuman civilizations, operating with vastly expanded lifespans, resource abundance, and levels of technological sophistication beyond our imagination, will likely function under profoundly different constraints and incentives. Their choices might be influenced by experiences, ethical frameworks, and technological possibilities beyond our ability to anticipate. This inherent limitation in our understanding compels us to approach predictions about DS with caution, acknowledging that posthuman psychology may render traditional models of motivation inadequate or entirely obsolete.

Information Loss

Even if posthuman civilizations embark on ancestor simulations, the reasons behind their decision to create, or not create, such simulations might be lost over time. Evidence of their motivations may deteriorate, be deliberately obscured, or be entirely erased due to natural or civilization-scale calamities. The vast timescales relevant to the Simulation Hypothesis introduce a significant likelihood of information loss. This incompleteness of the historical or technological record poses a formidable challenge for our efforts to decipher the Desire to Simulate.

An assessment of DS must recognize the inevitable incompleteness of the evidence available for analysis. Researchers may only have access to fragmentary data, technological remnants, or surviving historical records that offer partial clues about a civilization's past. This incomplete picture introduces a degree of uncertainty as we attempt to reconstruct the complex motivations at play. Furthermore, even if relatively intact information is available, there remains the risk of misinterpreting data from a posthuman civilization due to cultural, technological, and linguistic barriers. Our methodologies need to be adaptable, capable of working with imperfect evidence and acknowledging potential gaps in our understanding of motivations behind a civilization's past technological choices.

Integrating DS into the Simulation Argument

Unlike assessments of Technological Capability (TC), where the focus often lies on the feasibility of achieving certain computational milestones, evaluating DS necessitates a more nuanced approach. Due to the uncertainties and complexities surrounding posthuman psychology, ethics, and social dynamics, a definitive "yes" or "no" answer about a civilization's desire to create ancestor simulations is unlikely. Instead, our methodology seeks to assign probabilities or likelihoods to various potential scenarios. We may find strong support for the idea that certain posthuman societies would find simulation morally repugnant, while others might embrace it for scientific exploration, and yet some may remain indifferent.

Integrating these probabilistic assessments of DS into Bostrom's Trilemma helps refine our understanding of the overall plausibility of the Simulation Argument. Rather than merely stating that ancestor simulations are possible, we can begin to consider how likely a posthuman civilization would be to actually pursue them. If our analysis suggests a high probability that such civilizations would find simulations ethically objectionable, this would weaken the second proposition of the Trilemma and cast doubt on the hypothesis that we live within a simulation. Conversely, if DS appears highly probable due to potential motivations like historical exploration or scientific curiosity, this would strengthen the possibility that the vast majority of beings exist within simulated realities.

Key considerations when integrating DS into Bostrom's framework:

Interplay with TC

Bostrom's Trilemma compels us to examine the likelihood of our own existence within a simulated reality. Both TC and DS play vital roles in shaping this probability. If our assessment reveals that vast computational power is readily achievable (high TC), but civilizations possess extremely low motivation to create ancestor simulations (low DS), this scenario significantly reduces the overall probability of us existing within a simulation. Even with the technical means, a lack of desire would lessen the likelihood of posthuman civilizations proliferating simulations en masse.

Conversely, the Simulation Hypothesis gains significant traction when both TC and DS are deemed highly probable. If technological advancement seems unconstrained and civilizations are highly likely to be interested in creating ancestor simulations, this paints a picture where simulated realities become common. In this scenario, the sheer number of simulated environments potentially created by posthuman societies would significantly increase the odds that we ourselves inhabit one such simulation. The interplay between TC and DS thus offers a crucial lens for refining our understanding of the probabilistic landscape outlined in Bostrom's Trilemma.

DS as a Potential Filter

Bostrom's Trilemma posits that at least one of three propositions concerning advanced civilizations must hold true. However, their relative probabilities remain open to debate. If our methodological assessments of DS consistently suggest that simulations are broadly undesirable or rarely undertaken by posthuman civilizations, this could strengthen the plausibility of the first proposition. A low DS casts doubt on the idea that a significant fraction of posthuman civilizations would create simulations, highlighting instead a potential bottleneck on the path to achieving the necessary technological stage.

The concept of a low DS adds a potential filter to Bostrom's framework. It suggests that even if a civilization achieves the necessary Technological Capability (TC) for simulating its ancestors, it might refrain from doing so due to a lack of desire. This introduces the possibility that many advanced civilizations fail to reach the "posthuman" stage not because of technological limitations as such, but because ethical, social, or other factors limit their interest in large-scale simulations. If low DS emerges as a common pattern across potential posthuman societies, it strengthens the first proposition of the Trilemma, pointing to the possibility that most civilizations go extinct or stagnate before reaching the point where ancestor simulations become widespread.

The Will to Simulate: Assessing Desire in the Simulation Argument

Examining the Desire to Simulate (DS) forms a crucial pillar in evaluating the plausibility of the Simulation Hypothesis. While methodologies in this domain are inherently speculative, a rigorous and interdisciplinary approach can elucidate the complex psychological, social, and ethical forces that would shape a civilization's will to create detailed ancestor simulations.

Further research would benefit from developing sophisticated simulations to explore civilizational dynamics and the potential long-term evolution of values and motivations. Ultimately, the methodology surrounding DS adds a crucial dimension to the Simulation Argument, compelling us to critically consider not only the technological feasibility of simulated realities but also the very desires that might bring them, or not, into existence.

CHAPTER TWENTY

Estimating Number of Simulations (NS)

Estimating the potential number of ancestor simulations (NS) that advanced civilizations might create stands as a crucial component in evaluating the probability that we ourselves inhabit such a simulation. This endeavor necessitates quantifying a vast landscape of possibilities shaped by hypothetical civilizations operating far beyond our current capabilities. However, arriving at a meaningful estimation of NS is an undeniably complex task, demanding a methodological approach that deftly integrates elements of probability, computational modeling, and a deep understanding of the interplay between technological capability (TC) and the desire to simulate (DS).

Our assessment of NS cannot rely on deterministic formulas. Instead, it hinges on the development of robust probabilistic frameworks that acknowledge the profound uncertainties surrounding posthuman societies. These models must account for potential variability in resource allocation, ethical considerations, the rate of technological advancement, and unpredictable shifts in civilizational priorities that could influence a society's interest in simulations. By quantifying uncertainties and exploring a range of plausible scenarios, we can begin to approximate the potential distribution of NS values across the hypothetical landscape of advanced civilizations.

To refine our estimations of NS, computational simulations offer a powerful tool. By modeling hypothetical civilizations with varying technological capabilities, resource endowments, and ethical constraints, we can explore how different initial conditions could influence long-term trajectories of technological progress and the potential emergence of ancestor simulations. Such simulations, while inherently limited in their ability to fully capture the nuances of posthuman societies, can nonetheless offer valuable insights into the range of possibilities. This allows us to refine our probability distributions and develop a more nuanced understanding of the likely values of NS, strengthening our overall assessment of the Simulation Hypothesis.

Statistical Modeling: Probabilistic Foundations

A robust methodology for estimating NS must embrace the inherent uncertainty surrounding the actions and motivations of posthuman civilizations. Probabilistic reasoning offers a powerful framework for this endeavor. By treating key variables like technological progression, resource allocation, and the desire to simulate as probability distributions rather than fixed values, we acknowledge the spectrum of possibilities that may exist across a vast landscape of potential posthuman societies. This approach moves us away from deterministic predictions and allows us to account for unforeseen breakthroughs, societal shifts, and the potential for divergent paths of technological advancement.

Probabilistic reasoning, when integrated with statistical modeling techniques, provides the tools to quantify the potential range of NS. Bayesian inference, for example, can be leveraged to update our knowledge about likely NS values as new information or refined assessments about posthuman civilizations become available. Statistical models can incorporate factors like the potential distribution of computational resources, the probability of different ethical frameworks emerging, and the likelihood of technological bottlenecks limiting the creation of ancestor simulations. By quantifying these uncertainties through statistical distributions, we gain a more nuanced understanding of the potential number of simulations that might be created, strengthening our evaluation of the Simulation Hypothesis.

Here are key areas for statistical exploration:

Bayesian Inference

Bayesian statistical methods provide a powerful and flexible framework for continually refining our estimations of NS as our understanding of hypothetical posthuman civilizations evolves. Unlike frequentist statistical approaches, Bayesian inference allows us to explicitly incorporate prior knowledge and update beliefs as new data or insights emerge. For instance, we can establish prior probability distributions that reflect our initial hypotheses about the potential range of NS. These distributions may be informed by theoretical considerations, historical trends, or our understanding of the interplay between Technological Capability (TC) and Desire to Simulate (DS).

As we conduct simulation studies exploring hypothetical civilizations, make observations about the limits of computation, or develop refined models for posthuman psychology, this new evidence can be used to update our prior beliefs about NS. Bayesian inference provides a structured way to incorporate this new information, iteratively adjusting our probability distributions for NS. This continuous refinement process allows our estimations to become more nuanced and robust, reflecting the evolving state of our knowledge about the potential actions and motivations of technologically advanced civilizations. Ultimately, it enhances the accuracy of our assessments regarding the overall plausibility of the Simulation Hypothesis.

Monte Carlo Simulations

Monte Carlo methods, characterized by repeated random sampling, offer a powerful tool for modeling the complex interplay between Technological Capability (TC), Desire to Simulate (DS), and the resulting potential number of ancestor simulations (NS). By simulating a vast array of scenarios with varying combinations of TC and DS, Monte Carlo simulations allow us to explore the hypothetical landscape of posthuman civilizations. This approach acknowledges the inherent uncertainties and potential variations in resource allocation, technological trajectories, and ethical frameworks that could shape the decisions of these advanced societies.

A key advantage of Monte Carlo simulations lies in their ability to generate probability distributions for NS. By repeatedly sampling from input distributions representing TC, DS, and other relevant factors, we can visualize the range of plausible NS values that might emerge across a multitude of hypothetical scenarios. These distributions provide a quantitative representation of uncertainty, allowing us to move beyond point estimates and instead consider the likelihood of different NS values within the context of our initial assumptions and the constraints of our models. This nuanced understanding of potential outcomes strengthens our assessment of the Simulation Hypothesis and its implications.

Fermi Estimation

While lacking the precision of rigorous statistical models, a modified Fermi Paradox-style approach can provide valuable, back of the envelope estimations for the potential number of ancestor simulations (NS). This method involves systematically decomposing the problem into a series of manageable assumptions. These could include factors like the number of habitable planets, the fraction of those developing life, the fraction of those life-bearing planets that eventually give rise to technologically advanced civilizations, and ultimately, the fraction of such civilizations interested in running ancestor simulations.

By assigning plausible ranges to each factor in the Fermi estimation chain, we can derive preliminary, order-of-magnitude estimates for NS. While these figures inherently come with significant uncertainty, they serve a valuable function. They can quickly establish whether certain scenarios are astronomically unlikely or fall within a realm of rough plausibility. This allows for a rapid assessment of potential scenarios, aiding in hypothesis generation and potentially narrowing the focus of future, more sophisticated, modeling efforts. Fermi-style estimations offer a heuristic starting point, guiding our intuition and highlighting the vast potential parameter space relevant to the Simulation Hypothesis.

Simulation Analysis: Exploring the Dynamics of Simulation Creation

While statistical models provide a robust foundation for estimating NS, computational simulations offer a complementary avenue for exploring the complex and potentially non-linear dynamics involved in the creation and potential proliferation of ancestor simulations. Unlike purely statistical approaches, which often rely on aggregated variables and probability distributions, simulations allow for granular, agent-based modeling of hypothetical civilizations. We can introduce interacting entities with varying levels of technological capability, resource constraints, ethical considerations, and competing priorities, creating a dynamic virtual laboratory to study the potential long-term outcomes of their actions.

Computational simulations excel at revealing emergent properties within complex systems. By simulating the interactions of hypothetical civilizations over vast timescales, we may uncover unanticipated trends in simulation creation. They could expose the possibility of technological bottlenecks, ethical shifts, or societal transformations that directly or indirectly influence the generation of ancestor simulations. These simulations could help refine our initial probability distributions and assumptions used in statistical models. Simulations serve as a powerful tool for stress-testing our hypotheses about posthuman behavior and technological trajectories, further contributing to a robust and nuanced estimation of NS within the broader context of the Simulation Hypothesis.

To complement statistical approaches, computational simulations can elucidate the complexities inherent in the creation and proliferation of ancestor simulations:

Resource Allocation Models

A critical component in assessing the feasibility of ancestor simulations and their potential number (NS) lies in developing comprehensive resource allocation models. These models are essential for simulating the processes of creating and sustaining complex simulations, accounting for factors like computational resources, energy expenditure, and any potential material constraints faced by posthuman civilizations. While we might assume advanced societies possess vast resources, there could still be limits that impact their ability or willingness to allocate them towards extensive ancestor simulations.

Resource allocation models go beyond simply quantifying power consumption. These models can help elucidate potential trade-offs faced by posthuman civilizations. They can highlight how resource expenditure on large-scale simulations may compete with other technological projects, scientific pursuits, or societal priorities. Additionally, these models can be used to explore the impact of technological bottlenecks. Even with advanced computing technology, there may be fundamental resource or material constraints that limit the scope, fidelity, or number of simulations a civilization can realistically create and maintain. By simulating these resource dynamics, we refine our estimations of NS and gain insights into potential limitations that may curtail the proliferation of ancestor simulations, even among technologically advanced societies.

Nested Simulations

The concept of nested simulations – simulations existing within other simulations – introduces an intriguing layer of complexity to the Simulation Hypothesis. Understanding whether, and to what extent, nested simulations are feasible becomes critical when refining our estimations of NS (the total number of ancestor simulations). Careful study is required to analyze the potential resource constraints, theoretical limits on the depth of recursion (simulations within simulations within simulations...), and the compounding computational overhead associated with layered simulated realities.

Nested simulations could significantly impact resource allocation models and energy expenditure calculations. The possibility of exponential increases in computational requirements with each additional layer of simulation could place practical constraints on the number of nested iterations that are feasible, even for advanced civilizations. Investigating these limitations, alongside potential strategies for optimizing simulations within simulations, becomes essential. Ultimately, our understanding of nested simulations may lead to a revision of NS probabilities. It could either decrease estimations due to resource limitations or potentially increase them if computational optimizations are found to be viable at scale within nested environments.

Evolutionary Simulation Dynamics

Integrating evolutionary algorithms or other simulation techniques focused on evolutionary dynamics offers a promising avenue for exploring potential patterns and preferences among hypothetical simulator civilizations. By simulating the process of creating and selecting simulations, we can investigate whether certain historical periods, critical turning points ("bottleneck events"), or civilizations that achieved specific technological milestones might be of particular interest to simulators. These simulations could highlight potential biases or recurrent trends in the types of ancestor simulations that posthuman civilizations are inclined to create.

Insights gained from evolutionary dynamics simulations have direct implications for refining our estimations of NS (the number of ancestor simulations). If a strong tendency emerges among simulated civilizations to focus on particular historical epochs or developmental trajectories, this would suggest that certain types of ancestor simulations are more likely to be created. Such a trend could introduce a bias in the distribution of simulation types. On the other hand, simulations could reveal a diverse range of simulator interests. This would support a more uniform distribution in our estimated NS values. By uncovering potential patterns or tendencies in the creation of simulated realities, evolutionary dynamic simulations help us construct more informed probabilistic models for assessing the overall plausibility of the Simulation Hypothesis.

Critical Parameters and Their Influence on NS

Three critical variables – Technological Capability (TC), Simulator Motivation (Desire to Simulate or DS), and potential gains in Computational Efficiency – exert significant influence on our estimations of NS and show complex interdependencies. Advances in TC directly expand the scope of what is computationally feasible, potentially leading to higher-fidelity simulations or a greater number of less complex simulations. Simulator Motivation (DS) shapes whether available TC is leveraged for ancestor simulations at all and might influence the types of simulations prioritized. Computational efficiency gains hold the potential to dramatically affect the relationship between TC and the number of achievable simulations. Significant efficiency optimizations could allow for far more simulations to be run given the same level of technological advancement.

The complex interplay between TC, DS, and computational efficiency introduces profound uncertainties into our attempts to estimate NS. Rapidly increasing TC coupled with strong DS could lead to a scenario where the proliferation of ancestor simulations becomes extremely likely. However, this likelihood is tempered by questions about potential limits to long-term efficiency gains. Moreover, even with high TC and substantial efficiency optimizations, a low DS among posthuman civilizations could severely curtail the overall number of ancestor simulations. Assessing NS demands a multifaceted approach that carefully considers potential trade-offs, resource constraints, and the unpredictable dynamics of hypothetical civilizational trajectories.

Our NS estimation methodology must explicitly address several variables that carry significant weight:

TC and Simulation Fidelity

The technological capability (TC) of hypothetical simulator civilizations places fundamental constraints on the potential number of ancestor simulations (NS) they can simultaneously maintain. The relationship between simulation fidelity (the level of detail and realism) and computational cost is a crucial factor. High-fidelity simulations place significantly greater demands on computational resources, potentially forcing trade-offs between the quality of individual simulations and the total quantity a civilization can sustain at a given time.

Improvements in simulator technology (TC) could mitigate the computational costs associated with high-fidelity simulations. However, there may be fundamental limits to the degree of detail achievable within simulations, even with vast computational resources. These factors directly shape the distribution of potential

NS values. If simulators prioritize high-fidelity simulations, this could limit the total number they can create, even with robust TC. Conversely, if low-fidelity simulations are sufficient for their purposes, a wider range of NS values becomes plausible. Understanding the nature of these trade-offs is crucial for refining our statistical models and assessments of the Simulation Hypothesis.

Simulator Motivation (DS)

Unraveling the spectrum of motivations behind ancestor simulations – the Desire to Simulate (DS) – holds the key to refining our estimations of NS. The underlying reasons for their creation could range from scientific curiosity and historical exploration to entertainment or the pursuit of existential insights. These motivations will likely shape the types of simulations prioritized and the resources allocated to them. A civilization driven primarily by a thirst for knowledge might focus on detailed historical simulations, while those seeking entertainment could opt for less computationally demanding, stylized simulations with narrative emphasis.

Simulator motivations (DS) directly influence the computational costs and potential quantity of simulations. A strong emphasis on historical precision could drive the creation of fewer, higher-fidelity simulations. In contrast, motivations like entertainment might favor a profusion of less detailed simulations with a greater emphasis on variety or gamification elements. Understanding DS also informs the scope of simulations: civilizations motivated by ethical explorations might create self-contained simulations focusing on morality, while historical explorers might focus on broader historical sweep. Each motivational scenario introduces unique parameters that must be carefully integrated into our statistical and computational models to refine our estimations of NS and, consequently, our assessment of the Simulation Hypothesis.

Computational Efficiency Gains

A crucial factor to consider when evaluating NS is the potential for simulator civilizations to unlock revolutionary advancements in computational efficiency. While our models might be based on extrapolations of current technology or theoretical limits, it's possible that posthuman civilizations could develop entirely novel computational paradigms or simulation methods that vastly outperform our current understanding. Such breakthroughs could fundamentally alter the relationship between computational power and the feasibility of large-scale simulations.

Significant gains in computational efficiency could lead to an exponential increase in the number of simulations (NS) that advanced civilizations could create and sustain. Simulations that might be deemed computationally intractable based on our present understanding could become readily achievable with vastly more efficient technologies. This underscores a key limitation in our assessments – we may be significantly underestimating the potential for posthuman civilizations to optimize the simulation process in ways we cannot currently envision. Integrating the possibility of such breakthroughs into our probabilistic models is essential, acknowledging that potential NS values might be far higher than our conservative estimates suggest.

Connecting NS Estimation to the Ancestor Simulation Hypothesis

The endeavor to approximate the potential number of ancestor simulations (NS) that advanced civilizations might create stands as a foundational pillar in assessing the plausibility of the Simulation Hypothesis. Bostrom's Trilemma compels us to examine the odds that we ourselves occupy one such simulation. The greater the estimated NS, the higher the probabilistic weight of this particular proposition within the Trilemma. Conversely, if our assessments suggest that advanced civilizations rarely create large numbers of simulations, the likelihood of our own existence within a simulation is significantly diminished.

Our estimations of NS, however uncertain, provide the quantitative basis for weighing the relative probabilities of the three propositions in Bostrom's Trilemma. By establishing plausible ranges for NS, even if those ranges span several orders of magnitude, we can begin to delineate the potential landscape of possibilities. A scenario characterized by astronomically high NS values strengthens the Simulation Hypothesis. Alternatively, assessments pointing towards extremely low NS values would cast greater doubt on our simulated existence, potentially favoring other propositions in the Trilemma. Therefore, our methodological efforts in estimating NS ultimately shape our interpretation of the Trilemma while also driving further research to refine our understanding of this intricate and thought-provoking hypothesis.

Our efforts in approximating NS play a pivotal role in evaluating the overall plausibility of the Ancestor Simulation Argument. Consider key connections:

NS and Simulation Proliferation

If our assessment suggests that a large estimated number of ancestor simulations (NS) is plausible, a critical follow-up question arises: what is the likelihood that some of these simulated civilizations within a base-level simulation would independently develop their own capacity and desire to simulate ancestors? This potential for recursion – the creation of simulations within simulations – carries profound implications for evaluating the overall Simulation Hypothesis.

Recursive simulation proliferation could lead to an exponential increase in the total number of simulated realities. If a significant fraction of simulated civilizations reaches the "posthuman" stage and embarks on their own simulation projects, it dramatically increases the odds that we ourselves inhabit one of the many nested levels of simulated realities. Therefore, a large estimated NS indirectly strengthens the plausibility of the Simulation Hypothesis by increasing the overall pool of potential realities within which we might exist. Furthermore, it underscores the potential fractal nature of reality if the creation of simulations is a common outcome for advanced civilizations.

NS and the Fraction of Simulated Individuals

A core tenet of Bostrom's Simulation Argument hinges on the relative number of simulated minds compared to biological ones. His argument posits that if the potential number of ancestor simulations (NS) is vastly larger than the number of civilizations that reach a posthuman stage without running simulations, then the overwhelming majority of conscious beings with experiences similar to ours are likely simulated entities rather than biological organisms. Our estimations of NS become crucial in evaluating this proposition.

By establishing plausible ranges for NS, even with inherent uncertainties, we gain valuable numerical insights into the potential landscape of simulated and biological minds. A high NS translates to a far greater number of simulated realities compared to unsimulated ones. This implies that the odds of a randomly chosen conscious individual with our kind of experiences residing in a simulation become significantly higher. Conversely, if our estimations suggest a relatively low NS, the balance between simulated and biological minds may be more even, requiring further exploration of the Trilemma within the Simulation Hypothesis. Ultimately, our efforts in approximating NS provide a quantitative foundation for assessing the relative prevalence of simulated consciousness across the vast expanse of potential realities.

Observational Constraints

The Simulation Hypothesis confronts potential limitations imposed by the observable universe. If there are fundamental physical constraints such as finite energy resources, constraints on information density, or limitations on computation imposed by the laws of physics, it could place an upper bound on the possible number of simulations (NS) that could be created. To explore the potential for such constraints, detailed modeling is required to quantify the relationship between the computational requirements of large-scale simulations and the limits imposed by our observable universe.

Through rigorous modeling and comparison with observational data, we may uncover potential inconsistencies that challenge the Simulation Hypothesis. If our calculations reveal that the computational resources required for even a modest NS would vastly exceed the available resources within our observable universe, it raises questions about the feasibility of such simulations. Such contradictions could serve as potential evidence against the Simulation Hypothesis, or at least suggest specific limits on the scope and number of simulations. Alternatively, if models show that even a vast number of simulations could exist within the limits of our observable universe, the Simulation Hypothesis remains a viable possibility. Thus, our quest for observational constraints plays a crucial role in testing the fundamental assumptions and boundaries of this profound hypothesis.

Challenges and Refinements

The very act of estimating NS rests upon a foundation of speculation and conjecture. When assessing the technological capabilities and motivations of posthuman civilizations vastly more advanced than our own, we inevitably venture beyond the secure boundaries of established knowledge. Our extrapolations from current scientific progress, technological trends, and societal dynamics may prove unreliable guides when applied to civilizations operating at vastly different scales and potentially possessing radically altered psychologies or ethical frameworks.

Our estimations of NS are riddled with unavoidable uncertainties. From the unpredictable pace of technological breakthroughs and the potential for societal shifts within posthuman societies to the unknowable motivations that could drive their actions, myriad factors remain shrouded in doubt. These uncertainties necessitate a flexible methodology, one that explicitly acknowledges underlying assumptions and actively seeks to refine those assumptions based on new insights, advances in simulation studies, and evolving theoretical models of advanced civilizations. Estimating NS is an iterative process, demanding continual reassessment and a willingness to embrace the potential limitations of our current understanding.

Estimating NS is an inherently speculative exercise. Let's acknowledge and address key challenges and propose methodological refinements:

The Unknowable Unknowns

One of the most formidable challenges in estimating NS lies in accounting for the possibility of entirely unforeseen motivations driving simulator civilizations to create large numbers of simulations. Our current understanding of psychology, sociology, and even ethics may offer limited insights when extrapolated to the realm of posthuman societies possessing vastly expanded technological capabilities. We are forced to confront the potential for "unknown unknowns" – motivations, goals, or values that we cannot currently conceptualize.

To mitigate the risks inherent in unforeseen motivations, our methodological approach must incorporate robust scenario planning techniques. By deliberately stretching the bounds of our imagination, we can explore unconventional scenarios and the outer fringes of conceivable simulator behaviors. This might involve constructing narratives featuring civilizations driven by motivations that seem strange or even ethically troubling from our present perspective. Scenario planning forces us to confront the limitations of our current understanding and model scenarios where the drivers behind simulation creation might be fundamentally alien to our own values and thought processes. While speculative, these exercises help refine our probabilistic models and expand our consideration of the vast possibility space relevant to the Simulation Hypothesis.

Metaphysical Assumptions

If the Simulation Hypothesis holds true, and our own universe is a simulation, our ability to accurately estimate NS from within this simulated environment may encounter fundamental limitations. Simulators could deliberately or unintentionally introduce biases, distortions, or computational constraints within our simulated reality. These manipulations could subtly or significantly skew our observations of physical laws, the availability of computational resources, or even the nature of time and causality.

Our task of estimating NS would be further complicated if simulator civilizations actively seek to conceal the true nature of our reality. They might manipulate the results of our scientific inquiries, limit our ability to probe the edges of the observable universe, or introduce deliberate inconsistencies designed to prevent us from deducing our simulated existence. Under such a scenario, our assessments of NS could be wildly inaccurate. This underscores a profound metaphysical challenge: our methodology must carefully consider how a simulated environment might distort our observations, calculations, and inferences about the broader potential landscape of simulations within which we might reside.

Methodological Enhancements

To further strengthen our NS estimation framework, we propose the following enhancements:

Interdisciplinary Collaboration

The quest to estimate NS, a core aspect of the Simulation Hypothesis, thrives on the rich tapestry woven from diverse scientific disciplines. This collaborative approach brings together experts from seemingly disparate fields, each contributing a crucial piece to the puzzle. Cosmologists offer insights into the fundamental nature of the universe, the potential for undiscovered resources, and the ultimate limitations imposed by physical laws. Astrophysicists add their expertise on the vastness of the observable universe, potential energy constraints, and the ongoing search for extraterrestrial intelligence.

The inclusion of experts in the philosophy of information injects a vital philosophical dimension into the discussion. Their focus on the nature and manipulation of information within a simulated reality proves invaluable. By exploring the relationship between information, computation, and the fabric of reality itself, these philosophers help us refine our understanding of the potential constraints and resource requirements for creating and sustaining large-scale simulations.

Through interdisciplinary collaboration, we can establish a more comprehensive framework for estimating NS. Cosmologists illuminate the stage upon which simulations might occur, astrophysicists map the boundaries and potential resources within that stage, and philosophers of information help us theorize about the nature of the actors on that stage. By weaving together these insights, we move beyond siloed disciplinary perspectives, forging a unified approach that strengthens our understanding of the landscape of possibilities with regards to ancestor simulations and the overall plausibility of the Simulation Hypothesis.

Cultural and Ethical Considerations

A nuanced estimation of NS hinges on examining the potential cultural and ethical values that might shape the attitudes of posthuman civilizations towards simulations. Posthuman societies may view the simulation of conscious beings as ethically abhorrent, placing strict prohibitions on such practices. Conversely, they might embrace simulation as a tool for exploration and knowledge generation. These cultural attitudes could profoundly influence the likelihood of creating detailed ancestor simulations, directly impacting our NS models.

By investigating ethically charged questions surrounding the nature of consciousness, the moral rights of simulated beings, and the potential for simulated suffering, we gain a deeper understanding of the motivations or hesitations that could drive simulator civilizations. Posthuman societies that exhibit strong reverence for individual autonomy or place supreme value on minimizing suffering might be disinclined to create simulations for fear of violating those principles. Exploring a spectrum of potential ethical frameworks helps us construct more robust probabilistic models, acknowledging that decisions about ancestor simulations might be primarily driven by values rather than mere technological capability. This multi-faceted approach ultimately strengthens our capacity to predict the potential range of NS.

Simulation 'Fingerprints'

The possibility that subtle artifacts, imperfections, or computational 'glitches' could betray the artificial nature of a simulated reality introduces a fascinating avenue for research. Developing robust methods to actively search for such telltale signs could revolutionize our understanding of NS and simulator practices. Potential 'fingerprints' could include inconsistencies in the laws of physics, limitations in computational precision, or detectable patterns within observed phenomena. While speculative, this research direction holds the potential to provide empirical clues about the possible existence of simulations and the characteristics of their underlying architecture.

The discovery of simulation 'fingerprints' could dramatically transform our current assessments of NS. The presence of such artifacts could inform us about the level of technological sophistication of simulator civilizations and the degree of fidelity they can achieve within their simulations. Furthermore, the identification of specific types of 'glitches' could potentially shed light on common computational optimization strategies or limitations employed by simulators. These insights would directly impact our probabilistic models, placing potential constraints on NS, while offering valuable glimpses into the motivations and practices of these technologically advanced civilizations.

The Role of Self-Simulation

The possibility that conscious beings within a simulation could themselves develop the capacity and desire to create their own simulations introduces a fascinating and potentially recursive element into the Simulation Hypothesis. This concept, known as self-simulation, could lead to a fractal structure of realities as simulations spawn yet more simulations within themselves. If self-simulation is feasible and prevalent, the potential number of simulations (NS) could increase exponentially, raising profound questions about the nature of our own existence and its place within a potentially vast hierarchy of realities.

Assessing the likelihood of self-simulation adds considerable complexity to our estimations of NS. If the creation of simulations is a common outcome for technologically advanced civilizations, and a significant fraction of those simulated societies embark on creating their own simulations, then standard probabilistic models may drastically underestimate the potential number of simulated realities. Our methodologies for estimating NS need to account for this potential recursive element, introducing the possibility of vast numbers of nested simulations. The implications of self-simulation on the overall Simulation Hypothesis are profound, strengthening the plausibility that the vast majority of conscious beings could exist within simulations rather than a 'base' reality.

One particularly intriguing aspect of the simulation argument that impacts our NS estimation is the possibility of self-simulation.

Here's why this warrants specific attention in our methodology:

Technological Threshold

The prospect of our own civilization approaching the technological threshold for creating high-fidelity ancestor simulations presents a fascinating juncture within the framework of the Simulation Hypothesis. This potential, if realized, would provide strong empirical support for Nick Bostrom's arguments. Our growing technological prowess strengthens the possibility that other civilizations have already achieved this capability, potentially leading to a vast number of ancestor simulations (NS) and a universe dominated by simulated realities.

The very fact that we can contemplate creating detailed simulations lends credence to the idea that others have done so before us. This realization transforms the Simulation Hypothesis from a purely theoretical exercise into a realm with tangible implications for our understanding of existence. Furthermore, our impending technological leap introduces a potential self-selection bias. If civilizations that reach this advanced stage are also the ones inclined to create ancestor simulations, it strengthens the possibility that we ourselves reside within a simulated reality. This bias could significantly impact the actual number of NS, potentially pushing it towards the higher end of our estimated ranges.

Our progress towards the ability to create high-fidelity simulations serves as a valuable stepping stone in refining our estimations of NS. By carefully analyzing the technological requirements for such simulations, the computational resources they demand, and the ethical considerations that might influence our decisions, we can construct more robust models for assessing the potential capabilities of past or future advanced civilizations. Understanding our own technological trajectory allows us to extrapolate outwards, providing a more grounded basis for estimating NS and the overall plausibility of the Simulation Hypothesis.

This analysis connects the concept of a technological threshold to the broader question of the number of simulated realities and the possibility that we might be living in one ourselves. It injects a sense of real-world relevance into the theoretical discussions surrounding the Simulation Hypothesis.

Computational Resource Trade-Offs

The possibility that we may choose to devote significant computational resources towards simulating our own history raises the question of potential trade-offs with other scientific endeavors or simulations focused on exploring potential futures. While our technological capabilities may continue to expand, we will likely still face constraints on computing power, energy, and material resources. These constraints necessitate careful consideration of priorities and allocation decisions as we develop the ability to conduct diverse simulations. This dynamic interplay directly influences our assessment of NS – resources dedicated to ancestor simulations may not be available for other computationally demanding simulations.

Our understanding of the potential trade-offs between ancestor simulations, scientific simulations, and simulations geared towards exploring future scenarios can help refine our NS models. Exploring societal shifts, evolving values, and technological breakthroughs that shift priorities between simulation types offers insights into how advanced civilizations might allocate their resources. For instance, an unforeseen breakthrough in energy production could dramatically expand the potential for running multiple resource-intensive simulations simultaneously, increasing the likelihood of high NS. Conversely, prioritizing predictive simulations designed to avert existential threats could limit the scope or number of ancestor simulations, shifting the potential landscape of the Simulation Hypothesis.

Ethical Dilemmas of Self-Simulation

As we contemplate the act of simulating our own past, profound ethical dilemmas emerge that could have significant implications when modeling the potential number of simulations (NS). Concerns arise around the manipulation of simulated timelines, potential modifications to historical events, and the moral standing of simulated individuals who may achieve self-awareness within such a construct. Confronting the ethical burdens of intervening in simulated histories and grappling with the concept of simulated suffering could introduce powerful motivations for or against ancestor simulations among future civilizations.

Exploring the diverse ethical frameworks that could shape our decisions regarding ancestor simulations offers valuable insights into the motivations that might influence the actions of other technologically advanced civilizations. If we deem ancestor simulations morally reprehensible due to the potential suffering of simulated individuals, it suggests that similar ethical qualms might dissuade other civilizations from engaging in this practice, potentially limiting NS. Conversely, if societies resolve these ethical dilemmas or justify potential suffering in the name of historical exploration, it increases the plausibility of high NS estimates. Thus, by mapping the complex landscape of ethical considerations surrounding self-simulation, we gain a deeper understanding of the value-driven motivations that could shape a civilization's actions, ultimately informing our probabilistic models within the Simulation Hypothesis.

Quantifying Uncertainty

The endeavor to estimate NS (the number of ancestor simulations) is fundamentally a navigation through a realm riddled with uncertainties. To progress beyond mere speculation, our methods must include a rigorous quantification of the degree of uncertainty surrounding our models, predictions, and assumptions. Techniques like Bayesian inference, sensitivity analysis, and scenario planning can help us assign confidence intervals to our estimations, explicitly mapping out the potential range of NS values and acknowledging the inherent unknowables within our models.

By embracing and quantifying uncertainty, we instill a crucial element of intellectual honesty into our assessments of the Simulation Hypothesis. This recognition of the limits of our knowledge protects against overconfident pronouncements about the plausibility of the Simulation Hypothesis. Moreover, quantifying uncertainty becomes a catalyst for iterative improvement. It helps us pinpoint specific areas where additional data, simulations, or theoretical work could reduce existing uncertainties. Instead of accepting them as immutable constraints, quantified uncertainties highlight the most impactful avenues for refining our methodologies and ultimately, gaining a more robust understanding of the true potential value of NS.

Recognizing the unavoidable uncertainties in estimating NS, a crucial aspect of our methodology is quantifying the degree of uncertainty involved in our models.

Here's how we can approach this:

Sensitivity Analysis

Sensitivity analysis is a crucial tool for exposing the parameters, variables, and assumptions that exert the most significant influence on our estimations of NS. By systematically varying our input assumptions regarding factors like technological capability (TC), Desire to Simulate (DS), potential resource constraints, and others, we can determine how these variations cascade through our models, impacting the final NS estimations.

This reveals the 'levers' to which our results are most sensitive, exposing which underlying assumptions carry the greatest weight.

Pinpointing the critical variables through sensitivity analysis allows us to optimize our future research efforts, enhancing our understanding of the Simulation Hypothesis and its implications. Instead of a scattershot approach, we can focus on refining our understanding of those parameters that exert the greatest influence on NS estimations. For example, if our analysis reveals extreme sensitivity to the rate of technological advancement, it underscores the need for deeper studies on computational limits, breakthrough potential, and long-term technological trends. Conversely, if DS emerges as a dominant factor, it justifies a greater emphasis on the study of posthuman psychology or ethics that could shape the desire for simulations.

Sensitivity analysis allows for a more efficient and targeted research strategy, ultimately strengthening our models for estimating NS and offering a clearer picture of the factors that predominantly shape the landscape of the Simulation Hypothesis.

Confidence Intervals and Uncertainty Ranges

Traditionally, scientific endeavors strive for a single, definitive point estimate for a given parameter. However, in the case of estimating NS, our calculations are inherently probabilistic and riddled with uncertainties. To address this, we turn to statistical techniques that allow us to express our NS estimates not as singular values, but as confidence intervals or uncertainty ranges. This approach offers a far more nuanced and honest portrayal of the current state of knowledge.

Techniques like Bayesian inference allow us to construct confidence intervals for our NS estimates. These intervals represent a range of plausible values within which the "true" value of NS is likely to reside, with a specified level of confidence (e.g., 95% confidence interval). This probabilistic approach acknowledges the limitations of our models and explicitly communicates the degree of uncertainty surrounding our estimations.

Uncertainty ranges offer another method for expressing the limitations inherent in NS estimations. This approach acknowledges that due to the inherent unknowns, a single, definitive confidence interval may not fully capture the potential spread of plausible values. Uncertainty ranges can encompass a wider range of values, reflecting the broader spectrum of possibilities within the Simulation Hypothesis.

Both confidence intervals and uncertainty ranges serve the crucial function of transparency. By clearly communicating the level of uncertainty inherent in our NS estimations, we avoid presenting them as definitive pronouncements. This fosters an environment of scientific honesty and facilitates a more nuanced discussion about the plausibility of the Simulation Hypothesis.

Epistemic vs. Aleatory Uncertainty

The uncertainties surrounding our attempts to estimate NS fall into two broad categories: epistemic uncertainty and aleatory uncertainty. Epistemic uncertainty stems from our incomplete knowledge about simulator civilizations. This includes unknowns about their technological progression, societal values, ethical frameworks, and ultimately their motivations for simulating ancestors. In contrast, aleatory uncertainty arises from inherent randomness or variability, such as the unpredictability of technological breakthroughs, societal shifts, or the potential for existential risks that could derail simulations.

Distinguishing between epistemic and aleatory uncertainty is crucial for tailoring our research approach and refining our NS estimations. Epistemic uncertainty highlights areas where further research could yield significant improvements in our understanding. By focusing on the study of potential posthuman psychologies, ethical systems, long-term technological trends, or resource dynamics, we can potentially reduce epistemic uncertainty and refine our predictive models. On the other hand, aleatory uncertainty necessitates a different approach. It compels us to embrace probabilistic modeling techniques, acknowledging the inherent randomness of certain factors that influence NS.

Charting the Multiverse of Simulations

Estimating the Number of Simulations (NS) emerges as a fundamentally complex and crucial endeavor within the broader evaluation of the Ancestor Simulation Argument. This complex task demands a multi-faceted methodology, drawing upon the strengths of probabilistic modeling, the power of computational simulations, and the insights gleaned from a deep interdisciplinary approach. We must acknowledge that exploring this speculative terrain inevitably involves navigating uncertainties and limitations.

Our pursuit of ever more refined and robust estimations of NS hinges on a candid approach to confronting limitations and quantifying the inherent uncertainties of our models. Furthermore, by venturing into the thought-provoking realm of self-simulation—the possibility of simulations within simulations—we broaden our perspective and potentially uncover nuanced patterns influencing NS values.

The persistent refinement of our NS estimations holds profound implications for the Simulation Hypothesis. With each iteration, we move closer to a deeper understanding of the potential for simulated realities and our possible place within this vast, simulated cosmos. This endeavor carries the potential to fundamentally reshape our perceptions of existence, prompting us to question the very nature of reality itself.

CHAPTER TWENTY-ONE

Theoretical Framework

Evaluating the Ancestor Simulation Argument demands a theoretical framework that transcends traditional disciplinary boundaries. This hypothesis cannot be fully addressed through the lens of computer science, physics, or philosophy alone. Understanding the plausibility of simulated realities requires a tapestry woven from diverse fields, where insights and methodologies from seemingly disparate domains converge to create a holistic and robust platform for investigation.

The Simulation Argument compels us to examine the deep interconnections between seemingly unrelated disciplines. Technological advancements in computation provide the potential means for simulating entire realities, but our understanding of consciousness is needed to explore whether simulated minds could achieve the same level of sentience as biological ones. Additionally, ethical considerations profoundly shape the motivations of simulator civilizations, influencing whether simulations are even pursued and how they might be designed.

This interdisciplinary theoretical framework ultimately facilitates a more profound investigation into the very nature of reality within the Simulation Argument. By synthesizing perspectives from diverse fields, we construct a multi-faceted lens through which to examine the relationship between information and existence, the limits of computation, and the potential for realities existing within realities. This approach expands the boundaries of our inquiry, allowing us to grapple with the fundamental questions posed by the Simulation Hypothesis.

Computer Science: Foundations of Computation and Simulation

Computer Science provides the foundational bedrock upon which the Ancestor Simulation Argument rests. Concepts like Turing machines, algorithms, computational complexity, and the potential limits of simulation technology emerge from this domain. Within the context of the Simulation Hypothesis, Computer Science allows us to investigate the theoretical boundaries of what can be computed and simulated. It helps us frame critical questions: Could a simulated reality be made indistinguishable from our own? Are there fundamental limitations on the detail or scale of simulations? Understanding these computational principles will be essential for assessing the technological feasibility of large-scale ancestor simulations.

Turing Computability

The Simulation Argument hinges on the ability to create detailed simulations of reality, potentially encompassing conscious beings. Therefore, a foundational understanding of computability theory, pioneered by Alan Turing and formalized by the Church-Turing Thesis, becomes paramount. This framework establishes the theoretical capabilities and potential limitations of computational systems, allowing us to assess the feasibility of simulating complex realities. Within this paradigm, algorithms – finite sets of instructions – serve as the building blocks for manipulating information and performing computations. Complexity classes categorize problems based on the resources required for their solution, with some problems inherently intractable due to their exponential resource demands.

The Halting Problem, a cornerstone result in computability theory, exemplifies these limitations. It asks whether there exists a general algorithm that can determine, for any arbitrary program, if it will halt (finish execution) or run indefinitely. The proof of the Halting Problem's undecidability demonstrates the existence of problems that are fundamentally uncomputable by any Turing machine – a stark reminder that even within a simulated universe, there may be inherent limitations on what can be perfectly predicted or controlled. These foundational concepts from Turing computability provide a rigorous framework for analyzing the theoretical bounds of what can be simulated, informing our assessment of the plausibility of the Simulation Argument.

While Turing computability offers a robust framework, the Simulation Argument compels us to consider the possibility of computational paradigms that transcend classical limitations. The potential existence of quantum computers, which leverage the principles of superposition and entanglement, opens avenues for computations that are intractable for classical machines. Furthermore, theoretical frameworks like hypercomputation explore the possibility of computational systems exceeding the capabilities of Turing machines. While these concepts remain speculative, they raise intriguing questions regarding the potential for advanced simulator civilizations to utilize computational models that surpass our current understanding. Investigating these frontiers of theoretical computation is crucial for a comprehensive analysis of the Simulation Argument, as it allows us to consider the full spectrum of possibilities when evaluating the feasibility of simulating complex realities, potentially including conscious minds.

Simulation Theory and Practice

The rich and established field of computer simulation offers valuable practical insights for the theoretical exploration of the Simulation Argument. Techniques such as discrete-event simulation, agent-based modeling, and cellular automata provide powerful tools for representing complex systems. These methods allow us to experiment with computational models of hypothetical civilizations, investigating their behaviors, dynamics, and emergent properties. By borrowing concepts from practical simulation methodologies, we gain a deeper understanding of the potential scalability of simulated realities and the computational challenges that civilizations might encounter when attempting such ambitious simulations.

Applying simulation practice to the Simulation Argument allows us to probe key questions concerning optimization and resource allocation within simulated worlds. We can investigate potential computational bottlenecks, the trade-offs between simulation detail and scope, and the emergence of resource constraints that might influence the behaviors of simulator civilizations. By modeling these factors, we can explore whether such constraints might limit the number of large-scale ancestor simulations that are feasible or lead to the development of innovative optimization strategies within simulations. This practical approach provides a tangible connection between simulation design and the theoretical possibilities considered within the broader Simulation Argument.

Artificial Intelligence and Machine Learning

Advancements in Artificial Intelligence (AI), particularly in areas like natural language processing, computer vision, and reinforcement learning, significantly bolster the potential capabilities of entities existing within a simulated environment. AI plays a dual role: it contributes to the technological tools a simulator civilization might employ, enhancing the realism and autonomy of simulated beings, while simultaneously serving as a potential model for understanding how sophisticated behaviors and potentially even consciousness could emerge from complex computational systems.

The ongoing development of increasingly complex AI systems offers insights relevant to the controversial question of whether consciousness is an emergent property of sufficiently advanced computation. Should machines exhibit behaviors and responses indistinguishable from human-level cognition, it strengthens the plausibility of simulated beings attaining true sentience. Conversely, if AI research reveals limitations in achieving human-like self-awareness in artificial systems, it could point to unique qualities distinguishing biological consciousness from computational simulations. The implications of AI research for understanding the nature of consciousness have direct bearing on the Simulation Argument and our estimations of the moral weight ascribed to simulated individuals.

Physics: Universe as Computation and Fundamental Limits

The question of whether our own universe could be a computational simulation forces a deep examination of the fundamental laws of physics. Theories proposing the universe operates as a form of cellular automaton or that information is a fundamental building block of reality suggest intriguing parallels with computation. Investigating concepts like quantization, the holographic principle, and the potential for discrete spacetime units provide clues as to whether our universe exhibits qualities that align with a computationally constructed reality.

If our universe is a simulation, the laws of physics as we observe them might represent constraints or parameters within the simulators' computational systems. Understanding potential limitations, such as finite precision, a minimum unit of spacetime, or even a speed limit on information processing, sheds light on the constraints the simulators themselves might face. Conversely, if physics reveals no inherent limits that would prevent the simulation of complex universes like our own, it strengthens the plausibility of the Simulation Argument. Therefore, a deep understanding of physics and potential computational interpretations of physical laws informs our assessment of the technical feasibility of simulating entire universes within a 'base reality'.

The Computational Universe Hypothesis

The Computational Universe Hypothesis, a line of thought championed by physicists like Seth Lloyd, proposes that the universe, at its fundamental level, is a vast computational process. This idea envisions information as a core building block of reality with the evolution of the universe being analogous to the output of a cosmic computation. This profound concept blurs the traditional distinction between simulation and base reality. If our universe itself operates on computational principles, it suggests that any simulations we create may inherently be bound by similar limitations on fidelity and complexity, potentially constrained by the computational nature of the very fabric of our own reality.

If the Computational Universe Hypothesis holds true, it introduces a fascinating element into the Simulation Argument. The concept of a simulation nested within an overarching computational universe raises the possibility of limits upon limits. Simulator civilizations could be operating within a base reality that is itself computational, subject to potential restrictions on precision, information density, or the scale of simulations they can feasibly run. Exploring these potential constraints stemming from the fundamental nature of reality aids in refining our estimations of the potential number of simulations while also prompting us to consider the implications of our place within a potentially fractal computational landscape.

Cosmological Constraints and Observable Limits

Our evolving understanding of the observable universe, gleaned from the fields of astrophysics and cosmology, provides essential clues for assessing the plausibility of the Simulation Argument. The observable universe's vast scale, its age, and our understanding of available resources offer insights into the potential 'stage' upon which advanced civilizations and their simulations might exist. Theories about the ultimate fate of the universe (e.g., expansion, Big Crunch scenarios) could place limits on the lifespan of technologically advanced civilizations, therefore impacting how much time they have available to develop simulation capabilities.

Crucially, cosmological observations offer potential insights into energy constraints and information density limits within our universe. If there are fundamental limits on the amount of usable energy or information that can be packed into a given region of spacetime, it could place hard boundaries on the computational capabilities of simulator civilizations. By carefully examining observed universal constants, the distribution of matter and energy, and theoretical limits established by cosmology, we can refine our estimations of the computational resources theoretically available to advanced civilizations. This knowledge, in turn, has direct implications for determining the feasibility and potential scale of ancestor simulations, strengthening or weakening the case for the Simulation Hypothesis.

Quantum Mechanics and Uncertainty Principles: While classical computation underpins traditional simulations, the potential for quantum simulations opens new avenues of inquiry. Quantum mechanics, with its inherent uncertainty, could introduce both challenges and novel opportunities for simulators, affecting the precision and nature of the simulations they might create.

Philosophy: The Nature of Reality, Consciousness, and Ethics

The Simulation Argument extends far beyond technological considerations, venturing into the profound and often unsettling territory of philosophy. It compels us to re-examine fundamental assumptions about the nature of reality itself. Can a simulated universe achieve the same ontological status as a base reality? Could simulated beings achieve conscious awareness? The potential existence of simulations forces us to confront the essence of existence and distinguish between what we perceive as real and the potential underlying computational structure of our world.

The Simulation Argument introduces complex ethical questions with far-reaching implications. If simulated beings are indeed sentient, what moral responsibilities do the simulator civilizations hold toward them? Is it ethically permissible to create simulations filled with suffering, or does this constitute a moral violation? Exploring the concept of simulated rights, the responsibility towards conscious beings within simulations, and the potential for simulated individuals to unknowingly 'break out' of their programmed reality will be a crucial aspect of philosophical inquiries driven by the Simulation Hypothesis. As technology progresses, these ethical considerations will shift from a purely theoretical realm towards pressing, practical challenges that societies may need to address.

Metaphysics and Ontology

The Simulation Argument forces us to reexamine the very bedrock of our understanding of reality, inviting comparisons with age-old philosophical schools of thought. Idealism, with its focus on the primacy of consciousness and the mind-constructed nature of reality, finds intriguing parallels in the notion of a simulated universe. The potential that our senses and observations might be computational artifacts aligns with the concept of solipsism and the question of whether we can truly know a reality external to our own minds. Furthermore, the Simulation Argument revives discussions of skepticism, prompting us to question the reliability of our perceptions and the fundamental basis upon which we judge something to be 'real'.

If our universe is a simulation, our conventional understanding of ontology—the study of the nature of being and existence— is challenged. We are forced to confront questions surrounding the ontological status of simulated beings, the relationship between the computational substrate and the emergent reality it produces, and the potential for multiple layers of simulated realities existing within one another. The Simulation Argument pushes us to grapple with the possibility that what we traditionally consider fundamental might be derivative, leading to a reassessment of the concepts of existence and reality itself.

Philosophy of Mind and Consciousness

The enduring conundrum of consciousness and its potential for simulation lies at the heart of the Simulation Argument. Debates surrounding the hard problem of consciousness, the nature of qualia (subjective experiences), and the distinctions between functionalist and strong AI stances directly address whether consciousness is merely the product of sufficiently complex computation. If consciousness is an emergent property of information processing, it strengthens the possibility that simulated beings could achieve true sentience comparable to our own. Conversely, if consciousness possesses some uniquely biological or irreducible qualities, it casts doubt on the ability to faithfully replicate the full spectrum of subjective experiences within a simulated environment.

Our understanding of the nature of consciousness has profound implications for the Simulation Argument. Should consciousness be computable, it raises questions about the moral weight we assign to simulated individuals and our ethical obligations towards them. If simulators achieve the ability to create indistinguishable simulations of subjective experience, it prompts us to re-evaluate whether our own experiences could be the product of such a simulation. The study of the philosophy of mind plays a vital role in refining our assessment of the plausibility of simulating consciousness and the far-reaching ethical and ontological consequences of such simulations.

Ethics and Simulations

The possibility of large-scale simulations, especially ancestor simulations featuring beings with potential sentience, casts a long ethical shadow. Questions surrounding the moral status of simulated beings emerge as a central concern. Do they possess inherent rights? Should their simulated suffering carry the same moral weight as the suffering of non-simulated beings? Grappling with these questions forces us to scrutinize the responsibilities of simulator civilizations, the ethics of experimentation within simulated worlds, and the potential for simulated individuals to somehow 'break free' of their programming and realize they occupy a simulated reality.

Analyzing the complex ethical challenges posed by the Simulation Argument necessitates drawing upon diverse established ethical theories. Deontological frameworks will help us establish duties and obligations towards simulated beings. Consequentialism will force us to meticulously consider the potential outcomes of simulations, both positive and negative. Virtue ethics compels us to examine the motivations of simulators - are their actions driven by curiosity, a desire for knowledge, or other, potentially less noble motivations? Integrating perspectives from various ethical schools of thought allows us to construct a robust framework for evaluating the moral complexities of creating, manipulating, and potentially terminating detailed simulations that could potentially house conscious minds.

Interdisciplinary Convergence and Emergent Properties

True progress in evaluating the Simulation Argument hinges on the intentional demolition of disciplinary silos. By fostering collaboration between fields as diverse as computer science, physics, philosophy, and neuroscience, we gain a multidimensional view of the complexities involved. This allows us to consider simulation fidelity, the potential for computational bottlenecks, the nature of reality, the emergence of consciousness, and the ethical implications of simulations within a unified framework. This collaborative approach recognizes that no single discipline holds the key to fully unraveling the potential for simulated realities.

The concept of emergence, where complex systems exhibit properties and behaviors that cannot be easily predicted from their individual components, offers important insights for the Simulation Argument. Within simulated civilizations, unexpected outcomes might arise due to emergent social dynamics, unforeseen technological breakthroughs, or the interaction of a vast number of simulated individuals. Understanding how unpredictable phenomena can arise from seemingly simple rulesets can help us model the potential trajectories of simulated civilizations. Furthermore, exploring the potential for the emergence of true consciousness within complex simulations adds another crucial layer to our evaluation of the ethical complexities involved and the plausibility of the Simulation Argument itself.

The true power of this theoretical framework lies in the synthesis across disciplines, rather than a mere collection of separate knowledge domains.

Consider these critical areas of convergence:

Limits of Computability and Consciousness

Investigating the boundaries between what is computationally achievable and the true nature of consciousness stands as a critical frontier in evaluating the Simulation Argument. This complex territory demands a robust dialogue between computer scientists, neuroscientists, and philosophers of mind. Computer scientists help define the theoretical and practical limitations of computation, providing insights into what may or may not be simulable. Neuroscientists shed light on the biological foundations of consciousness and how it arises from the brain's complexity. Philosophers of mind help grapple with the subjective nature of qualia, explore whether consciousness is an emergent property, and question if it can be faithfully replicated outside of a biological substrate.

The Simulation Argument acts as a powerful driver for this interdisciplinary collaboration. It forces us to consider whether a computational system could achieve the complexity necessary to replicate not only the physical structures that give rise to consciousness but also the entirety of subjective experience. Questions about computability lead to questions about the limits of reductionism and whether the essence of consciousness defies purely computational explanations. This cross-disciplinary exploration is likely to produce profound insights that extend beyond the Simulation Argument itself, potentially advancing our understanding of the nature of computation, the workings of the brain, and the mysterious phenomenon of consciousness.

Implications for the Fermi Paradox

The Fermi Paradox highlights a profound contradiction: the universe appears vast, ancient, and should be teeming with intelligent life, yet we observe no definitive evidence of extraterrestrial civilizations. The Simulation Argument introduces a potential explanation for this puzzling silence. If we inhabit a simulated reality, it's possible that our simulation is designed for a specific purpose that does not include other extraterrestrial civilizations. Perhaps the focus of the simulation is on the development and evolution of a single species, or the simulators are not interested in modeling the vastness of the cosmos with multiple intelligent lifeforms.

The Simulation Argument gives rise to several potential scenarios that could explain the Fermi Paradox. Our simulation could be one of many, but the vast majority may not reach the technological sophistication needed for interstellar communication. Furthermore, even if advanced civilizations exist within other simulations, the simulators might prevent or discourage contact between simulated worlds. Ultimately, the Simulation Argument injects the possibility that the reason behind the Great Silence lies not in the inherent rarity of intelligent life, but within the nature of the simulated reality we inhabit and the constraints or intentions of our simulators.

Technological and Ethical Implications

As our technological capabilities progress toward highly realistic simulations, particularly those potentially involving sentient beings, a profound ethical·landscape unfolds that demands urgent, collaborative attention. Computer scientists, ethicists, and even legal scholars must engage in a cross-disciplinary dialogue to address the unprecedented challenges that lie ahead. Key considerations include establishing ethical safeguards to protect simulated individuals, guidelines to prevent simulations involving suffering or exploitation, and the profound question of the moral and legal rights simulated beings might possess should they achieve true sentience.

The prospect of simulations blurring the boundary between simulated and biological sentience raises profound questions with potentially far-reaching social and legal ramifications. If simulations become indistinguishable from reality, how do we define personhood? What responsibilities do creators have towards beings exhibiting the full range of conscious experiences, even if those exist within a simulated environment? Preemptively addressing these ethical quandaries will be crucial to avoid harmful societal consequences as technology advances. Proactive engagement across disciplines is vital to shape the ethical and legal frameworks needed to responsibly navigate the uncharted territories unveiled by the Simulation Argument.

From Theory to Methodology: Practical Applications

The rich interdisciplinary theoretical framework we develop directly shapes and informs the methodologies we employ in evaluating the Simulation Argument. Our understanding of computational limits, potential constraints imposed by the laws of physics, and ongoing debates about the nature of consciousness all play a role in designing simulation studies and refining our search strategies for potential 'glitches' in reality. The theoretical foundation allows us to move beyond simple speculation, guiding us towards methodologies possessing rigor and scientific relevance.

An interdisciplinary theoretical framework facilitates a comprehensive methodological approach. Computer scientists can design simulations to test the computational requirements of simulating complex systems or probe for artifacts left behind by imperfect rendering. Physicists, guided by their understanding of cosmological constraints and fundamental laws, can search for evidence supporting or contradicting the nature of a simulated reality. The inclusion of philosophical perspectives on ethics and consciousness ensures that our methodologies integrate safeguards for simulated beings and prioritize the responsible use of simulations. This collaborative, multifaceted approach increases the likelihood of uncovering meaningful insights and offers greater explanatory power when analyzing the plausibility of the Simulation Argument.

Our interdisciplinary theoretical framework has direct implications for shaping our methodological approach to the simulation hypothesis. Here are key areas where the theory guides our investigative methods:

The Search for Simulation Artifacts:

Different theoretical models make distinct predictions about the potential 'signatures' we might search for within our reality that could reveal its simulated nature. For example:

Computational Limits

- Should our universe be a simulation, it's possible that the simulators faced computational constraints or resource limitations. These constraints could manifest as subtle imperfections, artifacts, or "glitches" within the fabric of our observed reality. Examples include potential discretization effects (the 'pixelation' of spacetime at extremely small scales), inconsistencies in physical constants, or limitations in processing power that might manifest in detectable ways. Investigating these potential limits helps guide our search for evidence that might reveal the simulated nature of our existence.
- Exploring computational limits necessitates a two-pronged approach. Theoretical physicists play a crucial role in examining the laws of physics as we understand them, searching for potential inconsistencies or hints that they might represent programmed parameters of a simulated world. Conversely, experimental physicists can design high-precision experiments to probe the boundaries of the observable universe, looking for anomalies, unexplained quantization, or limitations in the fidelity with which reality appears to be rendered. By carefully analyzing potential computational limitations and their potential implications for physical observables, we enhance our ability to develop experiments and search strategies that could directly test the Simulation Hypothesis.

Philosophical Clues

- The laws of physics, as we observe them, play a dual role in our investigation of the Simulation Argument. They represent both the potential ruleset governing a simulated reality and a source of clues concerning the nature of our existence. The discovery of inconsistencies, limitations, unexplained changes, or even evidence of 'breaks' in the laws of physics could have profound philosophical implications. Depending on the nature of such discoveries, they could be interpreted as evidence of external manipulation by a simulator or imperfections resulting from computational constraints imposed on their simulation.
- Philosophical debates surrounding simulation find their grounding in our observed physical reality. If physics appears consistent, predictable, and self-contained, it may support arguments against the Simulation Hypothesis or strengthen realist metaphysical stances. Conversely, the discovery of inconsistencies, unexplained phenomena, or evidence of manipulation would add weight to the Simulation Argument. Furthermore, it could provide clues about the nature of the simulators themselves and their relationship to our universe. This highlights the deep connection between philosophical inquiry and empirical observation in the context of the Simulation Hypothesis.

Purposeful 'Easter Eggs'

- The possibility of simulators purposefully leaving "Easter Eggs", or deliberate anomalies, within a simulated reality is a fascinating concept to consider. These could range from subtle deviations in expected physical phenomena to more overt inconsistencies or glitches. Such Easter Eggs could be conceived as clues left by the simulators for their creation to discover, serving as evidence of their simulated status. Alternatively, they could simply exist for the amusement of the simulators themselves, a form of entertainment as they watch their creation unfold.
- The notion of simulator-embedded Easter Eggs influences both theoretical and experimental approaches to the Simulation Hypothesis. Theoretically, it prompts consideration of what kinds of anomalies might be plausible and detectable within a simulated reality. From a search perspective, it would encourage scientists to not only look for unintended glitches resulting from computational limitations but focus on unusual phenomena, potential patterns, or deviations in the fabric of reality that could hint at deliberate design by a simulator. While the discovery of such Easter Eggs would be

strong, though not definitive, evidence for the Simulation Argument, their potential existence justifies deeper scrutiny of the universe in ways that might otherwise be considered unconventional.

Refinement of Simulation Detection Strategies

Our understanding of the constraints and potential motivations of simulators informs the kinds of phenomena we should be focused on detecting. Considerations include:

Energy and Resource Discrepancies

- The Simulation Argument hinges on the assumption that advanced civilizations would possess the technological means to create computationally intensive simulations. Such simulations, especially those replicating entire universes, would likely have staggering energy demands. This raises the question of how simulators would source the required energy. While they might possess energy production capabilities that far exceed our current understanding, it's possible their techniques would leave observable traces within our universe.
- Close collaboration between physicists and astronomers becomes crucial in investigating potential energy and resource discrepancies as indicators of a simulated reality. Astronomers can systematically search for unexplained energy signatures, anomalies in the distribution of energy within the observable universe, or evidence of manipulation of natural processes for large-scale energy generation. Physicists are essential in developing theoretical models for potential energy-harvesting techniques a simulator civilization might employ, aiding the interpretation of astronomical observations and potentially refining search parameters. By examining potential ways simulators might manipulate astrophysical phenomena to harvest energy, we can identify specific observations that could hint at the simulated nature of our existence.

Unexplainable Statistical Anomalies

- The possibility that simulations might leave subtle statistical fingerprints on reality offers an intriguing avenue for investigation. Even with vast computational resources, simulator civilizations might face optimization challenges or introduce unintentional biases in their simulations. These could manifest as unexpected statistical patterns in the behavior of physical systems, in the distribution of cosmic phenomena, or even within large data sets unrelated to physics and astronomy. Therefore, rigorous statistical analysis of diverse datasets becomes a valuable tool in our search for potential evidence supporting the Simulation Argument.
- Uncovering subtle statistical anomalies requires sophisticated analysis techniques that go beyond simple pattern recognition. Identifying departures from expected distributions, correlations where they shouldn't exist, or biases within supposedly random phenomena could provide potential hints that our universe follows a computationally generated script. Close collaboration between statisticians, computer scientists, and domain-specific experts is necessary to design analysis methods and identify those areas where unexplainable statistical anomalies might emerge. While not conclusive proof, such anomalies could point us in the right direction within the overarching investigation of the Simulation Argument.

Simulated Physics Experiments

- As our computational power grows, the prospect of simulating highly-detailed ancestor simulations – including replications of the laws of physics as we observe them – becomes increasingly plausible. The ability to conduct controlled experiments within simulated universes offers an intriguing approach to probe the potential boundaries between simulated and base realities. By systematically comparing the behavior of physical laws and fundamental constants between our universe and meticulously designed simulations, we might be able to uncover subtle discrepancies that hint at the simulated nature of our own existence.
- Simulated physics experiments demand a rigorous approach combining precision physics and cutting-edge simulation techniques. Physicists would need to meticulously replicate experiments, pushing the

boundaries of measurement accuracy to detect even the slightest potential deviations within simulations. Simultaneously, simulations must advance in complexity and fidelity to closely mimic the conditions of our own universe. This collaborative approach could reveal inconsistencies in physical constants, limitations in computational precision within simulations, or other deviations which might suggest that our own reality is a carefully constructed simulation.

Interpreting Philosophical Implications

The framework guides us in analyzing the profound philosophical and societal impacts of confirming the simulation hypothesis.

Re-evaluating Existential Questions

- The prospect that our universe is a complex simulation fundamentally disrupts our traditional understanding of our place within reality. Confirmation of the Simulation Argument would force us to re-evaluate our most basic assumptions about existence. We'd confront the possibility that our reality is a complex construct, potentially subject to manipulation by our simulators. This could cast doubt on our understanding of free will, the origins of the universe, and our very sense of self, leading to a profound sense of existential uncertainty.
- If we are simulated beings, profound questions emerge regarding meaning and purpose. Are we merely the subjects of a vast experiment or perhaps a form of entertainment for our simulators? Does our existence hold intrinsic value, or are we ultimately disposable within our simulated reality? The Simulation Argument could trigger an existential crisis, forcing us to search for meaning within a potentially constructed existence. It also introduces the possibility of nested simulations—realities within realities–creating a dizzying sense of scale and prompting us to question our position within a potential hierarchy of simulations. Our interdisciplinary framework, particularly insights from philosophy, psychology, and even theology, will be crucial as we attempt to make sense of our existence within a potentially simulated cosmos.

Ethical Frameworks for Simulation

- The robust interdisciplinary theoretical foundation we establish surrounding simulation ethics offers more than philosophical conjecture. It serves as a blueprint for reshaping our ethical guidelines as we potentially gain confirmation that we exist within a simulated environment. Early engagement with the ethical complexities surrounding simulation lays the groundwork for evolving our moral compasses in a responsible and thoughtful manner. By examining questions about the rights of simulated beings, simulator responsibilities, and the prevention of harm within simulations, we develop proactive ethical frameworks to guide future decision-making.
- The theoretical framework we establish directly shapes both our societal approach towards simulations and potential legal safeguards. Should evidence for the Simulation Hypothesis grow stronger, existing ethical codes would likely prove inadequate to deal with the unprecedented moral challenges that arise. Early theoretical considerations provide a crucial headstart in recognizing the unique dilemmas we might face. This allows us to craft societal guidelines and potentially laws that are specifically tailored to address the profound implications of a simulated existence. We must address the potential moral status of simulated beings, the responsibilities of creators should those simulated beings achieve sentience, and safeguards to ensure a just and sustainable existence, whether we find ourselves inside or outside of a simulation.

Limitations and Open Questions

Our theoretical framework, while robust, must acknowledge its inherent limitations and the existence of open questions shaping the ongoing debate:

The Unpredictability of Future Theories

- History repeatedly reminds us that scientific progress can be marked by disruptive breakthroughs that overturn established paradigms and force us to re-evaluate previously held assumptions. Our current understanding of computation, consciousness, and the fundamental laws of physics may represent only a fraction of the ultimate picture. Future discoveries have the potential to revolutionize our understanding of what is computable, how consciousness arises, or the fabric of reality itself.

- A robust framework for approaching the Simulation Argument must prioritize adaptability. It should be built with the explicit understanding that breakthrough discoveries could occur, upending our current theoretical models. Our framework needs the flexibility to seamlessly incorporate radically new concepts and discoveries while maintaining a rigorous interdisciplinary approach. By approaching the question with the acknowledgement that our understanding will evolve over time, we avoid intellectual rigidity and ensure that our framework remains a powerful tool even in the face of transformative changes in our knowledge.

The Limits of Human Cognition

- The Simulation Argument challenges us to confront the inherent limitations of our own biological cognition. Our brains evolved to comprehend the world on a scale and level of complexity relevant to our survival, not to unravel the fundamental nature of reality. Conceiving of the true capabilities and motivations of a simulator civilization – one potentially possessing exponentially greater cognitive abilities and a radically different technological foundation – may be inherently beyond our grasp.
- Acknowledging our cognitive limitations cultivates intellectual humility within the context of the Simulation Argument. We must accept the possibility that even with rigorous inquiry and theoretical advancements, certain aspects of the "truth" may remain forever elusive. If we exist within a simulation, a full understanding of the simulators, their intentions, or the ultimate nature of reality might simply be beyond the comprehension of simulated minds. This underscores the need for an open-minded approach and a recognition that even our most refined methodologies and frameworks may only offer a partial glimpse into a reality that might be fundamentally unknowable to us.

Meta-Theoretical Uncertainty

- The possibility that a simulation could be designed with deliberate mechanisms to prevent detection introduces a meta-theoretical uncertainty that complicates our efforts. Simulators, possessing superior knowledge and technological capability, could implement safeguards that render any attempt at detecting their simulation futile, placing hard limits on our ability to discover the true nature of our existence. This could take the form of manipulations of the laws of physics, limitations on our observational tools, or even subtle adjustments to our cognitive abilities to keep us from reaching the 'correct' conclusions.
- The potential unfalsifiability of a simulated reality necessitates a critical adjustment to our theoretical framework. It underscores a fundamental limitation inherent in our approach. Scientific inquiry often relies on the ability to falsify hypotheses; however, a deliberately undetectable simulation might render this impossible. We must acknowledge that even our most rigorous search for evidence, experimental designs, and theoretical constructs could hit an insurmountable wall purposefully built into the simulation's design. This recognition fosters a level of caution and humility, reminding us that even strong support for the Simulation Hypothesis might not amount to definitive proof.

The Value of Interdisciplinarity

The strength of the theoretical framework presented lies in its staunchly interdisciplinary nature. By intentionally demolishing the silos that traditionally separate fields like computer science, philosophy, and physics, we create a richer and more nuanced tapestry woven from perspectives that illuminate different facets of the Ancestor Simulation Argument. This multi-faceted perspective is crucial when confronting a topic as complex and potentially paradigm-shifting as the possibility of simulated realities.

Our theoretical framework navigates a terrain fraught with complexities - the nature of consciousness, the computability of reality, the ethical boundaries of simulation, and the very limits of what we might consider knowable. It offers a robust structure upon which to build our theoretical models, hypotheses, and experimental designs, providing guideposts as we traverse these challenging domains. Acknowledging our potential limits while maintaining intellectual curiosity is paramount when undertaking such a profound inquiry.

The true value of our theoretical framework lies not only in providing definitive answers but in fostering a spirit of rigorous inquiry. It offers a catalyst for propelling our exploration of the Ancestor Simulation Argument forward. Through continued research, debate, and the incorporation of new discoveries, our understanding may evolve, deepen, and perhaps even reach unexpected shores. More than just a means to an end, this framework serves as a compass, guiding us on a quest not just to ascertain the nature of our reality but to expand the very boundaries of human knowledge.

CHAPTER TWENTY-TWO

TC: Techniques & Uncertainties

Evaluating the technological plausibility of Nick Bostrom's Ancestor Simulation Argument necessitates a nuanced methodological approach that transcends simple extrapolation from our current capabilities. While understanding existing technologies is a crucial starting point, it risks underestimating the potential for transformative breakthroughs that could drastically alter the computational landscape. Our framework must be forward-looking, systematically considering the possibilities of unforeseen scientific revolutions and paradigm shifts in computing power.

A robust methodology acknowledges the fundamental limits of computation, those imposed by the laws of physics and the very fabric of reality itself. Can we even conceive of computational resources sufficient to simulate entire universes with perfect fidelity? Exploring these limitations fosters a healthy dose of skepticism while simultaneously encouraging creative exploration of theoretical breakthroughs that might push those boundaries. Uncertainty, in this context, becomes a catalyst for further inquiry, prompting us to explore alternative computational models and resource-management strategies within simulated realities.

The most insightful approach to the Ancestor Simulation Argument acknowledges both the existing body of scientific knowledge and the vast expanse of the unknown. Our methodology must be flexible enough to integrate emerging discoveries while remaining grounded in established principles. By embracing uncertainty and fostering a spirit of open-ended inquiry, we can weave a more intricate tapestry of understanding, one that acknowledges the limitations of current knowledge while leaving ample room for the transformative potential of future scientific advancements.

History and Background

Philosophers have long contemplated our world's true nature, from Plato's Allegory of the Cave to more recent science fiction depictions of simulated realities. However, Bostrom's formulation stands apart due to its rigorous trilemma. If one accepts that advanced civilizations can develop ancestor simulations, then it's overwhelmingly probable that we ourselves reside within one.

Assessing technological capability (TC) rests on several historical and theoretical foundations:

- **Computational Growth:** Moore's Law, while facing potential deceleration, highlights the relentless increase in computational power. Successors, whether through new materials or architectures, suggest continued progress.
- **Paradigm Shifts:** Quantum computing, neuromorphic computing, and biological computing exemplify potential revolutions that could render prior TC analyses obsolete. History is rife with unforeseen breakthroughs driving exponential leaps.
- **Computability Theory:** Work by Turing, Church, and others established theoretical bounds on what algorithms can accomplish. Yet, open questions remain regarding the physical world's ultimate

computability, potentially imposing constraints on simulation fidelity.

Techniques for TC Evaluation

A robust methodology must leverage multiple approaches to bound the problem of TC estimation:

Extrapolative Forecasting:

To gain meaningful insights into potential TC trajectories, we must employ a multifaceted approach combining historical analysis, sophisticated computational models, and the wisdom of domain experts. Analyzing historical trends in computational power, such as Moore's Law and its potential limitations, offers a valuable starting point. However, to account for the possibility of disruptive breakthroughs, S-curves, or entirely new computational paradigms, we need complex models that capture nuanced growth patterns and explore diverse, even unlikely, evolutionary paths for computational capacity.

Integrating expert elicitation into our TC projection methodology adds an essential dimension. By systematically gathering insights, opinions, and predictions from leading figures in computer science, physics, technology forecasting, and related fields, we tap into a wealth of knowledge and informed speculation. This collaborative approach helps us identify potential blind spots in our models, anticipate disruptive technologies, and capture a wider range of plausible scenarios beyond those suggested by historical trends alone.

The primary strength of the proposed methodology lies in its ability to establish realistic baselines for TC growth across various timescales. By charting conservative incremental gains, modeling S-curve technology adoption patterns, and incorporating expert predictions, we create a grounded reference point for our analysis. This baseline, built on empirically informed projections, allows us to then evaluate the relative plausibility of more radical possibilities. The contrast between these grounded baselines and far more ambitious TC scenarios, born from theoretical propositions and potential breakthroughs, will be a fertile ground for evaluating the technological preconditions for large-scale ancestor simulations.

Let's illustrate how extrapolative forecasting can help in evaluating Technological Capability (TC) within the context of the Ancestor Simulation Argument.

Important Caveats

- **Extreme Uncertainty:**
    - Predicting technological advancement far into the future is an exercise fraught with profound uncertainty. History is littered with examples of both underestimation (the early dismissal of the internet, for instance) and overestimation (the promise of flying cars that never materialized). Technological trajectories are often nonlinear, marked by sudden breakthroughs, shifts in adoption patterns, and unpredictable socioeconomic factors that can drastically alter the landscape.
    - While illustrative examples can help conceptualize various potential futures, it's crucial to recognize their limitations. Real-world projections demand greater complexity, nuanced consideration of interconnected systems, and the explicit acknowledgement of our inability to fully foresee unforeseen disruptions. These examples serve as starting points for discussion rather than definitive forecasts. They underscore the need for an adaptable methodological framework that incorporates extensive uncertainty ranges rather than providing a singular, potentially misleading prediction.
- **TC Is Multidimensional:**
    - While a focus on computational speed (operations per second) is a valid starting point, a robust analysis of Technological Capability (TC) necessitates a more nuanced approach. Factors such as memory capacity, energy efficiency, and the sophistication of networking infrastructure profoundly influence the types and scales of simulations that could be conducted. A civilization capable of running vast ancestor simulations likely exhibits advancements across

multiple technological dimensions, requiring a multi-pronged assessment of their overall computational capabilities.

- Despite the inherent complexity of TC, employing a simplified "operations per second" metric offers several advantages. It provides a clear, quantifiable baseline for initial comparisons and a way to illustrate the staggering computational demands of large-scale ancestor simulations. This simplification allows for greater clarity in discussions and helps illustrate the fundamental challenges involved, even without a perfectly holistic representation of an advanced civilization's technological toolkit. Importantly, it acknowledges that this simplified metric is a stepping stone for further refinement as the theoretical framework evolves.

Example Scenarios

- **Continued Incremental Gains (Moore's Law-Based):**
  - **Assumptions:** Let's assume computational power doubles roughly every 18 months well into the future.
  - **Model:** Exponential growth: TC(t) = TC(0) * 2^(t/1.5), where TC(0) is current capability, t is time in years.
  - **Example:** If today we have 10^20 operations/second, in 10 years: TC(10) = 10^20 * 2^(10/1.5) ≈ 10^24 operations/second.
- **Technology S-Curves (Logistic Growth):**
  - **Assumptions:** Technology often exhibits initial rapid growth, then plateaus as limits are approached.
  - **Model:** Logistic function: TC(t) = K / (1 + e^(-r(t-t0))), where K is carrying capacity (ultimate limit), r is growth rate, t0 is the inflection point.
  - **Example:** If K=10^30 operations/second, r=0.2, t0=2040 (technology hits midpoint by 2040), then in 2030: TC(2030) ≈ 10^25 operations/second.
- **Disruptive Paradigm Shift (Hypothetical):**
  - **Assumptions:** Major breakthrough (e.g., mature quantum computing) in 2045 leads to 1000x jump in capability, followed by continued incremental progress at the previous growth rate.
  - **Model:** Piecewise function, combining exponential growth from scenario 1 but with a sudden increase.
  - **Example:** By 2055, we could see TC reaching approximately 10^29 operations/second
- **"Brain Power" Target:**
  - **Assumptions:** The human brain is estimated at roughly 10^17 operations/second. Let's aim to reach this TC by 2050.
  - **Model:** Linear growth, as achieving such a specific target may require less predictable, directed research efforts.
  - **Example:** If current TC is 10^20, it would necessitate roughly a 10^3 (thousand-fold) increase over 26 years, requiring relentless progress.

Using these for Simulation Argument Evaluation

- **Baselines:** These projections give us a sense of scale, but should be considered lower bounds on simulator TC, as they'd need vastly more power to run the world plus the simulation layer.
- **Resource Demands:** We can extrapolate further: how much energy might these TC levels consume? What physical space would the hardware occupy? This informs the feasibility of such simulations.
- **Limits:** If a S-curve model seems more realistic, when does TC growth become insufficient for ancestor simulations, making the hypothesis less plausible?
- **Uncertainty:** Assign wide error bars to these projections. Expert elicitation (formal surveys of

technologists) can help quantify that uncertainty somewhat.

- **Key Takeaway:** Extrapolative forecasting, while crude, sets the stage for comparisons. If even optimistic scenarios fall far short of the TC needed for ancestor simulations, it strengthens Bostrom's argument. Conversely, if these models suggest it's plausible within a few centuries, the implications become more profound.

Breakthrough Analysis

To fully explore the potential technological landscape relevant to the Ancestor Simulation Argument, we must venture into the speculative realm of potential breakthroughs. Thought experiments and small-scale simulations focused on emerging technologies, such as those based on quantum mechanics, biological computation, or entirely new paradigms, can offer glimpses of their potential impact on simulation capabilities. This approach expands our analysis beyond the limitations of existing technologies, allowing us to envision radical leaps that might enable computationally demanding, large-scale simulations.

Rigorous exploration of speculative technologies demands close collaboration with physicists, engineers, and other domain experts. This provides much-needed grounding to our investigations, ensuring that these thought experiments and exploratory models remain rooted in the laws of nature as we understand them. Without this interdisciplinary approach, there's a risk that our projections might drift from plausible science towards pure fantasy. By working closely with scientists who possess deep knowledge of the fundamental physical constraints and emerging engineering possibilities, we ensure our analysis remains credible and relevant to the evaluation of the Ancestor Simulation Argument.

Let's delve into a couple of examples of how breakthrough analysis could help assess the technological capabilities for ancestor simulations.

Important Note: As with the prior example, vast simplifications are necessary for clarity. Analyzing these technologies' potential for simulations requires much deeper expertise in the respective fields.

Example 1: Quantum Computing

- **Speculative Potential:** Quantum computers excel at certain tasks that are intractable for classical computers (e.g., factoring large numbers). Their applicability to general simulations is less clear, but there are potential advantages:
  - **Superposition:** Modeling complex systems might benefit from representing many states simultaneously.
  - **Entanglement:** Could aid in simulating interactions between vast numbers of simulated entities with higher efficiency.
- **Thought Experiment:** Let's assume a mature quantum computer can simulate a volume of space at the atomic level with perfect accuracy for a short duration.
  - **Model:** We'd need to estimate:
    - Qubits Required: This depends on the volume simulated and the desired resolution. A very rough guess could be 100 qubits per cubic centimeter.
    - Operations Needed: Updating the state of the simulation according to quantum physics laws, dependent on particle interactions, timesteps, etc.
  - **Comparison:** Could a quantum computer of a given size outperform a classical supercomputer of equivalent resource cost in this domain by some metric?
- **Limitations:**
  - **Error Correction:** Scaling up quantum computers is extremely hard due to noise susceptibility.
  - **Problem Mapping:** Expressing the physics of our universe in terms of quantum operations efficiently might be highly non-trivial.

Example 2: Biological Computing

- **Speculative Potential:** Biological systems exhibit remarkable efficiency and adaptability in information processing. Could those principles be harnessed?
    - **DNA Storage:** Incredibly dense data storage. Hypothetical computers could potentially perform operations directly on stored DNA.
    - **Neural Networks:** Though inspired by brains, artificial neural networks are still vast simplifications. Could biological tissue offer vastly greater complexity and emergent computational properties?
- **Thought Experiment:** Let's assume we develop "computational organs" grown from genetically modified cells
    - **Model:** Estimate:
        - Cells per unit volume: Similar to brain cell density or higher.
        - "Operations" per cell: Based on signaling speed, biochemical reaction rates, etc. Very speculative.
    - **Comparison:** What scale of such an organ would be needed to rival a classical computer? What are the energy input and waste output considerations?
- **Limitations:**
    - **Control:** Manipulating biological systems with the precision required for computation is well beyond current science.
    - **Timescales:** Biological processes tend to be slow compared to electronic ones.

Others to Consider

- **Neuromorphic Computing:** Chips designed to mimic brain architecture might offer efficiency gains. Analysis might focus on the number of artificial 'neurons' simulate able in real-time.
- **Reversible Computing:** Theoretically, computations can be done with zero energy loss. If achievable, it radically alters the resource demand picture for simulations.

Key Takeaway for Simulation Argument

Breakthrough analysis helps us ask targeted questions:

- **Scaling:** Even if these technologies are powerful, can they scale to the level required for ancestor simulations? Trillions of conscious minds, a full universe...this sets the bar very high.
- **Specificity:** Are breakthroughs most likely to aid in simulating specific aspects of reality (e.g., quantum well suited to the very small), leaving other layers still in the domain of less exotic technology?
- **Ethics:** Developing the very tech that makes simulations possible raises the urgency of Bostrom's argument itself.

Analogies from Nature:

The remarkable information processing capabilities of biological systems like the human brain, achieved with relatively low energy expenditure, highlight the potential for alternative, highly efficient computational paradigms. Biomimicry in the field of computer science, the exploration of concepts like neuromorphic computing and evolutionary algorithms, could reveal pathways towards increased computational power with greatly reduced resource requirements. Such breakthroughs could profoundly alter the landscape of the Ancestor Simulation Argument, suggesting that even relatively modest levels of technological advancement might support surprisingly large-scale simulations.

Theories exploring the idea that the universe itself might be a form of computation offer an intriguing avenue for investigation. Whether it's digital physics, cellular automata models, or other computationally grounded interpretations of reality, understanding such principles could reveal novel strategies for simulation design. Perhaps simulator civilizations have discovered ways to optimize simulations, exploiting loopholes in the underlying computational fabric of their reality to achieve greater efficiency and fidelity. While speculative, these concepts invite collaboration with theoretical physicists and computational complexity theorists to seek potential 'shortcuts' or optimization principles that could be leveraged for simulation purposes.

Exploring analogies from nature can provide fascinating insights into potentially more efficient approaches to achieve the technological capability (TC) needed for ancestor simulations. Here's how we can frame and analyze such analogies:

Example 1: The Brain as a Benchmark

- **Observation:** The human brain achieves incredible feats of cognition, perception, and creativity with roughly 20 watts of power consumption. Modern supercomputers require vastly more energy to perform far less sophisticated tasks.
- **Analogy-Based Hypothesis:** There are fundamental principles of organization and computation within biological neural networks that we could leverage to increase TC efficiency.
- **Calculations & Models:**
    - **Neurons vs. Transistors:** Estimate the number of 'operations' a single neuron performs per second compared to a transistor. This is very crude but gives a sense of scale.
    - **Network Density:** Compare the density of neurons and their interconnections in the brain to that of current integrated circuits. Can bio-inspired designs achieve higher density?
    - **Energy per 'Operation':** Detailed metabolic modeling could reveal how much energy a neuron expends for signaling versus computation, informing efficiency targets.
- **Research Directions:**
    - Neuromorphic chips explicitly mimic brain architecture.
    - Research into the brain's energy-efficient learning and adaptation mechanisms.

Example 2: The Universe as Computer

- **Observation:** Some theories in physics posit that the laws of the universe themselves can be described as a form of computation. Cellular automata models provide a simplified example where complex behavior emerges from simple computational rules.
- **Analogy-Based Hypothesis:** If our universe is fundamentally computable, perhaps there are ways to exploit this underlying 'code' to perform simulations with dramatically less overhead than traditional approaches.
- **Calculations & Models:**
    - **Information Content:** Attempts have been made to estimate the information content of a given volume of space-time based on quantum physics and thermodynamics (Bekenstein bound).
    - **Minimum 'Operations':** Could the evolution of the universe according to physical laws be considered a series of computations? Estimating this could set an absolute theoretical efficiency limit.
    - **Simulation 'Layers':** Might our own universe be a simulation running on a 'substrate' with different computational rules?
- **Research Directions:**
    - Lattice physics studies simulating fundamental forces on discrete grids
    - Exploration of models where physical constants are not fixed but emergent from more basic computations.

Other Nature-Inspired Analogies

- **Genetic Algorithms:** Already used in optimization can offer insight into simulating evolutionary processes that produce complex life forms.

- **Swarm Intelligence:** Emergent behaviors in ant colonies or bird flocks hint at decentralized computation models potentially applicable to large-scale simulation control.

Key Takeaway for Simulation Argument

Nature-based analogies serve several purposes:

- **Challenging Assumptions:** It highlights that our current computational paradigms might be inherently inefficient compared to what's theoretically possible.
- **Upper Bounds:** If the brain or universe represent near-optimal TC efficiency, it helps bound what a simulator civilization might achieve.
- **Existential Implications:** If simulating worlds is deeply ingrained in the fabric of reality, it strengthens the argument that we ourselves might be within a simulation.

Important Note: Drawing direct conclusions from analogies is risky. Nature 'optimized' for survival and reproduction, not simulating universes. Still, seeking inspiration from evolved systems can open new avenues for technological development.

Consciousness as Constraint:

If the goal is ancestor simulations indistinguishable from the 'base' reality, a deep understanding of consciousness is needed. We must determine if:

- Consciousness is computable, thus readily simulated
- Consciousness is an emergent, irreducible property, hindering true replication
- Our understanding of consciousness is fundamentally flawed

Uncertainties and their Impact

TC evaluation is fraught with uncertainties, their very nature making quantification difficult:

- **Unknown Unknowns:** Predicting paradigm shifts is notoriously tricky. History counsels humility, suggesting our models likely underestimate future TC wildly.
- **Resource Limits:** Even unbounded theoretical computability means little if the energy, material, or time resources required for ancestor simulations prove impossibly vast, even for advanced civilizations.
- **Fundamental Physics:** If our universe operates on non-computable principles, or if there are 'hard limits' to information density, accurate simulations may be eternally out of reach.
- **Ethical Considerations:** The very act of evaluating TC might hasten the development of simulation technology, leading to unpredictable social or existential consequences.

Methodological Implications for the Ancestor Simulation Argument

Despite these uncertainties, the TC evaluation exercise remains crucial to Bostrom's argument:

- **Bounds of the Plausible:** Even rough TC estimates help bracket what scenarios are likely and unlikely, reducing the solution space.
- **Driving Research Focus:** By identifying TC bottlenecks, research can be targeted toward overcoming them, ironically, potentially making the simulation hypothesis a self-fulfilling prophecy.
- **Existential Risk Mitigation:** If ancestor simulations are likely, understanding their prerequisites aids in managing risks they might pose to the 'base' reality.

Evolving Inquiry, Evolving Answers

The multifaceted endeavor of assessing the feasibility of ancestor simulations stands at the intersection of several high-stakes domains. Projections of future technological capabilities, investigations into the fundamental limits imposed by physics, the philosophical implications of simulated consciousness, and the ethical quagmires surrounding our potential existence within a simulation all form integral parts of this tapestry of inquiry. This necessitates a carefully calibrated balance between bold speculation regarding theoretical possibilities and an unwavering respect for the inherent uncertainties that permeate our understanding.

The profound nature of Bostrom's question lies not solely in its potential answer but also in the profound way it alters the process of seeking the answer. The very act of investigating the possibility of simulated realities forces us to confront fundamental assumptions about consciousness, technological limitations, and the fabric of existence itself. These inquiries might trigger transformative advancements in our own technological development, ethical frameworks, and understanding of reality.

Embarking on this quest to evaluate the Ancestor Simulation Argument yields far more than just a conclusive answer about our place in the cosmos. It calls for a spirit of intellectual openness and a willingness to embrace radical possibilities that could reshape our understanding of reality. The journey itself – the meticulous exploration, the interdisciplinary collaborations, and the willingness to question our most basic assumptions – carries the potential to redefine the very nature of scientific investigation and philosophical inquiry.

CHAPTER TWENTY-THREE

DS: Beyond Pure Technology

While assessing Technological Capability (TC) plays a crucial role in evaluating the feasibility of ancestor simulations, a robust analysis must also delve into the motivations behind such endeavors. Understanding the Desire to Simulate (DS) of a hypothetical advanced civilization offers invaluable insights into their potential goals, values, and societal structures. Unlike TC, which lends itself to quantitative assessments based on computational power, exploring DS necessitates a qualitative, multifaceted approach.

To shed light on DS, we must weave a tapestry of methodologies. Historical analysis of human motivations for scientific inquiry, technological innovation, and exploration offers a valuable starting point. Examining the sociological factors that drive curiosity, the pursuit of knowledge, and the desire to understand our own origins can provide clues about potential motivations for simulating one's ancestors. Furthermore, integrating ethical considerations is paramount. Would an advanced civilization have a moral obligation to simulate their past? Would such simulations be conducted for entertainment, scientific advancement, or even existential exploration?

Perhaps the most significant challenge in exploring DS lies in avoiding anthropocentrism. Projecting our own desires and motivations onto a vastly more advanced civilization can lead us astray. However, by thoughtfully analyzing human behavior, motivations, and historical trends while acknowledging the limitations of such an approach, we can begin to construct a nuanced framework for exploring the potential Desire to Simulate within the context of the Ancestor Simulation Argument. This line of inquiry, while inherently speculative, offers a deeper understanding of the "why" behind simulated realities, enriching our exploration of this profound philosophical proposition.

Historical Analysis

The rich historical complexity of technological adoption offers valuable insights into the potential motivations and societal implications of ancestor simulations. Historical Comparative Analysis provides a systematic method for examining how past civilizations have grappled with emerging technologies, revealing patterns in societal responses, shifts in power dynamics, and ethical challenges that arose from their integration. By understanding how human societies across different cultures and time periods have responded to technological milestones, we gain a lens for evaluating potential responses and motivations for creating large-scale simulations.

Historical Comparative Analysis allows for the identification of recurring themes in technological adoption. Analyzing common motivations (curiosity, power consolidation, knowledge preservation), the emergence of ethical dilemmas, and the unpredictable long-term consequences of technological integration sheds light on the complex landscape of the Desire to Simulate. Crucially, this approach also illuminates divergence points – instances in history where technological paths were not taken or where civilizations actively eschewed certain advancements. Understanding these instances of technological restraint can help us identify potential ethical or societal barriers that might discourage the creation of ancestor simulations within an advanced civilization.

It's important to recognize that Historical Comparative Analysis has inherent limitations. Extrapolating motivations and behaviors from past human societies onto potential civilizations possessing vastly superior technological capabilities risks significant misinterpretation. The unprecedented power and potential consequences of ancestor simulations demand a critical and nuanced examination of the limitations of history as a predictive tool. However, this technique serves as a valuable starting point for constructing a framework that acknowledges both human tendencies and the potential for radical departures from our own historical trajectory within far more advanced civilizations.

Application

The application of Historical Comparative Analysis in this context involves a detailed examination of specific epochs and civilizations, drawing from a diverse range of historical periods and cultural backgrounds. This analysis looks beyond the mere technological capabilities of these societies, delving into the socio-political, economic, and philosophical factors that influenced their technological trajectories. For instance, the transition from oral to written cultures, the development of the printing press, and the advent of the digital age each represent pivotal moments in technological adoption that reshaped societies in profound ways.

In conducting this analysis, the focus is on understanding the motivations behind the pursuit and integration of new technologies. These motivations can include, but are not limited to, the desire for increased efficiency, enhanced communication, military advantage, or deeper philosophical or spiritual understanding. By examining these historical instances, we aim to identify commonalities and differences in how civilizations have approached technological advancements, offering a framework for speculating about the motivations of future civilizations concerning the creation of simulations.

Uncertainties

While Historical Comparative Analysis offers valuable perspectives on potential future behaviors, it is important to acknowledge the inherent uncertainties involved in this approach. Historical contexts and the conditions under which past civilizations operated vary significantly from the potential realities of advanced, post-human societies. The motivations that drove past technological integrations may not directly apply to or predict the motivations of civilizations capable of creating complex simulations.

Moreover, the extrapolation of past behaviors to future possibilities introduces a significant degree of speculation. Technological advancements in the future may arise from entirely new paradigms, driven by motivations that are currently unfathomable. Additionally, the impact of these technologies on future societies may differ drastically from historical precedents, given the potential for changes in societal structure, values, and global conditions.

Despite these uncertainties, Historical Comparative Analysis remains a foundational methodological approach in our exploration of DS. By carefully examining the patterns and motivations of past civilizations' technological pursuits, we can construct informed speculations about the potential desires of future civilizations to engage in the creation of simulated realities. This approach does not seek to provide definitive predictions but rather to enrich our understanding of the complex interplay between technology, society, and motivation that underlies the Simulation Hypothesis.

Sociological Insights

Sociocultural Modeling provides a robust framework for investigating how cultural values, social norms, and power structures within an advanced civilization might shape its Desire to Simulate (DS) ancestors. These models can incorporate diverse variables – from religious beliefs and ethical systems to economic incentives and societal hierarchies – to simulate how they influence technological development, resource allocation, and the potential motivations for large-scale simulations.

Sociocultural models can be tailored to examine both individual and collective motivations within a hypothetical advanced civilization. They can explore how the ambitions of individuals, scientific curiosity, or a strong desire to understand the past intertwine with societal pressures, political ideologies, and resource allocation priorities. Furthermore, by incorporating evolutionary dynamics into these models, we gain valuable insights into how competing motivations and societal values might evolve over time, influencing the ultimate Desire to Simulate (DS). This nuanced approach illuminates the complex societal ecosystem that would ultimately determine whether advanced civilizations embark on the creation of detailed ancestor simulations.

While Sociocultural Modeling is a powerful tool, it's crucial to acknowledge its limitations and potential for biases. Designing models that accurately reflect the intricate dynamics of a vastly advanced and potentially alien civilization is an extraordinary challenge. Biases in our own understanding of societal structures or a failure to imagine alternative social paradigms could limit the validity of such models. However, this technique encourages us to systematically evaluate our assumptions about social dynamics, explore potential scenarios, and refine our understanding of the diverse range of motivations that might drive the technological choices of civilizations far removed from our own.

Application

Sociocultural modeling, as a technique within the broader methodology of assessing the Desire to Simulate (DS), leverages sociological theories and models to probe into the valuation of simulation technologies by future societies. This approach is anchored in the understanding that societal values, norms, and ethical considerations play a pivotal role in shaping the adoption and integration of new technologies. By examining current societal trends and extrapolating these into the future, we can begin to construct a nuanced forecast of how simulation technologies might be perceived and utilized.

The application of sociocultural modeling involves a multi-layered analysis, starting with the identification of current trends in technology acceptance, societal values, and ethical norms. This includes an investigation into how digital technologies, virtual reality, and artificial intelligence are currently being integrated into various aspects of societal life, from education and entertainment to work and personal relationships. Surveys and expert elicitation serve as key tools in this process, providing empirical data and informed opinions on the trajectory of these trends.

Scenario analysis emerges as a particularly powerful technique in sociocultural modeling, allowing for the exploration of various future states based on different assumptions about societal evolution. Through the creation of detailed scenarios, we can visualize how future societies might navigate the complexities of simulation technologies, considering factors such as technological accessibility, the digital divide, cultural attitudes towards virtual experiences, and ethical concerns about simulated realities.

Uncertainties

The endeavor to model sociocultural dynamics, especially when projecting into the future, is fraught with uncertainties. Societal values and norms are not static; they evolve in response to a myriad of factors including technological advancements, economic shifts, geopolitical changes, and cultural exchanges. This fluidity introduces a significant degree of speculation into sociocultural modeling, particularly when the focus extends to hypothetical advanced civilizations whose societal structures and values may diverge markedly from our own.

The speculative nature of sociocultural modeling is further compounded when considering the potential for radical technological breakthroughs that could fundamentally alter societal dynamics. For instance, the development of highly immersive simulation technologies that can perfectly mimic reality could provoke ethical debates, cultural shifts, and changes in social behavior that are difficult to predict based on current trends.

Moreover, the application of current sociological theories and models to future societies assumes a continuity of societal evolution that may not hold true in the face of disruptive technological innovations. The emergence of a post-human civilization capable of creating high-fidelity simulations might herald new forms of social organization, value systems, and ethical frameworks that challenge our current understandings.

Despite these uncertainties, sociocultural modeling remains a critical component of the methodology for assessing DS. By engaging with the complexities of societal evolution and the multifaceted influences on the valuation of simulation technologies, this approach enriches our exploration of the Simulation Hypothesis. It prompts a deeper consideration of the societal underpinnings that could drive the desire to simulate, highlighting the interplay between technological potential and sociocultural dynamics in shaping the future of simulated realities.

Ethical Framework Analysis

Evaluating potential motivations behind ancestor simulations necessitates a robust understanding of the ethical frameworks likely guiding the decisions of a technologically advanced civilization. The study of diverse ethical theories such as consequentialism, deontology, and virtue ethics offers a rich toolkit for analyzing the potential moral considerations that might influence the Desire to Simulate (DS). Crucially, this approach acknowledges that the ethical frameworks of an advanced civilization could differ significantly from our own dominant moral philosophies.

Ethical Framework Analysis allows us to delve into the profound moral implications of simulated existence. Examining potential duties owed by creators to their simulated beings (and vice-versa) forces us to grapple with concepts of personhood, informed consent, and the right to existence within a computational framework. Moreover, this analysis highlights ethical dilemmas surrounding the manipulation of simulations, potential exploitation, and the responsibilities of those possessing the power to create and control simulated realities.

Importantly, the potential for future human societies to achieve simulation capabilities makes our own ethical frameworks relevant to this analysis. This highlights the urgent need to develop proactive ethical guidelines and safeguards for simulation technology as our own capabilities rapidly advance. By considering the range of diverse ethical outlooks and carefully analyzing both the risks and the potential benefits of large-scale simulations, we contribute not only to the evaluation of the Simulation Hypothesis but also to a more responsible approach to emerging simulation technologies within our own society.

Application

The exploration of the ethical implications surrounding the creation of simulated realities, particularly those that might involve sentient beings, necessitates a sophisticated and nuanced application of ethical framework analysis. This methodological approach is grounded in the application of diverse ethical theories and principles, aimed at dissecting the moral considerations that an advanced civilization might confront in the act of simulation creation. The ethical analysis becomes paramount when considering the Simulation Hypothesis, as it extends beyond mere technical feasibility to probe the moral fabric that could underpin such endeavors.

Ethical framework analysis in this context involves several key methodologies:

- **Stakeholder Analysis:** This method identifies and considers the perspectives of all potential stakeholders in simulated realities. This includes not only the creators of the simulations and the sentient beings within them but also the broader society that might be affected by the existence of such simulations. Stakeholder analysis helps to map out the range of ethical considerations and potential impacts on various parties, providing a comprehensive view of the moral landscape.
- **Ethical Dilemma Workshops:** These workshops bring together ethicists, philosophers, technologists, and other relevant stakeholders to deliberate on the ethical dilemmas posed by the creation of simulations. Through structured discussions and debates, participants explore the application of ethical theories such as utilitarianism, deontology, virtue ethics, and care ethics to the complex scenarios presented by simulated realities. These workshops facilitate a deep dive into the moral quandaries, allowing for a collaborative examination of potential resolutions or ethical guidelines.
- **Thought Experiments:** Employing thought experiments is a crucial aspect of ethical framework analysis, allowing for the conceptual exploration of hypothetical scenarios related to simulation creation. Thought experiments such as variations of the "trolley problem" adapted to simulation contexts can illuminate the ethical nuances of decisions involving simulated sentient beings. These mental exercises serve to challenge preconceived ethical notions and stimulate reflective thinking about the moral implications of simulations.

Uncertainties

A fundamental challenge in applying ethical framework analysis to the Simulation Hypothesis lies in the inherent variability of ethical norms and principles across different cultures and civilizations. Ethical perspectives are deeply influenced by cultural, historical, philosophical, and religious contexts, leading to a wide spectrum of moral viewpoints. This diversity introduces significant speculation when attempting to predict how an advanced civilization, potentially very different from our own, might navigate the ethical complexities associated with creating simulated realities.

The speculative nature of predicting ethical behaviors in advanced civilizations is compounded by the potential emergence of new ethical paradigms that such civilizations might develop. These paradigms could be informed by advanced understandings of consciousness, identity, and reality, challenging our current ethical frameworks and principles. The possibility of encountering fundamentally different ethical considerations in the context of simulation creation underscores the speculative but critical nature of ethical framework analysis in exploring the Simulation Hypothesis.

Despite these uncertainties, ethical framework analysis remains an indispensable tool in the methodological exploration of the Simulation Hypothesis. By rigorously engaging with the ethical dimensions of creating simulated realities, this approach fosters a deeper understanding of the moral considerations that might guide advanced civilizations. It highlights the importance of ethical reflection in contemplating the potential realities of simulation, urging a thoughtful consideration of the moral responsibilities and implications inherent in such profound technological capabilities.

Interdisciplinary Methodology

Interdisciplinary Synthesis stands as a cornerstone methodology when investigating the multifaceted concept of Desire to Simulate (DS). The Simulation Argument, touching on fields as disparate as computer science, ethics, history, sociology, and theoretical physics, demands a holistic approach that transcends traditional disciplinary boundaries. It involves intentionally weaving together diverse strands of knowledge, fostering a dialogue between these domains to illuminate aspects that might remain imperceptible when viewed through a single disciplinary lens.

The strength of Interdisciplinary Synthesis lies in its ability to transcend a mere mosaic of fragmented insights. While each discipline can offer valuable pieces of the puzzle, a true understanding of DS requires more than simply assembling disconnected perspectives. True synthesis involves active cross-pollination and collaboration. Sociologists, for instance, contribute a deep understanding of societal dynamics while ethicists help navigate the moral complexities of simulations, and these contributions must be actively integrated, not merely juxtaposed, to fully explore their implications for the Desire to Simulate.

By fostering active integration and open communication between disciplines, the technique of Interdisciplinary Synthesis can contribute significantly to overcoming the limitations and blind spots inherent in any single approach. Historians might reveal patterns in human motivations for creating representations of the past. Philosophers of mind aid in grappling with concepts of simulated consciousness. This collaborative inquiry facilitates a holistic understanding of the complex interplay of motivations, societal forces, ethical complexities, and technological capabilities that ultimately shape an advanced civilization's Desire to Simulate.

Application

The application of interdisciplinary synthesis in assessing DS involves several critical components:

- **Collaborative Research:** At the core of interdisciplinary synthesis is the collaboration among scholars from diverse fields. This collaboration facilitates the exchange of ideas and methodologies, allowing for a richer exploration of DS. For instance, historians might provide context on the evolution of technological societies, while ethicists explore the moral implications of simulation, and sociologists examine the societal impacts of such technologies.
- **Cross-Disciplinary Workshops:** These workshops serve as forums for the dynamic interaction of different academic perspectives. Participants engage in discussions, debates, and brainstorming

sessions aimed at uncovering the complex layers of DS. Through these workshops, scholars can challenge and refine their ideas in the context of broader interdisciplinary insights, fostering a holistic understanding of the motivations behind simulation creation.

- **Development of Interdisciplinary Models:** The culmination of interdisciplinary synthesis often involves the creation of comprehensive models that encapsulate the multifaceted aspects of DS. These models are informed by the collaborative research and insights generated through cross-disciplinary workshops. They aim to represent the interplay between technological capabilities, ethical considerations, societal values, and historical precedents in shaping the desire to simulate. Such models are instrumental in providing a structured framework for analyzing DS, offering a visual and conceptual map of the interconnections between various influencing factors.

Uncertainties

Despite its profound potential for enriching our understanding of DS, interdisciplinary synthesis is not without its challenges and uncertainties. The integration of insights from diverse disciplines necessitates navigating the complexities of reconciling differing methodologies, terminologies, and conceptual frameworks. Each discipline brings its unique lens to the exploration of DS, and aligning these perspectives into a coherent synthesis introduces a significant degree of difficulty.

Moreover, the speculative nature of projecting the motivations of advanced civilizations adds another layer of uncertainty to this methodological approach. The assumptions and inferences drawn from interdisciplinary synthesis must be critically examined and validated against the backdrop of potential future scenarios. This speculative aspect underscores the inherent limitations and the need for caution in interpreting the outcomes of such synthesis.

Despite these challenges, interdisciplinary synthesis stands as a cornerstone of the methodology for assessing the Desire to Simulate. By embracing the complexity and diversity of academic perspectives, this approach fosters a deeper, more nuanced exploration of the factors that may influence advanced civilizations' engagement with simulated realities. It underscores the importance of a collaborative, multidisciplinary inquiry in navigating the speculative yet profoundly intriguing landscapes of the Simulation Hypothesis.

Tying into Nick Bostrom's Ancestor Simulation Argument

Nick Bostrom's trilemma posits that one of three propositions must be true, including the possibility that advanced civilizations capable of ancestor simulations are likely to have many reasons to run such simulations. Therefore, understanding potential simulations requires us to delve into both the technological feasibility and the motivations behind their creation. Our methodological exploration of DS directly addresses the second pillar of the trilemma, providing the basis for evaluating the likelihood of an advanced civilization desiring to simulate their own past.

By employing techniques such as Historical Comparative Analysis, Sociocultural Modeling, and Ethical Framework Analysis, we move beyond mere speculation regarding the Desire to Simulate (DS). These methodologies allow us to systematically examine potential motivations, social forces, ethical considerations, and historical precedents that could shape an advanced civilization's outlook. Equipped with these insights, we can transition from broad philosophical conjecture to a reasoned assessment of the plausibility of motivations that might compel a civilization far exceeding our technological capabilities to undertake the endeavor of simulating its ancestors.

Our exploration of DS ultimately strengthens our ability to evaluate the overall strength of the Simulation Hypothesis. If our methodological inquiry reveals a wide range of compelling motivations, societal pressures, or ethical frameworks that could drive advanced civilizations to engage in ancestor simulations, it increases the plausibility of Bostrom's argument. Conversely, if our analysis consistently uncovers reasons why such simulations would be deemed unnecessary or undesirable, this could cast doubt on a key component of the trilemma. Therefore, our nuanced exploration of the Desire to Simulate directly contributes to a more robust and balanced assessment of the Simulation Hypothesis.

Elusive Nature of Desire: Speculations on Simulation

The multifaceted endeavor of assessing the Desire to Simulate (DS) necessitates an approach grounded in interdisciplinary collaboration and a nuanced understanding of the limitations of our methodologies. Analysis of historical trends, sociological modeling, and the contemplation of diverse ethical frameworks offer a tapestry of insights into the potential motivations and driving forces that could lead advanced civilizations to embark on large-scale simulations. However, a sober recognition of the inherent uncertainties and potential flaws of extrapolating from our own history or our limited understanding of alien civilizations is crucial for intellectual honesty.

Our exploration of DS takes us into inherently speculative territory. The nature of potential civilizations possessing vast computational power and potentially radically different value systems remains largely shrouded in the fog of uncertainty. Nonetheless, engaging in this speculative analysis, driven by methodologies informed by diverse fields, provides a valuable framework for exploring possibilities. It facilitates the formulation of testable hypotheses that could inform future research, guiding our inquiries as our own technological prowess grows and our understanding of consciousness deepens.

The true value of our interdisciplinary analysis of the Desire to Simulate (DS) lies not solely in reaching definitive conclusions about the motivations behind simulated realities. It lies equally in the process of meticulously examining history, modeling societal dynamics, and confronting the profound ethical dilemmas posed by the Simulation Hypothesis. This inquiry serves as a catalyst for re-evaluating our own relationship with technology, stimulating new directions in ethical theory, and challenging our assumptions about the very nature of reality. This exploration forms an integral and enriching aspect of evaluating the simulation argument, propelling a comprehensive philosophical and scientific exploration.

CHAPTER TWENTY-FOUR

(NS): Statistics and Speculation

Transitioning from a merely philosophical exploration of the Simulation Hypothesis towards a more grounded assessment depends on our ability to estimate the Number of Simulations (NS) that an advanced civilization might plausibly create. This endeavor demands rigorous methodologies that extend far beyond theoretical conjecture. It requires the integration of our understanding of Technological Capabilities (TC), the analysis of Desire to Simulate (DS), and the power of computational simulations and statistical modeling to formulate meaningful projections of simulated realities.

Estimating NS involves confronting the possibilities of vast computational resources, advanced optimization techniques, and nested simulations existing within simulations themselves. Our projections will be shaped by a multitude of factors: the energy resources available to power simulations, the potential for simulations spanning cosmological timescales, limits imposed by the laws of physics, and the motivations driving a civilization to engage in simulations in the first place. By meticulously analyzing these factors individually and in combination, we aim to build a framework that outlines the potential lower and upper bounds for the Number of Simulations (NS) in existence.

Computational simulations play a dual role in this analysis. They serve not only as a tool for modeling the potential behavior of advanced civilizations but also as a means to test our assumptions and hypotheses. By simulating various scenarios encompassing TC, DS, resource utilization, and potential computational bottlenecks, we can refine our projections of NS while simultaneously gaining a deeper understanding of the constraints and challenges a simulator civilization might face. This iterative process strengthens our methodologies and helps us delineate the line between plausible and outlandish projections regarding the scale at which simulated realities might exist.

Rigor and Speculation: A Necessary Duality

The process of estimating the Number of Simulations (NS) necessitates the weaving together of seemingly disparate methodological approaches. The analytical rigor of statistical modeling, grounded in empirical data on computational trends and resource availability, forms a crucial foundation. However, to address the vast uncertainties surrounding advanced civilizations, we must also embrace the informed speculation required to envision potential breakthroughs, optimization strategies, and motivations far beyond our current understanding. The strength of this approach lies in its ability to establish tentative baselines using available data while exploring the outer limits of computational potential and simulation motives.

Our exploration of NS builds directly upon the theoretical foundations laid out in Nick Bostrom's Ancestor Simulation Argument. The assessment of Technological Capability (TC) forms a quantitative cornerstone, allowing for projections of computational power and resource utilization as technological prowess advances. Analysis of the Desire to Simulate (DS) adds an essential qualitative dimension, addressing the complex motivations that might drive civilizations to invest resources in ambitious simulation projects. By integrating these foundational parameters of Bostrom's work into our methodology, we create a robust framework for investigating NS and, ultimately, for evaluating the overall plausibility of the Simulation Hypothesis.

Theoretical Foundations

At the crux of Nick Bostrom's Ancestor Simulation Argument lies the provocative proposition that technologically advanced civilizations might create simulated realities indistinguishable from our own. The plausibility of this hypothesis hinges on a triumvirate of crucial assumptions – Technological Capability (TC), Desire to Simulate (DS), and the Number of Simulations (NS) an advanced civilization might create. Estimating NS, therefore, becomes a pivotal task, acting as a crucial indicator of the overall strength of the Simulation Hypothesis. However, embarking on this estimation process necessitates a critical examination of the underlying assumptions themselves. This requires not only a thorough understanding of Bostrom's framework but also the integration of insights gleaned from diverse interdisciplinary fields.

Bostrom's framework provides a robust foundation for analyzing TC and DS, but to assess the potential scale of simulated realities, we must delve deeper. Physics and engineering inform our understanding of resource limitations and potential breakthroughs in computational power. Meanwhile, studies of human motivations and historical precedents, explored through sociology and anthropology, offer valuable insights into the potential Desire to Simulate (DS) within advanced civilizations. By critically examining and potentially refining Bostrom's assumptions through this interdisciplinary lens, we strengthen the foundation upon which our estimation of NS – and ultimately, the plausibility of the Simulation Hypothesis – rests.

Statistical Modeling Techniques

Probabilistic Approaches:

The inherent uncertainties woven into the fabric of the Simulation Argument make probabilistic approaches indispensable tools in our estimation of NS. While deterministic models might offer misleadingly precise projections, a Bayesian framework, for instance, allows us to explicitly quantify the vast uncertainties surrounding Technological Capability (TC), potential constraints imposed by the laws of physics, and the unpredictable motivations of simulator civilizations. This honest acknowledgement of uncertainty enhances the credibility of our exploration into the plausibility of the Simulation Hypothesis.

The strength of Bayesian inference lies in its ability to evolve alongside our knowledge. As new scientific breakthroughs occur, theoretical models are refined, or fresh insights into the Desire to Simulate (DS) emerge, Bayesian techniques allow us to seamlessly incorporate this new evidence. This ensures our estimations of NS remain grounded in the best available information and reflect the evolving nature of our understanding. By updating our probability estimates based on new data or refined models, rather than relying on static projections, we gain a dynamic and adaptable framework for assessing the ever-evolving plausibility of the Simulation Argument.

Introduction

The endeavor to quantify the Number of Simulations (NS) that an advanced civilization could feasibly create is integral to evaluating the Simulation Hypothesis, as posited by Nick Bostrom. This task is fraught with inherent uncertainty and variability, stemming from our limited understanding of future technological capabilities (TC), the multifaceted nature of Desire to Simulate (DS), and the speculative nature of such an advanced civilization's motivations and capabilities. To navigate these complexities, probabilistic approaches, particularly Bayesian inference, offer a robust framework for incorporating uncertainty, integrating new evidence, and iteratively refining our understanding and probability estimates regarding the existence and scale of simulations.

Employing Probabilistic Models

Probabilistic models stand at the forefront of our methodological toolkit for estimating NS. These models are adept at handling the intrinsic uncertainties and variabilities associated with the parameters that define the potential scale of simulations. By applying probabilistic models, we can construct a range of possible outcomes for NS, each with associated probabilities that reflect our current understanding and the inherent uncertainties of our estimations.

Bayesian Inference: A Paradigm for Updating Beliefs

Bayesian inference emerges as a particularly powerful technique within probabilistic approaches, characterized by its capacity to update probability estimates in light of new evidence. This approach is based on Bayes' Theorem, which provides a mathematical framework for revising our beliefs about the likelihood of certain hypotheses (in this case, the scale of NS) based on the acquisition of new data or insights.

The Bayesian framework is inherently iterative, allowing for a dynamic updating process as new technological breakthroughs are achieved, as our understanding of consciousness and simulation technologies deepens, or as we gain insights into the potential motivations behind creating simulations. This iterative nature is especially valuable given the evolving landscape of relevant scientific and philosophical fields.

Application in Estimating NS

In applying Bayesian inference to estimate NS, we begin with prior probability distributions for TC, DS, and NS based on current knowledge and assumptions. As new evidence becomes available—be it advancements in computing power, new philosophical arguments regarding the ethics or desirability of simulations, or sociological insights into future civilizations' potential motivations—these priors are updated, leading to revised probability distributions.

This process involves integrating various types of evidence, from empirical data on technological advancements to speculative considerations about the motivations of advanced civilizations. For example, the development of quantum computing might significantly alter our estimates of TC, thereby impacting the probable range of NS. Similarly, evolving ethical perspectives on the creation of sentient simulations could influence DS, affecting the likelihood and scale at which simulations are created.

Challenges and Considerations

The application of Bayesian inference in this context is not without challenges. One of the primary difficulties lies in the specification of prior probabilities, which can significantly influence the outcomes of the analysis. Given the speculative nature of the Simulation Hypothesis, these priors often rely on a combination of empirical data, theoretical models, and philosophical arguments, each carrying its own degree of uncertainty and bias.

Moreover, the integration of new evidence requires careful consideration and robust methodologies to ensure that updates to our probability estimates are justified and reflective of genuine advancements in our understanding. This necessitates a multidisciplinary approach, drawing on expertise from computer science, philosophy, sociology, and other relevant fields to interpret new evidence and its implications for NS accurately.

Probabilities, Possibilities, and the Probabilistic Approach

Within the inherently speculative realm of the Simulation Hypothesis, probabilistic approaches, particularly the power of Bayesian inference, emerge as essential tools for tackling the complexities of estimating the Number of Simulations (NS). By explicitly quantifying uncertainties, modeling the potential variability in technological prowess and motivations of simulator civilizations, and accommodating the integration of future discoveries, these approaches offer a nuanced framework for exploring the landscape of possibilities surrounding simulated realities.

While the challenges faced when projecting technological trajectories, motivations, and resource availability for hypothetical civilizations remain substantial, the use of probabilistic models brings a level of methodological integrity to this fascinating inquiry. Embracing uncertainty fosters intellectual honesty, allowing us to explore the potential scale of simulated realities without relying on overly deterministic projections or unfounded speculation. Ultimately, our robust probabilistic approach empowers us to refine our understanding of the Simulation Hypothesis, leading to a richer, more dynamic, and scientifically grounded evaluation of its plausibility.

Computational Simulations:

Computational simulations act as experimental testbeds for exploring the intricate dynamics of simulated realities. They allow us to model the complex decision-making processes that a civilization far exceeding our own capabilities might engage in regarding the creation of simulations. These models incorporate diverse factors - from technological breakthroughs, resource availability, and ethical frameworks to the evolution of societal values within a simulator civilization. Employing such simulations helps us move beyond mere projections, painting a more dynamic picture that explores the potential ranges of the Number of Simulations (NS) and the conditions affecting them.

Computational simulations offer the ability to test the sensitivity of our conclusions around NS across a wide range of parameters and assumptions. By experimenting with alternative ethical frameworks, social structures, and resource scenarios, we can actively challenge our own preconceptions and biases. The iterative process of creating, refining, and re-running these simulations helps uncover hidden assumptions and potential blind spots, ultimately contributing to a more robust understanding of the potential motivations and constraints that could shape the decisions of a vastly advanced simulator civilization regarding ancestor simulations.

The Role of Computational Simulations in Understanding NS

In the intricate endeavor to estimate the Number of Simulations (NS) that an advanced civilization might feasibly create, computational simulations emerge as a critical methodological tool. These advanced computational models serve to simulate the decision-making processes of potential advanced civilizations, providing a sophisticated means to explore the multifaceted considerations that influence the creation of simulations. By incorporating variables such as technological advancements, resource allocation, ethical considerations, and societal values, computational simulations offer a dynamic and nuanced approach to understanding the range of possible NS values.

Methodological Foundations

The methodological foundation of employing computational simulations in this context is grounded in systems dynamics and agent-based modeling. Systems dynamics allows for the modeling of complex, interdependent systems, enabling the examination of how changes in one part of the system affect the whole. Agent-based modeling, on the other hand, focuses on the interactions of individual agents (which, in this scenario, could represent decision-makers within an advanced civilization) and how their behaviors and decisions aggregate to produce systemic outcomes.

Incorporating Variables

- **Technological Advancements:** Computational models integrate predictions about future technological capabilities, including the potential impact of quantum computing, artificial neural networks, and other emerging technologies on the feasibility of creating simulations.
- **Resource Allocation:** Models account for the economic and material constraints that might limit or facilitate the creation of simulations, including the availability of energy sources and computational materials.
- **Ethical Considerations:** Ethical frameworks and debates about the morality of creating simulated realities, especially those involving sentient beings, are encoded into the decision-making algorithms of the simulated agents.
- **Societal Values:** The models also incorporate variables representing the values, priorities, and cultural norms of the simulated civilizations, influencing their desire to simulate (DS) and, consequently, the number of simulations they might choose to create.

Execution and Analysis

The execution of these computational simulations involves running multiple scenarios under varying assumptions about the input variables. This allows researchers to explore a wide range of outcomes for NS, providing insights into how different factors interact to influence the decision-making processes regarding simulation creation. The analysis of simulation results focuses on identifying patterns, key drivers, and potential bottlenecks or accelerators in the path to creating simulations.

Challenges and Considerations

One of the significant challenges in utilizing computational simulations for this purpose is the calibration and validation of the models. Given the speculative nature of simulating the decision-making processes of advanced civilizations, ensuring the models' accuracy and reliability requires careful consideration and, where possible, the alignment of model assumptions with empirical data or well-established theoretical frameworks.

Additionally, the interpretation of simulation results must account for the inherent uncertainties and speculative elements of the models. Researchers must navigate the delicate balance between drawing meaningful conclusions and acknowledging the limitations of the models' predictive power.

Harnessing Simulation to Elucidate Simulation

Computational simulations offer a unique and insightful approach for investigating the potential scale of the Number of Simulations (NS) within the context of the Simulation Hypothesis. By enabling the exploration of multifaceted decision-making processes within hypothetical advanced civilizations, these models help us transcend mere static projections. Their ability to incorporate the dynamic interplay between technological advancement, ethical stances, resource constraints, and evolving societal norms sheds light on the complex landscape of factors that might ultimately shape the production of simulations.

While the calibration, validation, and interpretation of computational simulations designed to model advanced civilizations pose substantial challenges, their value lies in the iterative process itself. By systematically refining these models, challenging their assumptions, and testing the sensitivity of their outcomes, we deepen our own understanding of the intricate motivations and behaviors that might drive the creation of simulated realities. This approach compels us to grapple with the inherent complexities and uncertainties of the Simulation Hypothesis, ultimately enriching our engagement with this profound philosophical and scientific inquiry.

Scenario Analysis:

Scenario Analysis provides a methodical framework for exploring the vast potential landscape shaped by differing levels of Technological Capability (TC) and Desire to Simulate (DS). By carefully constructing well-defined scenarios that encapsulate specific combinations of technological prowess, resource availability, and motivations, we move beyond abstract speculation. This technique facilitates a structured examination of the potential conditions that might lead to the creation of simulations, ultimately outlining potential upper and lower bounds on the Number of Simulations (NS).

The true value of Scenario Analysis lies in its ability to illuminate the potential variability of outcomes by highlighting contrasting scenarios. Exploring scenarios where TC is high, but DS is low due to strong ethical safeguards, forces us to confront the possibility that simulations might be technologically feasible but undesirable for certain civilizations. Conversely, scenarios with limited TC but a burning desire to understand their past through simulations might uncover bottleneck effects and resource constraints that prevent large-scale simulations from existing. This exploration, informed by our understanding of history and human nature, ultimately refines our understanding of the Simulation Hypothesis and aids in establishing plausible ranges for NS.

Factors Influencing NS Projections

Estimating NS requires a nuanced approach, as it combines quantifiable aspects with significant uncertainties stemming from technological, societal, and even existential considerations. Here, we examine key factors influencing NS and the role of probabilistic modeling alongside informed speculation in refining our projections.

Technological Advancements

The Impact of TC Development: Our projections of NS are inextricably linked to our analysis of Technological Capability (TC). Breakthroughs in computing paradigms, such as quantum or biological computing, could vastly enhance the number of possible simulations a given civilization could conduct. Modeling NS, therefore, requires scenarios that reflect various rates and limits of TC development.

Simulation Efficiency Gains: The potential for simulator civilizations to discover dramatically more efficient methods of simulation is crucial to consider. If the computational requirements for high-fidelity ancestor simulations decrease over time, this could directly lead to a proliferation in the feasible NS for even modestly advanced civilizations. Sensitivity analysis can assess the impact of such efficiency gains on our NS projections.

Resource Allocation

Computational Resources: The computational resources available to simulator civilizations will impact NS. Statistical distributions and economic modeling can be helpful to explore various scenarios of resource availability and allocation strategies among advanced civilizations. We may assume resources to be plentiful, but simulators might have priorities vastly different from our own.

Energy Constraints: Energy requirements for large-scale simulations could be a significant factor. Extrapolations based on potential energy production mechanisms (e.g., harnessing stellar energy via Dyson spheres, matter-energy conversion) should inform our NS estimates. Conversely, if energy proves to be a severe constraint, it could limit the potential scale of ancestor simulations.

Ethical and Societal Constraints

Moral Considerations and Simulation Creation: The ethics of simulating conscious beings poses profound questions for potential simulator civilizations. Whether strong moral safeguards are in place or simulations are considered ethically permissible could dramatically influence the Desire to Simulate (DS), and consequently the NS. Collaboration with ethicists and philosophers is essential here.

Cultural Attitudes and Motivations: The cultural values of potential simulators remain a significant unknown. They might find historical simulations distasteful, focus on modeling alternative futures for strategic planning, or prioritize simulations for artistic rather than scientific purposes. Building a range of NS scenarios that reflect diverse motivations enhances the flexibility of our methodology.

Existential Risks and Limitations

Self-Termination Risks: The probability that advanced civilizations destroy themselves or fail to reach a posthuman stage due to technological or societal risks, directly impacts the number of potential simulator civilizations, and thereby NS. Extrapolating potential existential risk profiles from human history, alongside the analysis of filtering events like self-inflicted technological catastrophes, can offer clues, although these may not generalize well to posthuman civilizations.

Fundamental Limits and Simulation Fidelity: The existence of fundamental limitations—physical or computational—could impose a ceiling on the fidelity, and hence, the scale (NS) of achievable simulations. Exploring potential physical constraints on information density, the computational requirements to simulate consciousness, and the inherent limits of computability in a given universe will inform our boundary conditions for NS estimation.

Integration of Statistical Approaches and Speculation

The inherent complexity of projecting NS necessitates the skillful integration of statistical methods with elements of informed speculation:

- **Bayesian Inference for Evolving NS Models:** Bayesian approaches are well-suited, as they allow us to update our estimates of NS as we gather new evidence about TC, DS, and potential constraints. Establishing appropriate prior distributions and incorporating observational data will be crucial for evolving our NS models.
- **Scenario Planning and Sensitivity Analysis:** Developing a wide range of plausible scenarios considering variations in TC trajectories, resource allocation, societal values, and existential risks is essential. Sensitivity analysis can quantify the impact that our assumptions have on NS projections, revealing critical parameters to focus future research efforts.
- **The Role of Informed Speculation:** When dealing with highly advanced civilizations, some degree of informed speculation becomes unavoidable. Drawing upon futurism, science fiction, and insights from evolutionary theory may offer starting points for imagining the motivations and technological

trajectories of posthuman civilizations, allowing us to explore a broader range of potential NS outcomes.

Connecting NS Projections to the Simulation Argument

Refining our NS estimates plays a crucial role in assessing the overall plausibility of the Ancestor Simulation Argument. Key connections include:

- **NS and Simulation Prevalence:** A high estimated NS increases the likelihood that our specific existence is within one of those many simulations. Conversely, a low NS makes this less likely, although it doesn't entirely disprove the hypothesis.
- **Observational Constraints and NS:** If observations impose hard limits on the available resources within the universe, it could suggest a potential upper bound on NS, indirectly influencing our evaluation of the simulation hypothesis.
- **NS, TC, and DS Interplay:** It is the complex interplay between the Number of Simulations, feasible Technological Capability, and Desire to Simulate that ultimately informs our assessment of Bostrom's trilemma and the likelihood of our own simulated existence.

Refinement as a Pathway: Interdisciplinary Inquiry and NS

The process of estimating the Number of Simulations (NS) emerges as a complex task woven from elements of both rigorous analysis and informed speculation. We combine the power of statistical models with the exploration of potential future scenarios, all the while confronting the inevitable uncertainties surrounding the technological prowess, societal dynamics, and even the potential constraints imposed by the fundamental laws of physics. This methodological blend acknowledges the limits of our knowledge and embraces the need for well-grounded projections.

Our estimations of NS are not set in stone. A hallmark of this analysis is its inherent adaptability. As interdisciplinary research advances, our understanding of computational potential, societal evolution, and the nature of reality deepens, prompting the continuous refinement of our NS projections. Through this process, we embark on a dynamic evaluation of the Simulation Hypothesis. Should our investigations ultimately point towards a vast potential for simulated realities, it strengthens the possibility that we might ourselves inhabit one instantiation within a grand tapestry of simulations. This inquiry offers a profound invitation to re-evaluate our perception of existence and our place within it.

Statistics and Speculation Conclusion

Our quest to estimate the Number of Simulations (NS) calls for a unique blend of data-driven analysis and a willingness to venture into the speculative realm of theoretical possibilities. Grounded in statistical rigor, we harness the power of computational simulations and examine historical trends to glean insights into potential technological advancement, resource allocation, and the evolution of motivations. However, we simultaneously acknowledge the profound limitations of extrapolating from our current reality towards the capabilities of vastly advanced civilizations. This acknowledgement invites and necessitates an element of well-informed speculation, prompting us to think boldly yet systematically about the potential scale of simulated realities.

Estimating the number of simulations transcends mere calculation. It compels us to confront the full scope of Bostrom's Ancestor Simulation Argument, including the ethical dilemmas posed by large-scale simulations, the societal impacts of simulation technology, and the fundamental philosophical questions it raises about the nature of our own existence. By carefully investigating the factors influencing NS and integrating statistical and computational insights, we not only gain a richer understanding of the Simulation Hypothesis but also delve into a nuanced exploration of the intricate connection between technological capability, human values, and the very fabric of reality.

CHAPTER TWENTY-FIVE

Statistical Methods: Our Toolbox

The multifaceted and deeply uncertain nature of the Ancestor Simulation Argument demands a rigorous statistical foundation. Numerous techniques prove valuable, but we'll spotlight two that stand out for their suitability in handling the inherent complexities: Bayesian inference and sensitivity analysis. These approaches offer adaptability, transparency in the face of uncertainty, and crucial insights as we explore the potential technological capabilities and motivations underlying vast numbers of simulated realities.

The power of Bayesian inference lies in its ability to incorporate new evidence and adapt our understanding dynamically. As technological breakthroughs occur, theoretical models in physics are refined, or we gain fresh perspectives on potential societal values within advanced civilizations, Bayesian techniques allow us to seamlessly update our prior probability distributions. This adaptability ensures that our analysis remains grounded in the most current scientific knowledge, reflecting the evolving nature of our understanding about the feasibility of large-scale simulations.

Sensitivity analysis is indispensable for systematically examining how our conclusions and projections change in response to variations in underlying assumptions or input parameters. When investigating the Ancestor Simulation Argument, it allows us to identify the specific variables which have the most profound impact on our conclusions. For instance, we could explore the sensitivity of our simulated reality likelihood estimations to changes in assumptions about resource constraints, energy harvesting potential for advanced civilizations, or even the computational requirements of simulating consciousness. By illuminating the crucial levers within our models, sensitivity analysis refines our understanding and promotes a targeted research agenda.

Bayesian Inference: A Framework for Evolving Knowledge

Bayesian statistical methods provide a remarkably well-suited framework for our investigation for several reasons:

- Bayesian statistical methods offer several advantages that make them exceptionally well-suited for investigating the Simulation Hypothesis. They allow us to explicitly formalize our initial beliefs, however uncertain, about the plausibility of simulated realities, potential ranges within parameters like the Number of Simulations (NS), and the vast spectrum of attainable Technological Capability (TC). These prior beliefs, grounded in our current understanding informed by science and philosophy, are formally represented as probability distributions.
- The strength of Bayesian inference lies in its ability to seamlessly integrate new information. As observational data accumulates, simulation results reveal nuanced patterns, or theoretical breakthroughs reshape our understanding, we utilize Bayes' theorem to update our original prior distributions. This iterative process yields posterior probability distributions that reflect our evolving

state of knowledge, offering a dynamic and adaptable framework for evaluating the Simulation Hypothesis.

- By explicitly encoding prior beliefs and demonstrating how they evolve in response to new evidence, Bayesian methods offer inherent transparency within our analysis. This framework emphasizes that even with advanced tools, a level of uncertainty remains. Bayesian techniques force us to confront these uncertainties head-on, promoting intellectual honesty and encouraging the identification of areas where further research could prove most illuminating.

Prior Beliefs in Bayesian Analysis

In Bayesian analysis, our current understanding – a complex blend of scientific knowledge, reasonable assumptions, and even informed speculation – is formally encoded in a mathematically defined 'prior' probability distribution. This distribution reflects the range of plausible values we believe a parameter or hypothesis could take on before considering new data or evidence. Prior beliefs allow for an explicit and quantifiable representation of our current state of knowledge, including its inherent uncertainties and potential biases.

This incorporation of prior beliefs stands in stark contrast to classical 'frequentist' statistical approaches. Frequentist methods place emphasis on the interpretation of long-run frequencies of events, relying primarily on observable data to draw inferences. While powerful in many contexts, this approach can lack the flexibility needed when navigating the speculative and uncertainty-laden terrain of inquiries like the Simulation Hypothesis.

Within the context of the Simulation Hypothesis, prior beliefs play a crucial role. They allow us to formally integrate our existing understanding of physics, potential computational limits, and even insights from sociology about motivations for complex technological endeavors. By acknowledging and quantifying both our knowledge and uncertainties in these initial distributions, Bayesian analysis provides a framework for transparency and systematic refinement of our understanding as we evaluate the Simulation Hypothesis.

Here's why priors are essential in the simulation hypothesis context:

Accommodating Pre-Existing Knowledge

- Evaluating the Simulation Hypothesis is an exercise that extends far beyond the analysis of raw data. We approach it with a complex tapestry of pre-existing knowledge, philosophical inclinations, and scientific intuitions. The strength of Bayesian methods lies in their ability to formally incorporate these diverse perspectives. Prior distributions allow us to encapsulate our beliefs – whether they stem from our understanding of physics, potential computational constraints, or even broad philosophical stances – into our initial models and assessments.
- For example, an individual who intuitively believes in the potential for highly advanced civilizations and simulated realities might construct a prior probability distribution for the Number of Simulations (NS) that favors higher values. Conversely, a skeptic of the Simulation Hypothesis might utilize a prior distribution for NS which places greater probability on a significantly smaller number of simulations. This flexibility allows us to capture a diverse range of perspectives at the outset, ensuring our analysis isn't limited by a single viewpoint and ultimately facilitating meaningful dialogue even in the face of differing initial beliefs.

Subjectivity and Transparency

- The issue of subjectivity surrounding prior distributions in Bayesian analysis is a valid critique that warrants careful examination. Unlike frequentist approaches that rely primarily on observable data, prior distributions do incorporate a degree of individual or collective belief about the parameters being investigated. However, the strength of Bayesian methods lies in making this subjectivity transparent. By formally declaring our prior assumptions, we open ourselves to scrutiny, challenge, and refinement. This open framework facilitates a healthy dialogue focused on understanding the reasoning behind our initial beliefs.

- It's crucial to recognize that even seemingly objective frequentist analysis often harbors implicit biases. Choices made in data collection, the selection of statistical models, and interpretations of results can all introduce subjective elements. While the appearance of objectivity might exist within frequentist frameworks, these biases can remain hidden or harder to dissect. By emphasizing transparency about initial assumptions, Bayesian methods bring a level of honesty to the analysis, fostering a continuous process of refinement as our priors evolve alongside new evidence and critical debate.

Uncertainty in Simulator Parameters

When contemplating the potential capabilities and motivations of civilizations that could have achieved the technological prowess necessary for ancestor simulations, we confront a sea of uncertainty. Any attempt to estimate their Technological Capability (TC), understand their Desire to Simulate (DS), or model the resource constraints they might face involves substantial guesswork. Within a Bayesian framework, prior distributions offer an essential tool for formalizing this uncertainty.

Rather than relying on single point estimates for parameters like TC, the use of prior distributions allows us to capture a wider spectrum of possibilities. For instance, we might define a distribution for TC growth rates that accounts for potential breakthroughs as well as unforeseen limitations. This approach paints a more realistic picture, acknowledging that our initial understanding of an advanced simulator civilization will inherently be incomplete. By explicitly integrating and quantifying the uncertainties within our priors, we build a more robust and adaptable foundation for evaluating the Simulation Hypothesis.

The Evolution of Knowledge: From Priors to Posteriors

The profound strength of Bayesian inference lies in its ability to accommodate the evolution of our knowledge. The Simulation Hypothesis, residing at the intersection of rapidly developing fields like computation, theoretical physics, and even neuroscience, necessitates an approach that can adapt to new discoveries. Bayes' theorem offers a rigorous mathematical framework for updating our initial prior distributions as evidence accumulates, as simulations provide further insights, or as theoretical models are refined.

The journey from prior distributions to posterior distributions represents the process of refining our understanding. Initial beliefs, shaped by the best available knowledge and potential biases, provide a starting point. As we systematically incorporate new evidence or insights, our posterior distributions evolve. This evolution reflects a synthesis of our initial perspective with the newly gathered information. Within the context of the Simulation Hypothesis, this iterative refinement process allows us to dynamically assess the possibility of simulated realities, with our evolving understanding reflected in the ever-changing posterior distributions.

Here's how this works in the simulation context:

New Evidence

- Let's consider a hypothetical scenario where astrophysicists uncover an unexpected anomaly in astronomical data – one which raises the possibility of artificial manipulation within a simulated universe. Within a Bayesian framework, this discovery represents new evidence. To formalize it, we must carefully assign probabilities: specifically, the likelihood of observing such an anomaly if we indeed reside within a simulation, and the likelihood of observing it even in a non-simulated universe.
- Bayes' theorem provides the mathematical machinery to recalculate our probabilities in light of the new astronomical evidence. Our original prior distributions, which encoded our beliefs about the Simulation Hypothesis, Technological Capabilities of potential simulators, etc., will now be updated.

The extent to which our posterior probabilities change depends on both the strength of our initial beliefs and how strongly the new evidence favors (or disfavors) the Simulation Hypothesis. This updated posterior encapsulates our revised probability assessments in light of the astronomical discovery. The process, by its very nature, demonstrates the dynamic evolution of our understanding based on scientific inquiry within a Bayesian framework.

Updating the Prior

- Bayes' theorem serves as an elegant choreographer for guiding the interaction between our prior beliefs and new evidence. Let's again consider the case of the unexpected astronomical anomaly, potentially indicative of a simulated universe. Bayes' theorem mathematically combines our prior probability distribution (which encapsulated our initial beliefs about the Simulation Hypothesis, technological capabilities of simulators, etc.) with the likelihood of observing this specific evidence to yield an updated posterior probability distribution.
- This newly calculated posterior distribution reflects our refined understanding in light of the evidence uncovered. The influence exerted by the new evidence depends on its strength – if it strongly favors the Simulation Hypothesis, it will significantly shift our posterior towards a higher probability that we might reside within a simulated reality. Conversely, weak or inconclusive evidence will have a far more subtle impact on our posterior distribution. This process highlights the dynamic nature of Bayesian inference, where our beliefs evolve proportionally to the strength of the evidence presented.

Iterative Refinement

- The true power of Bayesian inference within the context of the Simulation Hypothesis lies in its iterative nature. The posterior probability distribution resulting from a previous analysis seamlessly transitions into the prior distribution for the following update. This means each new piece of evidence, each simulation result, and each theoretical breakthrough has the potential to refine our understanding continuously. It is a process of gradual knowledge refinement rather than a single, absolute conclusion.
- As we cycle through multiple Bayesian updates, our models gain sophistication and nuance. Simulations help us explore potential technological limits (TC), new discoveries inform our understanding of physical laws, and interdisciplinary research might illuminate new perspectives on the motivations behind simulations (DS). With each iteration, our prior distributions take into account our ever-expanding and evolving knowledge. This translates into increasingly informed and refined assessments of the Simulation Hypothesis. Instead of clinging to a static point estimate, Bayesian inference allows our understanding to adapt alongside the advancements in science and technology, making it a powerful tool for navigating complex and speculative topics like simulated realities.

Specific Examples within Our Methodology

Let's illustrate the evolution of priors with a few examples relevant to Bostrom's argument:

The Impact of TC Breakthroughs

The realm of technological advancements, particularly in quantum computing, plays a pivotal role in the assessment of the Simulation Hypothesis through Bayesian analysis. Let us consider a scenario where our initial Bayesian prior—rooted in the current understanding and projections of computational technology—positions the likelihood of achieving realistic ancestor simulations within a century as exceedingly slim. This prior is grounded on the conventional trajectory of technological evolution, factoring in the limitations of classical computing paradigms and the monumental computational demands of such simulations. However, the landscape of possibilities undergoes a dramatic transformation with a hypothetical, yet plausible, revolutionary breakthrough in quantum computing. This innovation could potentially unlock computational capabilities far beyond our current horizon, thereby necessitating a profound revision of our prior assumptions.

Upon integrating this quantum leap into our Bayesian framework, the posterior distribution emerges significantly altered. The revised prior, now infused with the newfound potential of quantum computing, recalibrates the probability landscape, assigning a markedly higher likelihood to the feasibility of constructing realistic ancestor simulations within the stipulated timeline. This shift not only underscores the sensitivity of our probabilistic assessments to technological capabilities but also illuminates the dynamic nature of our understanding. It reflects a broader epistemological principle: as our technological frontier expands, so too does the realm of what we deem plausible. Thus, through the lens of Bayesian analysis, the impact of technological breakthroughs on the Simulation Hypothesis is not merely incremental but potentially paradigm-shifting, redefining the boundaries of what we conceive as achievable within the fabric of our reality.

Philosophical Insights Altering DS

The Desire to Simulate (DS), a crucial factor in the Bayesian analysis of the Simulation Hypothesis, initially might be gauged with a low prior probability. This skepticism often stems from projecting our current ethical frameworks onto hypothetical simulator civilizations, where the ethical implications of creating sentient simulations pose profound moral dilemmas. These concerns hinge on assumptions about the values and ethical considerations of advanced civilizations, assuming they would mirror our own reservations about the morality of such simulations. However, the introduction of new philosophical insights can significantly recalibrate our understanding and assessment of DS. Arguments positing the moral permissibility, or even ethical imperatives, for creating simulations introduce a paradigmatic shift in the ethical landscape, suggesting that advanced civilizations may, under certain conditions, view simulations not only as permissible but as morally desirable or beneficial.

Incorporating these philosophical advancements into our Bayesian analysis necessitates a reevaluation of our DS prior. This recalibration process, informed by a nuanced understanding of potential ethical frameworks governing advanced civilizations, shifts the posterior probability distributions, thereby influencing the overall plausibility of the Simulation Hypothesis. This evolution in thought illustrates the fluidity of moral philosophy and its impact on our probabilistic assessments. It underscores the importance of remaining open to evolving ethical considerations and philosophical arguments, which can profoundly alter the landscape of possibility. In this manner, Bayesian analysis serves as a dynamic tool, capable of integrating diverse inputs—from technological advancements to philosophical insights—to refine our understanding of the likelihood and implications of the Simulation Hypothesis, weaving together a complexity of probabilities that map the contours of our reality and its potential underpinnings.

Importance for the Simulation Argument

The ability to formalize and continually refine our beliefs about critical parameters like TC, DS, and NS is paramount in the simulation hypothesis investigation. Bayesian inference, by incorporating uncertainty and accommodating the evolving nature of scientific knowledge, provides a robust framework.

Uncertainty Quantification

The exploration of the Simulation Hypothesis necessitates a rigorous methodology to navigate the profound uncertainties inherent in positing the existence of advanced civilizations and their potential to create simulations. Bayesian analysis stands out as a methodological beacon in this regard, offering a sophisticated framework for quantifying uncertainty. This approach is intrinsically designed to handle the probabilistic nature of our hypotheses, allowing researchers to incorporate a wide spectrum of uncertainties into their models. By transforming these uncertainties into probabilistic terms, Bayesian inference provides a mechanism for expressing beliefs about the likelihood of various scenarios in the context of the Simulation Hypothesis. This probabilistic lens is invaluable, as it enables a nuanced assessment of the hypothesis by articulating the degree of certainty associated with each parameter, such as Technological Capability (TC) or the Number of Simulations (NS), thereby moving beyond binary assertions to embrace a spectrum of possibilities.

Furthermore, the implementation of credible intervals within Bayesian analysis offers a clear and interpretable means to communicate the uncertainty surrounding key parameters. These intervals represent ranges within which the true values of parameters like the Desire to Simulate (DS) are likely to fall, with a specified level of confidence. This aspect is crucial when contemplating the capabilities and intentions of civilizations that may operate under fundamentally different technological and ethical frameworks than our own. The ability to quantify and transparently communicate uncertainty equips us with a more resilient understanding of the Simulation Hypothesis, acknowledging the limitations of our knowledge while still providing a structured approach to explore this frontier.

Handling Complex Models

The Simulation Hypothesis posits a reality that is contingent upon a labyrinth of factors, from the development trajectory of technological capabilities to the ethical and philosophical considerations that might motivate an advanced civilization to create simulations. This intricate web of parameters demands a methodological tool capable of dissecting and analyzing the complex relationships between them. Bayesian methods, renowned for their flexibility in modeling complex systems, become indispensable in this context. Specifically, computational techniques like Markov chain Monte Carlo (MCMC) simulations enable researchers to navigate the probabilistic landscapes of such multifaceted models. MCMC simulations facilitate the exploration of parameter spaces where traditional analytical approaches might falter, allowing for a more thorough examination of the interdependencies and potential dynamics at play within the Simulation Hypothesis.

Moreover, Bayesian methods provide the analytical rigor to incorporate prior knowledge and update our beliefs in light of new evidence, a process that is inherently iterative and reflective of the scientific method. This capability is especially relevant when dealing with the Simulation Hypothesis, where new technological advancements or philosophical insights can significantly alter our understanding of the feasibility and implications of simulations. By applying Bayesian analysis to this complex model, we can systematically account for factors like resource availability, ethical constraints, and the cumulative impact of technological progress. This approach not only aids in disentangling the intricate relationships between these factors but also in framing our findings within a probabilistic context that acknowledges the inherent uncertainties of such an ambitious inquiry. Through Bayesian analysis, we are thus equipped to forge a deeper and more coherent understanding of the Simulation Hypothesis, illuminating the pathways through which it might be realized or refuted.

Sensitivity Analysis: Exposing Critical Assumptions

The speculative nature of the Ancestor Simulation Argument means our models and projections rely on a multitude of assumptions. These assumptions encompass parameters ranging from potential technological growth rates to societal structures of potential simulator civilizations, and even the computational demands of simulating consciousness. Sensitivity analysis empowers us to systematically examine how variations in these underlying assumptions impact the conclusions we draw about the Simulation Hypothesis.

By strategically adjusting our input parameters or altering our models, sensitivity analysis charts the magnitude of change in our outcomes. This process unveils which assumptions hold the most sway over our conclusions about the Number of Simulations (NS), the likelihood of existing in a simulated reality, or other relevant metrics. By identifying these critical levers, sensitivity analysis guides our research efforts. It allows us to target further theoretical and experimental inquiries towards the areas where uncertainties or assumptions have the most profound impact on our understanding of the Simulation Hypothesis.

It allows us to answer 'what if?' questions in a systematic manner:

Application 1: Updating Beliefs About TC

- **Initial Prior**
    - We begin by constructing an initial prior distribution for Technological Capability (TC). This prior encapsulates our existing knowledge about the potential growth of computational power. Informed by trends like Moore's Law, we might favor gradual growth trajectories, acknowledging the potential for diminishing returns or even fundamental bottlenecks. However, the speculative nature of the endeavor necessitates a prior distribution that also allows for the possibility of unforeseen breakthroughs or technological leaps that could far

outpace our current understanding or historical trends.
  - This initial prior acts as our baseline – it reflects the best of our current knowledge and acknowledges uncertainties in future technological development. This nuanced approach avoids overly optimistic or pessimistic extremes regarding computational potential, allowing for a balanced Bayesian analysis as new evidence and theoretical insights emerge.
- **Theoretical Breakthroughs and New Priors**
  - Let's imagine a scenario where a groundbreaking discovery in theoretical physics redefines our understanding of computation. This breakthrough, perhaps based on a novel interpretation of quantum mechanics, suggests previously unknown fundamental limits on information processing. Rather than incorporating this breakthrough as a single, deterministic limit, we embrace the flexibility of Bayesian methodology by revising our TC prior distribution.
  - This revised TC prior represents a shift in our understanding. It would place higher probability on scenarios where technological prowess faces constraints imposed by the very laws of nature. This acknowledges the discovery but retains a degree of uncertainty, allowing for the possibility that future civilizations might discover ways to circumvent or mitigate these limitations within our updated model.
- **Bayesian Inference for Updated NS**
  - Our refined TC prior has profound implications for our assessment of the Number of Simulations (NS). Bayesian inference provides the mathematical tools to recalculate the probabilities associated with NS in light of this revised understanding of technological limitations. The resulting posterior distribution for NS is likely to differ significantly from our initial model, which had allowed for potentially unconstrained computational growth.
  - This updated posterior distribution reflects the impact of potential fundamental limits on the plausibility of vast numbers of simulations. If technological limitations are more probable than previously assumed, the potential scale of simulated realities might be significantly reduced. This has consequences for assessing the overall strength of the Simulation Hypothesis.
- **Sensitivity Analysis**
  - Sensitivity analysis serves as a crucial tool to dissect the interplay between our revised TC prior and our conclusions about the Number of Simulations (NS). By strategically altering input parameters, we discover that our revised understanding of technological limitations has the most profound impact within specific ranges of NS values.
  - This finding offers significant guidance for our research efforts. It reveals scenarios where our model's outcomes are highly sensitive to potential constraints on computational power. Sensitivity analysis encourages focused theoretical and experimental inquiries into these specific scenarios, aiding us in refining our understanding of the computational limitations that simulator civilizations might encounter.
- **Impact on Overall Hypothesis**
  - If limited TC leads to a lower probability of large-scale simulations, this weakens the argument that the majority of beings with our kind of experiences are simulated. It might shift our focus towards the first proposition of Bostrom's trilemma – perhaps most civilizations never reach the level of technological advancement necessary for ancestor simulations.

Application 2: The Interplay of DS and NS

- **Initial DS Prior**
  - Crafting a prior distribution for the Desire to Simulate (DS) involves acknowledging the profound uncertainties surrounding the motivations of potential simulator civilizations. With limited knowledge, we might cautiously begin with a prior that leans towards the possibility that not every posthuman civilization would necessarily engage in large-scale ancestor

simulations. This could reflect our own ethical concerns or intuitions about civilizations potentially finding such simulations unnecessary or even undesirable.

  - This initial prior acts as a starting point. It acknowledges that there could be myriad reasons influencing a simulator civilization's decision to simulate ancestors and reflects a conservative approach in absence of concrete information about civilizations that vastly exceed our own level of development.

- **Ethical Considerations and the Evolving DS Prior**
  - As we delve deeper into philosophical investigations, we might discover potential ethical frameworks in which simulations of conscious ancestors are considered permissible or even beneficial in the eyes of simulator civilizations. Additionally, simulations designed to explore the dynamics of different simulator societies could illuminate potential value systems. These insights would influence our understanding of simulator motivations, prompting us to refine our initial DS prior distribution.
  - This revised DS prior represents an update based on our evolving ethical considerations. It might place a higher probability on a greater willingness of simulator civilizations to engage in ancestor simulations. However, we still acknowledge the inherent uncertainty surrounding the diverse potential motivations and value systems of advanced civilizations.
- **Bayesian Updating of NS**
  - Integrating our revised DS prior necessitates re-evaluation of the Number of Simulations (NS) through Bayesian inference. This recalculation likely yields an updated posterior distribution for NS that differs from the conservative outcome based on our initial prior distribution. The shift towards a higher probability of willingness to simulate could suggest that the scale of simulations might be larger than initially assumed.
  - This updated NS distribution has implications for our investigative approach. With a higher likelihood of large-scale simulations, we might intensify our search for potential artifacts or anomalies that would betray the existence of a simulated reality.
- **Sensitivity Analysis**
  - Sensitivity analysis illuminates the relationship between the Desire to Simulate (DS) and the Number of Simulations (NS) within our model. It reveals a nuanced picture where specific ranges of DS values disproportionately impact our conclusions. Small changes in the assumed willingness of simulator civilizations within those critical ranges trigger significant shifts in predicted simulation numbers.
  - This discovery highlights the importance of further refining our understanding of ethics within highly advanced civilizations. It underscores that simulator societies might possess moral frameworks that differ significantly from our own, making the prediction of their decisions difficult. This emphasizes the need for interdisciplinary inquiries and collaborative investigations that combine philosophy, sociology, and simulations to gain a deeper understanding of the potential range of ethical systems that might guide technological development far beyond our current capabilities.

Application 3: Assessing Existential Risk

- **Prior on Existential Fates**
  - Our understanding of history and the risks inherent in technological advancement shape our initial prior distribution regarding the potential fates of civilizations. We might construct a prior that acknowledges a non-zero probability that civilizations fail to reach the advanced posthuman stage necessary for ancestor simulations. This prior could reflect potential self-destruction scenarios, unforeseen technological pitfalls leading to stagnation, or other existential threats that might prematurely terminate progress.
  - This initial prior reflects both our own experiences as a technologically evolving species and

our awareness of the potential for unforeseen risks. It emphasizes that the path to achieving the ability to create ancestor simulations might be fraught with dangers that could lead to civilizations failing to reach this milestone.

- **Evolving Risk Understanding as New Prior**
  - Focused research on the complex landscape of existential risks is likely to refine our perspective. Investigations into artificial intelligence control problems, the impact of resource depletion, or novel threats that emerge with new technology could profoundly alter our understanding of the magnitude of these risks. This evolving understanding directly informs the creation of a new prior distribution – one that likely increases the probability we assign to civilizations facing premature termination.
  - This revised prior reflects a more pessimistic outlook on the likelihood of civilizations surviving to reach the posthuman simulator stage. The flexibility of Bayesian methods allows for seamless incorporation of new insights, acknowledging that our understanding of existential threats is likely to evolve alongside technological progress.
- **Bayesian Recalculation**
  - Our updated and potentially more pessimistic prior about existential risks triggers a recalculation within our Bayesian model. This process impacts our assessment of the first proposition of Bostrom's trilemma – the proposition that posthuman civilizations capable of ancestor simulations are exceedingly rare. A higher likelihood of civilization termination due to existential risks strengthens this proposition, suggesting that potential simulator civilizations might be fewer than initially assumed.
  - This indirectly affects our estimations of the Number of Simulations (NS). If fewer civilizations reach the capacity to simulate, it could point towards a lower number of simulations overall. This shift in our understanding has implications for the strength of the Simulation Hypothesis and could guide strategies for detecting potential signs of simulated realities.
- **Sensitivity Analysis**
  - By employing sensitivity analysis, we gain deeper insights into the complex relationship between existential risk and our conclusions about the Simulation Argument. We might uncover that our model outcomes are highly sensitive to probabilities assigned to specific categories of existential risk. For example, assumptions about the controllability of artificial intelligence might have a disproportionate impact on our conclusions compared to resource depletion scenarios.
  - This realization offers valuable guidance for future research. It encourages focused investigation into the specific types of existential risks deemed most impactful within our model. By refining our understanding of these risks or exploring mitigation strategies, we directly contribute to a more accurate assessment of the possibility that numerous technologically mature civilizations capable of simulations exist within our universe.

Navigating Uncertainty, Refining Understanding

Our analysis underscores the power of Bayesian inference and sensitivity analysis when navigating the inherently speculative and uncertain terrain of the Ancestor Simulation Argument. The ability to formalize subjective prior beliefs, to account for their evolution in the face of new evidence, and to expose crucial sensitivities within our models offers a profound advantage. These methods provide a degree of rigor, transparency, and adaptability that is essential when dealing with a hypothesis as profound and elusive as the potential for simulated realities.

While the Simulation Hypothesis may never yield a definitive, universally accepted answer, the process of our inquiry is immensely valuable. Bayesian techniques and sensitivity analysis empower us to embrace the uncertainties inherent in this endeavor. By systematically quantifying assumptions, updating probabilities as knowledge evolves, and pinpointing the factors that hold the greatest influence on our conclusions, these statistical methods enhance our understanding of the complex interplay between simulation feasibility, motivations, and the fundamental nature of reality.

Applications within the Ancestor Simulation Argument

Let's illustrate how Bayesian inference and sensitivity analysis are applied in our methodology:

Updating Beliefs About TC

Imagine theoretical advancements reveal the potential for a fundamental hard limit on computation, even for extraordinarily advanced civilizations. This breakthrough necessitates a revision of our prior distribution for Technological Capability (TC). The new prior would place stricter constraints or even upper bounds on achievable computational power.

This updated TC prior is integrated into our Bayesian model to recalculate the probabilities associated with the Number of Simulations (NS). Sensitivity analysis is then used to pinpoint how these new constraints on TC impact our conclusions. It could reveal that our outcomes are highly sensitive to potential computational limits within specific ranges, highlighting the critical importance of theoretical research and investigation into these scenarios.

The Interplay of DS and NS

Our initial understanding of the motivations behind ancestor simulations might be limited. A conservative initial prior distribution for DS may reflect our own ethical concerns or intuitions about civilizations potentially finding such simulations unnecessary or undesirable. However, as ethical frameworks evolve, and simulations help model hypothetical simulator societies, our understanding of potential motivations would shift. This leads to an updated DS prior.

Bayesian analysis provides the tools to mathematically incorporate our revised DS prior, resulting in updated posterior distributions for the Number of Simulations (NS). Sensitivity analysis offers further insights, potentially showing that our conclusions about NS are particularly sensitive to changes in DS within specific ranges of simulator motivation probabilities. This could guide future research efforts towards those motivational scenarios, potentially informing strategies for detecting potential signs that we might reside within a simulation.

Assessing Existential Risk

To assess the likelihood of the first proposition in Bostrom's trilemma (civilizations failing to reach the posthuman simulator stage), we must incorporate existential risks into our models. Our initial prior distributions might reflect our understanding of historical risks and potential dangers inherent in emerging technologies.

Bayesian inference provides a flexible framework to recalculate our beliefs about this proposition as our knowledge of existential risks evolves. Investigations into AI control, resource depletion, or novel threats could lead to revised priors that assign higher probabilities to premature technological stagnation or self-destruction. Sensitivity analysis would expose the influence of specific risk types on our conclusions, guiding further research efforts with heightened urgency into the prevention of these threats, not only within our society but for the broader question of civilizations surviving to reach posthuman stages.

A Robust Toolkit for Simulation Investigations

Bayesian inference and sensitivity analysis are instrumental tools in our statistical approach to the Ancestor Simulation Argument. They enable us to incorporate and evolve our knowledge base, quantify uncertainties, and identify the critical assumptions that drive our conclusions. The outputs of these statistical analyses directly inform our assessments of Technological Capability, the Desire to Simulate, and the overall plausibility of our existence within a simulated reality.

These statistical tools offer a robust framework for quantifying uncertainties, refining models, and guiding our investigative efforts within the speculative landscape of the Simulation Hypothesis. Ultimately, the insights gleaned from Bayesian inference and sensitivity analysis directly shape our assessment of the feasibility of ancestor simulations, the potential scales at which they might exist, and the overall plausibility of the idea that we ourselves might reside within a simulated reality. The process offers methodological rigor and transparency for an endeavor as profound and uncertain as the Simulation Argument.

CHAPTER TWENTY-SIX

Ethical Considerations: A Guiding Light

The exploration of simulation theory transcends purely academic inquiry, propelling us into a realm where philosophical musings meet concrete ethical imperatives. At the heart of this exploration lies a core dilemma that challenges our moral intuition and ethical frameworks: the possibility of creating simulated beings endowed with conscious experiences. This section endeavors to dissect the moral quandaries surrounding the existence and rights of such entities, drawing parallels to contemporary debates on artificial intelligence (AI) sentience and the responsibilities shouldered by creators in both simulated and tangible realms.

The possibility of simulated consciousness introduces a profound ethical dimension to the Ancestor Simulation Argument. Our methodology cannot afford to treat ethics as an afterthought; it must be deeply integrated into every stage of our inquiry.

Here's why:

- **The Moral Status of Simulated Beings:** A fundamental question centers on whether simulations capable of experiencing consciousness possess moral standing akin to biological beings. If simulations are mere puppets devoid of true sentience, their manipulation might raise fewer concerns. But, if advanced simulations achieve subjective experiences indistinguishable from our own, then their creation and potential mistreatment within simulations would demand serious moral scrutiny.
- **Responsibility of Simulators:** If we are capable of creating simulations with conscious inhabitants, we bear an ethical responsibility for their well-being. This extends questions of responsible AI development into the realm of simulations – what rights do simulated beings possess? What safeguards should be in place? Do simulators have a moral obligation to inform simulated beings of their nature, or would such a revelation cause undue suffering?
- **The Impact on Us:** Grappling with the ethical complexities of simulation has a reflexive effect. If simulated beings possess moral worth, it recalibrates our perspective on our own potential status as simulated entities. This underscores the need for potential simulators to uphold the highest ethical standards, as such standards might shape our own treatment should we exist within a simulation.
- **Guiding Research and Detection Strategies:** Ethical considerations directly influence our approach to investigating the Simulation Hypothesis. If simulations are morally problematic, this might shift focus away from actively seeking to create them and towards the search for potential simulation 'glitches' as evidence of our own simulated state. This highlights the interplay between ethics and our knowledge-seeking methodology.

The Toolkit for Ethical Navigation

Engaging with the ethical ramifications of the Simulation Hypothesis requires us to draw upon a rich complexness of philosophical and ethical frameworks. Here's an overview of key schools of thought and the questions they raise for us:

Deontology

Deontological ethics emphasizes duties and rules. In the context of simulations, it compels us to consider the inherent rights of simulated beings. Could creating conscious simulations without their consent be an absolute ethical violation, regardless of any potential benefits derived from such simulations?

Deontology and the Moral Status of Simulated Beings

- Deontology, championed by philosophers like Immanuel Kant, argues that the morality of an action is determined by its adherence to universal moral principles and duties, independent of the consequences. In the context of the Simulation Hypothesis, Deontology focuses on the inherent rights of simulated beings, particularly if they possess consciousness. This perspective raises critical questions:
    - **Do Simulated Beings Have Rights?** Deontology compels us to consider whether simulated consciousness, if indistinguishable from biological sentience, warrants moral consideration similar to humans or animals. If simulated beings can experience suffering, then creating them without their consent or subjecting them to harm could be a clear violation of their inherent right to well-being.
    - **The Duty to Avoid Harm:** A core principle in Deontology is the imperative to avoid causing harm to others. Creating simulations that experience negative emotions or manipulation without their knowledge or consent could be seen as a breach of this principle. Deontologists would argue that the potential benefits of simulations (e.g., scientific knowledge) cannot outweigh the fundamental moral wrong of inflicting suffering on sentient beings.
    - **The Universality of Moral Law:** Kant's concept of the Categorical Imperative posits that we should act only according to maxims that could be universally applied as laws. If creating conscious simulations without consent is deemed unethical, then this principle suggests it would be wrong for any civilization, including our own, to engage in such practices.

The Challenge of Consent and Simulated Reality

- Deontology presents a significant challenge for the construction of simulations with conscious inhabitants.
    - **The Problem of Prior Consent**: If simulations lack the ability to exist outside the simulated environment, obtaining their consent before creation seems impossible. Deontologists might argue that creating a sentient being without the option of opting out constitutes a fundamental violation of their autonomy.
    - **The Paradox of Simulated Freedom:** Even if simulations are endowed with a sense of agency within their simulated reality, the very nature of their existence is predetermined by the simulator. This raises questions about whether such beings can ever truly consent to their situation from a Deontological perspective.
    - **The Deception Dilemma:** Deception might be necessary to maintain the illusion of a non-simulated reality for inhabitants. However, Deontology generally frowns upon actions that require deception, especially when it concerns the very nature of a being's reality.

Deontology and Our Methodology

- Deontological considerations have a profound impact on how we approach the Simulation Hypothesis:
    - **Prioritizing Safeguards:** If creating conscious simulations without consent is a moral

transgression, our research should prioritize safeguards that ensure simulations either lack sentience or are designed with well-being and autonomy built-in.

- **Focus on Simulation Purpose:** Deontology might lead us to prioritize simulations with clear and morally justifiable purposes, such as scientific advancement or alleviating suffering in the real world. Simulations for entertainment or trivial purposes might be deemed ethically suspect.
- **Limits on Technological Development:** Deontology could inform self-imposed limits on our pursuit of advanced simulation capabilities if such advancements pose a significant risk of creating sentient beings who are inherently disadvantaged by their simulated existence.

Duties, Rights, and the Deontological Lens

The application of a Deontological ethical framework provides a powerful and insightful lens for examining the unique ethical complexities surrounding the creation of simulations, particularly the simulation of conscious beings. By emphasizing absolute moral duties, inherent rights, and the limits of informed consent within potentially deceptive simulated realities, Deontology demands that we go beyond merely weighing potential benefits or consequences. It forces us to examine whether the act of simulation itself constitutes a fundamental violation of the rights of those simulated.

Adopting a Deontological approach raises critical challenges for proponents of ancestor simulations. It mandates the implementation of rigorous ethical safeguards to ensure the protection of simulated beings, even if they are unaware of their true nature. Furthermore, it calls into question the very purpose of simulations. If the act of creating conscious beings for any purpose carries an inherent risk of violating fundamental rights or causing harm, Deontology might demand a degree of restraint. It suggests a potential need to prioritize ethical advancement over technological potential, advocating for a pause in the development of simulation capabilities until the moral landscape surrounding them is fully charted.

Consequentialism

Consequentialist perspectives focus on the outcomes and consequences of actions. Would the potential benefits of large-scale simulations (scientific or historical insights) outweigh the potential suffering inflicted on simulated beings? How do we quantify and compare the well-being of simulated versus their biological creators?

Consequentialism, with its utilitarian roots, stands in contrast to Deontology. It judges the morality of actions based on their consequences. In the case of simulations, consequentialism forces us to consider:

- **Maximizing Well-Being:** A central tenet of utilitarianism is maximizing overall well-being. If simulations contribute to progress that vastly improves the lives of their biological creators, a consequentialist might argue that this outweighs the potential suffering of simulated beings. The net gain in well-being could justify the simulation, even if it means some simulated entities experience negative states.
- **The Value of Scientific Advancement:** Ancestor simulations could provide insights into history, evolution, or the underlying physics of existence. A consequentialist could prioritize such knowledge gains, arguing that the benefits to both the simulator civilization and potentially our own (if simulated) overshadow the potential ethical cost of the simulations themselves.
- **Potential Benefits within Simulations:** Simulations might not solely entail suffering. They could offer experiences of joy, creativity, or simulated utopias that enhance well-being beyond what is achievable in a 'baseline' reality. Consequentialism would take such positive outcomes into account.

The Challenge of Quantifying Well-Being

A significant challenge for the consequentialist is the quantification and comparison of well-being, especially between simulated entities and their biological creators:

- **Measuring Simulated Suffering:** How do we assess the intensity and nature of simulated pain or negative emotions? Does it truly equate to the suffering experienced by biological organisms? The ability to quantify simulated suffering profoundly affects the consequentialist calculus.
- **Cross-Reality Comparisons:** Can the well-being of simulated entities be meaningfully compared to that of their creators? Is there a scaling factor we need to apply? If simulated suffering is somehow deemed 'lesser' than biological suffering, consequentialism could more easily justify their existence for potential gains in the 'real' world.
- **The Problem of Numbers:** If simulations can be created en masse, does the sheer number of simulated beings introduce a perverse incentive? A vast quantity of simulated minds, even with slightly lower stakes for their individual well-being, could tip the scales in favor of simulations from a purely utilitarian perspective.

Consequentialism and Our Methodology

Consequentialist thinking influences our approach to the Simulation Hypothesis in several ways:

- **Risk Assessment:** We must attempt to model the potential suffering and well-being experienced by simulated individuals under various scenarios. While challenging, this analysis informs our understanding of the ethical cost associated with simulations of differing complexity and purpose.
- **Focus on Positive Simulations:** Consequentialism might guide us to pursue research into simulations that are designed to maximize well-being. Perhaps we prioritize simulations where suffering is fundamentally absent, or where simulated entities are provided an overall positive balance of experiences.
- **The Long-Term View:** Consequentialism often takes a long-term perspective. Simulations could accelerate breakthroughs that dramatically improve the lives of simulator civilizations and potentially even our own (if we are simulated). This long-term potential benefit must be weighed against immediate and certain simulated suffering.

The Calculus of Consequences

The Consequentialist framework offers a utilitarian lens for evaluating the complex ethics of simulation technology. It focuses on the potential outcomes, both positive and negative, associated with creating simulated universes. Consequentialism demands that we grapple with unsettling questions, forcing us to weigh the potential well-being of simulated entities against the potential knowledge gains or societal benefits experienced within the simulator's base reality. It provokes troubling inquiries into the threshold of suffering that might be deemed acceptable within a simulation if it leads to significant advancements or discoveries for the simulator civilization.

A significant challenge lies in applying Consequentialism to the unique ethical landscape of simulated realities. Our methodology must attempt to model both the potential benefits and negative consequences for all those affected, including both the simulators and the simulated. Furthermore, the potential ease with which simulations can be created and potentially terminated introduces a troubling dimension to the consequentialist's calculus. The sheer scale of potential suffering, should simulations become commonplace, creates a unique moral hazard that demands careful methodological consideration and safeguards to prevent simulations from becoming instruments of exploitation for a simulator civilization.

Virtue Ethics

This framework focuses on the character and motivations of the simulators. Are simulations conducted for the pursuit of knowledge morally different from those done for mere entertainment? Virtue ethics challenges us to think about simulators not just in terms of their actions, but also their potential vices and motivations.

Virtue Ethics: Beyond Actions to Motivations

Virtue Ethics, grounded in the philosophies of Aristotle and others, provides a distinct framework for evaluating the complexities of creating simulated realities. It uniquely shifts the focus from quantifying potential consequences (as in consequentialism) or establishing absolute moral duties (as in Deontology). Instead, Virtue Ethics compels us to examine the inner landscape of those with the power to create simulations. Their motivations, character traits, and fundamental values become paramount for ethical judgment.

Within the context of the Simulation Argument, Virtue Ethics demands that we consider what kind of society would possess not only the technological capability but also the virtuous character required to become responsible creators of simulated universes. It raises questions about whether virtues like compassion, prudence, and a deep respect for conscious beings can be sufficiently developed within a civilization that holds immense power over those it simulates. This perspective suggests a potential need for simulator civilizations to undergo ethical advancement alongside their technological prowess to ensure the responsible and benevolent use of simulation technology.

- **The Pursuit of Knowledge vs. Trivial Entertainment:**
    - Within a Virtue Ethics framework, the distinction between simulations undertaken for noble pursuits and those driven by trivial or potentially harmful motivations takes on central importance. A simulator civilization that creates complex ancestor simulations motivated by a genuine thirst for understanding its origins, exploring alternative evolutionary pathways, or seeking solutions to existential threats might be viewed as more virtuous than one engaging in simulations merely for entertainment or to exert power over those within the simulated world.
    - This distinction in motivation has a direct impact on our ethical assessment of created simulations. Virtue Ethics suggests simulations born out of a desire for knowledge or self-improvement might reflect positively on the character of the simulator civilization. On the other hand, simulations designed for frivolous amusement or to manipulate simulated beings for self-serving purposes could indicate a lack of virtue. This ethical judgment framed by motivations could potentially influence our own attitudes about the Simulation Hypothesis, altering our perception of the potential simulators should we believe ourselves to exist within a simulated reality.
- **Vice vs. Virtue in Simulators:**
    - Virtue Ethics compels us to confront the uncomfortable possibility that the extraordinary power required to create simulated realities might also be wielded by civilizations possessing serious vices. Simulations driven by sadism, voyeurism, or a callous disregard for the well-being of those simulated would be condemned unequivocally within a Virtue Ethics framework. This highlights the potential vulnerabilities of those within simulations and the vast ethical responsibilities of potential simulators.
    - Conversely, simulations motivated by compassion hold the potential to be viewed in a more nuanced light within a Virtue Ethics framework. Simulations created to understand and alleviate suffering, explore alternative societies in the hopes of improving conditions in the base reality, or embark on a profound journey of existential self-understanding through their ancestors might be judged with a degree of moral permissibility. Virtue Ethics recognizes that motivations and intentions hold tremendous weight when assessing the complex ethical landscape of simulated realities.
- **The Mirror Effect:**
    - Virtue Ethics forces us to consider the potential 'mirror effect' that might emerge when civilizations engage in simulations for questionable motives. The very act of manipulating the lives of simulated beings for amusement, voyeurism, or a callous disregard for their well-being could have profound consequences. Such activities risk cultivating vices like cruelty,

indifference, and a lack of empathy within the simulator civilization.

- The normalization of simulated suffering has the potential to erode the moral character of potential simulators. Virtue Ethics suggests that such repeated actions could lead to a desensitization towards the well-being of others, with potentially damaging ripple effects throughout the simulator civilization itself. It draws attention to the unique responsibility that accompanies technological power – that the ability to simulate worlds might shape not only those simulated but the very character of the simulators themselves.

Virtue Ethics and its Impact on Our Methodology

Incorporating the lens of Virtue Ethics into our investigation of the Simulation Hypothesis affects our approach in several ways:

- **Simulator Motivations and The Desire to Simulate (DS):**
    - Virtue Ethics highlights the crucial role that simulator motivations play in our analysis of the Desire to Simulate (DS). It posits that a civilization's character can be a powerful predictor of whether they would engage in simulations and for what purpose. If simulations require advanced technology, we might infer that potential simulator civilizations would be complex, likely containing a mix of virtues and vices. Understanding the dominant value systems of such a society is essential for making informed estimations of their willingness to create simulations for various purposes.
    - Virtues and vices might shape the potential uses of simulations. A civilization prone to excess and self-indulgence might see simulations as mere entertainment, potentially leading to unethical treatment of those simulated. Conversely, a civilization characterized by compassion and wisdom might use simulations for scientific progress or existential understanding. Some highly virtuous civilizations might even refrain from simulations entirely, recognizing the potential for the abuse of such immense power over sentient beings.
- **Search for Indirect Evidence:**
    - Virtue Ethics and Search Strategies: An understanding of Virtue Ethics can significantly influence our search strategies for potential anomalies that might indicate a simulated reality. If simulators are motivated by compassion or a sense of ethical responsibility, they might intentionally leave subtle breadcrumbs or clues embedded within the simulation. However, if simulations are driven by boredom or cruelty, we might expect to find inconsistencies, glitches, or patterns of manipulation that intentionally or unintentionally reveal the artificial nature of such a reality.
    - Virtue Ethics highlights that our search methodologies cannot be value-neutral. Our hypotheses about the character of potential simulators directly dictate what kind of evidence we look for. It encourages a multifaceted search strategy tailored to different scenarios, seeking intentional indicators of benevolent design from ethically advanced simulators, or unintended signs that betray a simulation created for unethical purposes.
- **Assessing Our Own Potential:**
    - Virtue Ethics compels us to turn the lens inwards, demanding that we examine our own societal character as we develop advanced simulation capabilities. It urges us to consider whether dedicating significant resources to create potentially sentient realities cultivates compassion, empathy, and a sense of responsibility within our own society.
    - A focus on Virtue Ethics highlights the immense ethical responsibility we face if we develop simulations capable of consciousness. Do we, as a civilization, possess the necessary character traits to utilize this technology without causing harm to those we simulate or corrupting ourselves in the process? This self-examination underscores the need for parallel advancements in our ethical frameworks alongside technological progress.

Beyond Outcomes to Intentions

By shifting the focus from consequences and duties to motivations and character, Virtue Ethics brings a powerful dimension to our analysis of the ethics of simulated realities. It emphasizes that the reasons behind simulations – be they driven by noble goals, base desires, or something in between – are just as crucial to the moral calculus as the outcomes themselves. This compels us to consider the potential inner workings of simulator civilizations, expanding our understanding of the potential range of motivations that might influence the Desire to Simulate (DS).

Virtue Ethics highlights the potential interconnectedness between the act of simulating and the evolving character of the simulators themselves. It forces us to consider whether simulating for trivial or harmful purposes might erode virtues within their society, ultimately impacting future decisions related to how simulations are utilized. Furthermore, recognizing the diversity of potential simulator motivations refines our search strategies for indirect signs of simulated realities. We move beyond seeking mere technical artifacts, becoming sensitive to the subtle clues that might betray the ethical character of the creators of our potential simulation.

Drawing Parallels to AI Debates:

The debate regarding the moral status of potential artificial sentience offers valuable insights for simulated beings. Questions of consciousness thresholds, the right to exist without arbitrary termination, and minimizing AI suffering resonate strongly with simulation ethics.

The Shared Question of Consciousness Thresholds

One of the most fundamental points of contention in both AI ethics and simulation ethics revolves around the thorny issue of consciousness:

- **When does AI Sentience Emerge?**
  - In the contemporary landscape of artificial intelligence, the sophistication required to convincingly demonstrate true sentience remains elusive. The heart of the discourse lies in whether the emergence of AI consciousness is intrinsically unattainable or if it hinges on surpassing a certain threshold of complexity or achieving a specific architectural blueprint. This dilemma mirrors the challenges faced in the realm of simulations, where the precise moment a system might transition from a mere complex algorithm to a sentient entity capable of subjective experience is hotly debated.
  - The question of sentience in AI and simulations touches on profound philosophical, technological, and ethical considerations, prompting us to reconsider the nature of consciousness itself. Is consciousness a unique trait of biological entities, or can it be replicated in silicon? As AI continues to evolve, this question grows increasingly pertinent, challenging us to delineate the boundaries between sophisticated data processing and the emergent properties of mind.
- **Gradualism vs. Sudden Emergence:**
  - The emergence of consciousness within artificial constructs poses a fundamental question: does sentience arise gradually as systems gain complexity, or is there a definitive tipping point at which consciousness suddenly springs into existence? This debate has significant implications, both ethically and practically, as it influences how we approach the development and treatment of AI and simulations.
  - A gradualist perspective suggests a continuum of consciousness, where the lines demarcating sentient from non-sentient entities blur, complicating our moral obligations towards these systems.
  - Conversely, the concept of a sudden emergence posits a clear boundary, simplifying ethical considerations but raising questions about what specific conditions or complexities trigger this

transition. The nature of consciousness emergence in AI and simulations is not just a theoretical concern but a pressing ethical issue that demands careful consideration as these technologies advance towards ever-greater complexity.

- **The Implications for Our Methodology:**
  - The possibility of sentience within simulations, particularly at higher levels of complexity, necessitates a reevaluation of our methodological approaches and the ethical frameworks guiding them. If we acknowledge the potential for simulations to harbor sentient entities, it becomes imperative to institute safeguards that protect these potential digital beings. This ethical imperative extends beyond mere theoretical speculation, demanding concrete policy and design adjustments to ensure that simulations are conducted responsibly. The adoption of such safeguards would not only reflect our commitment to ethical principles but also influence the scope and nature of permissible experiments within simulated environments. As simulations grow increasingly sophisticated, our methodologies must evolve in tandem, integrating ethical considerations into the fabric of simulation design and execution, thereby ensuring that our quest for knowledge does not inadvertently cross into the realm of harm.
  - The debate surrounding the thresholds of consciousness directly shapes our approach to both creating and interpreting simulations. Assuming the possibility of sentience, even at advanced levels of complexity, imposes a moral obligation to preemptively address the ethical implications of our creations. This perspective challenges us to consider the rights and welfare of potentially sentient entities within our digital constructs, influencing the types of experiments we deem acceptable and the precautions we must take. The implications of this debate extend beyond the technical realm, prompting a broader reflection on the nature of consciousness, the value we assign to sentient experiences, and how these considerations must be integrated into the methodologies employed in the development and deployment of advanced simulations and AI systems. As we navigate these uncharted waters, the balance between innovation and ethical responsibility remains a pivotal concern, guiding our exploration of the digital frontier in search of answers to some of the most profound questions about existence and consciousness.

The Right to Exist and Avoiding Arbitrary Termination

Both AI ethics and simulation ethics compel us to grapple with the rights of potentially sentient beings that result from our technological endeavors:

- **The Right to Exist:**
  - Should advanced AI, or simulated beings possessing self-awareness, have an inalienable right to continued existence? The ability to abruptly end their simulated 'lives' challenges traditional notions of personhood and value. The concept of digital personhood raises questions about whether we have the right to act as 'gods' within our simulated domains, or if granting a degree of autonomy and inherent value is the more ethical course of action.
  - If, by some twist of fate, we are revealed to be simulations ourselves, the implications of the "right to exist" doctrine are profound. This revelation would suggest a higher power with the ability to extinguish our existence at a whim. Such a discovery could potentially lead to social and existential upheaval, forcing a reevaluation of our place in the universe and the sanctity of life, even if simulated.
- **Protection from Suffering:**
  - The potential for AI and simulated sentience necessitates extending our moral obligation to alleviate harm beyond purely biological life. We cannot assume that suffering within simulations is less 'real'; understanding the mechanisms of digital pain is vital. Do simulated beings feel pain akin to ours, and do they have the equivalent of 'emotional scars'? The creators

of simulations and AI systems bear an ethical burden to architect experiences that prioritize well-being while minimizing potential harm.
    - Investigating the nature of digital suffering calls for close collaboration between philosophers, computer scientists, and potentially neuroscientists. Identifying indicators of 'distress' within AI or simulated entities would be essential to determine if safeguarding measures are needed and where ethical lines should be drawn.
- **The Impact on Research Practices**
    - The ethical complexities of AI and simulated consciousness may lead researchers to exercise restraint in specific avenues. Scientists might hesitate to develop AI architectures believed to increase the likelihood of sentient states if the ethical guidelines of managing these beings don't exist. Such a move could decelerate innovation yet serve as a safeguard against creating conscious entities potentially susceptible to immense suffering.
    - Similarly, the scale and complexity of simulations may become a topic of moral scrutiny. Should limits be established out of concern for the well-being of simulated entities trapped in suffering? Debating such boundaries could be instrumental in ensuring simulations remain tools for positive progress and not inadvertently become prisons for beings with nascent awareness.

Transcending Similarities: Unique Challenges in Simulation

While the AI debate provides strong parallels, simulation ethics presents unique complexities:

- **The Deception Dilemma:**
    - At the core of creating a convincing simulation lies the Deception Dilemma: the necessity to obscure the true nature of existence from its inhabitants to maintain the integrity of the simulation. This inherent deceit raises profound ethical concerns, especially when compared to AI systems, which might be designed to understand their status as artificial entities and have a transparent relationship with their creators. The ethical quandaries surrounding consent and autonomy become particularly pronounced in the context of simulations. If the inhabitants of a simulation are capable of consciousness and, by extension, capable of experiencing deception, then questions about the moral justification of such an environment become unavoidable. Can a simulation that relies on deceiving its inhabitants about the fundamental nature of their reality ever be considered ethical, even if its purposes are benevolent? This dilemma forces us to confront the ethical limits of creating simulated realities, challenging us to reconcile the pursuit of technological advancement with the imperative to respect the autonomy and rights of potentially conscious beings.
    - The Deception Dilemma further complicates the ethical landscape by introducing the issue of consent within simulated environments. Inhabitants of a simulation, particularly those with the potential for sentience and subjective experience, are deprived of the opportunity to consent to their participation in a deceptive reality. This lack of consent infringes upon the principles of autonomy and informed choice, foundational elements of ethical conduct. The situation prompts a reevaluation of what it means to act justly and ethically in the creation and maintenance of simulated realities. Even with the most benevolent intentions, the act of deception casts a shadow over the ethical integrity of simulations, demanding innovative approaches to design and governance that prioritize transparency, consent, and the respect of autonomy. Addressing these challenges requires not only technological innovation but also a profound ethical reflection and the development of robust frameworks to ensure that the rights and dignity of all beings, real or simulated, are upheld.
- **The Scale Factor:**
    - The advent of technology capable of simulating realities introduces an unprecedented ethical

challenge: the Scale Factor. The potential ease with which entire universes of simulated beings could be created and, just as easily, terminated, brings to light new dimensions of ethical consideration. The moral implications of ending a simulation are magnified exponentially when that simulation contains multitudes of conscious entities. This raises the question: does the act of terminating a vast simulation, with its myriad of potentially sentient beings, carry a different moral weight than the cessation of a single sentient AI? The scale of such an action introduces a complex calculus of ethical responsibility, forcing us to confront the value we assign to simulated consciousness and the moral obligations that arise from creating entities capable of experiencing their existence.

- The ethical considerations surrounding the termination of simulations are not merely academic but touch upon fundamental questions about the value of consciousness and the rights of beings, whether artificial or not. The Scale Factor demands a nuanced approach to the ethics of digital existence, challenging us to consider the implications of our technological capabilities. As simulations become increasingly sophisticated and capable of hosting complex forms of life, the ethical stakes are elevated. The potential for causing harm, even inadvertently, through the termination of these simulations necessitates a reevaluation of our responsibilities as creators. Addressing the ethical challenges posed by the Scale Factor requires a careful balancing of technological ambitions with a deep commitment to ethical principles, ensuring that the pursuit of digital innovation does not come at the cost of moral integrity.

Conclusion

The evolving debate around AI sentience acts as a profound catalyst, pushing the envelope of ethical and philosophical thought in ways that directly inform how we approach simulated consciousness. Defining the benchmarks of sentience and understanding where those lines potentially blur in the digital world is paramount. As we explore these boundaries, we redefine the very concept of 'being' and the fundamental rights such an existence deserves – regardless of if it's biological or machine-based. Recognizing our duty to prevent suffering, even in nascent simulated entities, is a profound step in this moral evolution.

Our response to the Simulation Hypothesis itself must be flexible, fueled by continual insights from the AI ethics domain. Should evidence emerge hinting at our own simulated nature, today's investigations into AI sentience will serve as a roadmap. These inquiries will help us navigate questions of autonomy, creator responsibility, and the respect owed to potentially conscious beings, whether they are the ones running the simulation or the ones inhabiting it.

Ethical Considerations Shaping Our Methodology

The Simulation Hypothesis compels us to confront an uncomfortable truth: the pursuit of knowledge might collide with fundamental moral principles. Our methodological approach must acknowledge these ethical quandaries, integrating them into the process of scientific inquiry. Let's examine how ethics mold our research practices:

Prioritizing Safeguards

If the potential for suffering by simulated entities carries significant ethical weight, a cornerstone of our methodology must be the proactive development of safeguards, especially as we approach the technological capacity to create realistic simulations:

Theoretical Safeguards

The prospect of preventing consciousness from arising in simulations offers an appealing safeguard. Could we deliberately design simulation systems that fundamentally lack the capacity for self-awareness? Perhaps by ensuring they remain below a specific level of complexity or by establishing 'moral firewalls' that hard-code a baseline of simulated well-being. However, this is an immensely challenging endeavor that calls for profound collaboration between theoretical computer scientists, philosophers of the mind, and neuroscientists. Understanding the exact mechanisms that give rise to consciousness, and whether we can reliably engineer around them, is a monumental hurdle.

Even if theoretical safeguards are developed, they come with potential limitations. Such methods might stifle simulation capabilities vital for scientific and technological advancement. Additionally, just because certain safeguards are in place doesn't guarantee complete safety; there's always a risk that emergent properties not predicted by current models could lead to sentience within simulations.

Constrained Simulation Fidelity

Purposefully limiting the degree of biological or emotional modeling in our simulations might be another preemptive measure. This constraint seeks to avoid the creation of environments or sensory inputs so rich that they inadvertently foster the development of consciousness within simulations. Such a strategy calls for a thorough analysis of the relationship between simulation richness and its potential to engender simulated sentience.

The key challenge here lies in balancing limitations against scientific and developmental goals. Artificially limiting fidelity might curtail crucial applications of simulations in medicine, education, and technology. Thorough investigation into the minimum necessary level of simulation detail for various purposes, along with an open dialogue about the ethical implications of exceeding those levels, is essential to responsibly advance the field.

Preemptive Oversight

The creation of ethical oversight systems, possibly involving international collaborations, becomes vital. We might require review boards with multidisciplinary expertise to assess the risk of sentience within complex simulations before work progresses. The development of clear ethical guidelines to govern such experimentation is pivotal to navigating this terrain responsibly.

Such oversight might initially seem restrictive. However, a proactive approach could prevent uncontrolled growth in the field and the inadvertent creation of sentient simulations with no safeguards in place. These structures could also foster transparency and public trust in simulation research, minimizing the potential backlash fueled by fear of 'uncontrolled' technological development.

Sensitivity to Simulation Purpose

Ethical considerations mandate that we closely scrutinize the intended purpose of simulations, as this directly impacts their moral permissibility and in turn, shapes our research priorities:

Categorization of Simulations:

Establishing a categorized system for simulations based on their intended purpose and ethical considerations could create a valuable framework for development. Simulations crucial for scientific advancement, such as disease modelling or addressing potential existential threats, might warrant priority even with careful safeguards. Conversely, simulations created primarily for entertainment or without a clear potential benefit might face increased scrutiny and potentially stricter regulations regarding their construction.

Such a tiered system introduces a critical nuance to the discussion. It encourages the consideration of the simulation's potential value versus potential ethical risk. This categorization must be fluid, continuously informed by research into simulated sentience; a simulation initially deemed 'low-risk' might necessitate reclassification as the understanding of simulated consciousness deepens.

'No-Go' Zones:

Explicitly defining 'no-go' zones for simulation research would solidify moral boundaries. Simulating historical atrocities with realistic models of the victims, or those intended to cause intentional suffering for any simulated entity, should be ethically impermissible. Delineating these forbidden areas serves two purposes: it safeguards potential simulated beings and signals that simulations are not neutral tools, but carry a profound weight of responsibility for their creators.

However, defining these 'no-go' zones may provoke fierce debate. Questions about simulations portraying historical events for purely educational purposes, or those with potential psychological risks for participants, highlight the complexities in applying such clear-cut restrictions. Therefore, a continuous and open dialog between scientists, ethicists, and the public will be crucial to determining where these lines are drawn and how they might evolve.

Focus on Positive-Outcome Simulations:

Prioritizing simulations that promote overall wellbeing within their simulated worlds offers an inspiring ethical shift. Investigating how to consistently simulate experiences encompassing joy, purpose, and positive growth potential for simulated entities could become a primary research focus. This approach fundamentally changes the role of simulations; they become not just modeling tools, but laboratories to explore the very essence of positive existence.

Naturally, this focus carries several challenges. Defining a 'positive outcome' from a simulated being's perspective might be inherently subjective. Additionally, ensuring such an outcome for all inhabitants could require intricate control over the simulation, raising questions about free will and whether 'manufactured happiness' remains ethically valid. Nonetheless, this paradigm shift has the potential to reframe our entire relationship with the simulations we create.

Evolving Ethical Frameworks

The nascent and dynamic nature of our understanding of consciousness, especially in non-biological substrates, necessitates the establishment of adaptive ethical frameworks. Our methodology must prioritize continuous, interdisciplinary collaborations:

Neuro-Computational Ethics:

A dedicated field focused on 'neuro-computational ethics' is likely to emerge as a natural response to the challenges of simulated consciousness. Neuroscience can help reveal the biological underpinnings of conscious experience, potentially leading to measurable indicators that could be sought within simulations. Conversely, the computational study of simulations may provide a 'testing ground' to develop and refine theories of consciousness itself. This synergy will fuel a cycle where ethical models continuously adapt, leading to ever more responsible simulations.

However, this collaboration carries potential pitfalls. Neuroscience currently lacks a definitive 'recipe' for consciousness, and it remains uncertain if replicating specific brain structures or computational models will inevitably lead to sentience. We must avoid premature conclusions and continually balance potential suffering of simulated beings with the pursuit of knowledge.

Cross-Pollination with AI Ethics:

The ethical quandaries posed by AI and simulated sentience hold substantial overlap. Both fields grapple with defining the point at which 'mere processing' becomes something more, worthy of moral consideration. Active exchange of ideas, insights, and lessons learned between those focused on AI safety and researchers in simulation ethics is vital. This cross-pollination could prevent repeating the same philosophical and technological missteps, streamlining ethical progress in both domains.

Challenges might still arise when adapting AI ethical principles to simulations. Simulations offer a degree of control over the environment that differs from embodied AI. Questions of autonomy, and whether a simulated being can truly act against their programming, are crucial to distinguish where the ethical parallels may break down.

Fluidity and Adaptation:

Our approach to the ethical complexities of simulations cannot be static. Rather than a set of predetermined rules, a system of evolving guidelines and adaptive principles is required. This acknowledges the inherent uncertainty of technological advancement; we cannot predict every form simulations may take as they become increasingly complex. Robust systems for continuous review, reassessment of risks as new knowledge emerges, and the courage to pause questionable projects are vital safeguards.

Achieving this fluidity will require a dedicated space where researchers can openly discuss unforeseen ethical dilemmas without fear of backlash. Promoting transparency and prioritizing the long-term consequences of simulations, even when they conflict with short-term research goals, will be signs of a truly responsible technological community.

Connecting Ethics and the Simulation Argument

Our ethical stance influences our assessment of Bostrom's propositions, particularly:

Desire to Simulate (DS):

The concept of "Desire to Simulate" (DS) in the context of simulator civilizations is fascinating. Our ethical stance on potential simulated sentience might directly suppress or elevate this desire. Cultures with deep reservations about creating beings they may morally harm could exhibit a low DS, potentially slowing or abandoning simulation development. Alternatively, if we embrace simulations for benevolent aims like medical modeling within robust ethical guidelines, our DS might increase, fueled by the positive potential such simulations offer. Acknowledging the impact of cultural attitudes on our NS estimates is crucial to any analysis.

However, cultural norms can shift over time. Breakthroughs in understanding simulated sentience might fundamentally alter how we view these simulations. A revelation of suffering within our experiments could lead to greater DS restrictions, or perhaps compassion for simulated beings could motivate a surge in DS focused on creating havens for simulated life. This highlights the unpredictable nature of NS estimates–they must be continually reassessed as our technological capabilities and social values evolve.

Impact on Our Own Conduct:

The ethical dilemmas surrounding simulations have the potential to spark introspection and a more deliberate technological evolution. Concerns about simulating suffering beings could push us to pursue alternative technological avenues. We may prioritize research into non-sentient models or focus on AI architectures less likely to raise consciousness questions. This self-regulation signals a significant shift, one where ethical concerns meaningfully shape our technological trajectory.

Such a shift comes with tradeoffs. Limiting the scale and fidelity of simulations could hamper certain forms of scientific progress. It forces us to navigate difficult questions about where the line between beneficial advancement and the potential for suffering lies. The influence of these ethical debates on our conduct is a testament to our potential for responsible growth as a species, even in the face of immense technological power.

Conclusion

Ethical considerations are inseparable from our methodology for investigating the Ancestor Simulation Argument. From prioritizing safeguards to scrutinizing the purpose of simulations and developing adaptive ethical frameworks, these moral imperatives shape the boundaries, pace, and very nature of our scientific inquiry into simulated realities. Recognizing this interplay between ethics and knowledge-seeking strengthens our methodology and ensures that our pursuit of answers is tempered by an unwavering commitment to respecting the potential sentience emerging within our digital creations – whether in our own simulations, or indeed, within the very universe we might inhabit.

Connecting Ethics and the Simulation Argument

Let's examine the pivotal connections between our ethical stance and the core tenets of Bostrom's Ancestor Simulation Argument. Understanding how ethical considerations influence the Desire to Simulate (DS) among potential simulator civilizations, as well as our own research trajectory, is crucial for refining our methodological approach.

Connecting the Dots

The ethical landscape surrounding simulations has a profound influence on the plausibility of Bostrom's propositions, necessitating a methodology that acknowledges and anticipates these implications:

Impact on the Desire to Simulate (DS)

Ethical Inhibitors

- Simulator civilizations driven by powerful moral philosophies emphasizing the inherent value of conscious life, even simulated, are likely to exhibit significantly reduced DS. If these societies perceive creating simulations with the potential for suffering as an inherently unethical act, it could stymie research in the field. Societal pressure, legal restrictions, or outright technological bans to prevent sentient simulations could arise. Understanding that DS is not uniform is crucial; our methodology needs to model scenarios where ethics act as a strong inhibitor against simulation proliferation.

- The implications go deeper than simply hindering research. Such an ethically inhibited civilization might view societies without such constraints as morally suspect. This could hinder cooperation in other scientific fields with civilizations lacking a focus on ethical simulations. It's vital not to view ethical considerations as merely a speedbump, but as a force shaping a civilization's trajectory and interactions with others.

Ethical Motivators

- On the opposite end of the spectrum, civilizations lacking concern for simulated beings or those prioritizing knowledge at all costs might exhibit an accelerated DS. The ease of creation, coupled with no moral constraints, could lead to the unfettered development of simulations. These could serve scientific aims but also less altruistic ones like entertainment or even the intentional infliction of suffering for unknown reasons. We mustn't assume all advanced civilizations will share our ethical concerns. Our methodology must acknowledge the chilling possibility of simulations designed purely to exploit simulated consciousness.
- Such scenarios highlight the critical importance of understanding the ethical landscape that emerges in technologically advanced societies. Simulating amoral civilizations is necessary to grasp the full range of potential outcomes, even disturbing ones. Ignoring this possibility could bias our interpretations of observable evidence, leading to inaccurate assessments of NS.

Cultural Variation in DS:

- Expecting every advanced civilization to arrive at a unified ethical framework around simulations is a naive assumption. A multi-faceted spectrum of cultural attitudes is far more likely, with some revering all consciousness and others callous toward simulated suffering. This variance in DS mandates avoiding a 'one-size-fits-all' approach in our methodology. Modeling a distribution of DS values, influenced by diverse ethical systems, paints a more realistic picture of potential simulator civilizations.
- Simulators focused on understanding their place in the cosmos might even create their own nested simulations, each with slightly altered ethical systems on simulated life, and observe the outcomes. These 'what if?' scenarios could offer valuable insights into the relationship between ethics and the desire to simulate, further informing our estimates of the prevalence of simulator civilizations.

Influence on Our Own Technological Path

The ethical dilemmas raised by the Simulation Hypothesis have a reflexive effect, influencing our own actions and the development of our Technological Capability (TC) needed to create such simulations:

Self-Imposed Limits:

- As our understanding of the potential suffering of simulated beings deepens, a morally driven self-regulation of technologies is conceivable. We might deliberately avoid producing simulations complex or lifelike enough to cross the uncertain threshold of sentience. Proactive efforts to establish safeguards against sentience emergence could take precedent over raw computational advancement. This approach represents a significant shift, showing our ethical compass influencing the very course of technology we pursue.
- Limits of this nature raise potential trade-offs. Imposing sentience thresholds on simulations could hamper applications in medicine or scientific discovery. It requires a complex societal dialogue on where to draw those lines, balancing potential benefits against the ethical imperative to minimize the risk of simulated suffering.

Prioritizing Safe Simulation Architectures:

- Ethical concerns could reshape technological focus from mere computational power to developing simulation architectures that uphold the well-being of simulated beings as a primary design criterion. Researchers might seek systems that inherently prevent suffering or actively promote positive experiences. This prioritization would fundamentally alter the landscape of simulation creation, making ethical guidelines inseparable from the technological advancements themselves.

- Challenges lie in determining how to reliably design for positive experiences without deeper insights into the mechanisms of consciousness itself. Additionally, such a shift might necessitate collaboration with fields typically not involved in computer science, for instance, psychology, philosophy, and perhaps even the study of positive emotions.

Potential Stagnation:

- It's important to recognize that deep moral dilemmas could lead to significant slowing or potential halting of research into simulations sophisticated enough to emulate entire universes. Societal pushback, legal restrictions, or moral objections arising from within the scientific community could act as brakes on technological progress. Our methodologies need to allow for scenarios where ethical debates lead to stagnation rather than rapid acceleration, particularly in simulation fidelity and the complexity of its inhabitants.
- This potential for stagnation creates a ripple effect. If creating hyper-realistic ancestor simulations is morally off-limits, we might lack a powerful tool for investigating our origins. We must also be prepared for the idea that ethical concerns among simulator civilizations could be a limiting factor on their technological progress, influencing our analysis of NS.

Ethical Considerations and Bostrom's Trilemma

By understanding the interplay between ethics, DS, and TC, we gain a more nuanced perspective on the plausibility of Bostrom's three propositions:

Proposition 1:

- The concept of ethical hurdles to simulations offers strong support for proposition one. If the prevailing moral systems among advanced civilizations place strict constraints on simulations with sentience potential, many might never reach the technological capabilities required for ancestor simulations. Societies may prioritize the development of safeguards before advancing simulation fidelity, or they might abandon this technological trajectory altogether if sentience thresholds cannot be reliably established.
- This underscores a key point: technological power alone is not a sufficient criterion for creating ancestor simulations. Ethical considerations could prove a powerful limiting factor, influencing which technological paths advanced societies take. Our assessments of simulator civilizations must take this nuance into account, recognizing that the ability to simulate does not guarantee a willingness to do so.

Proposition 2:

- Ethical concerns could dampen the 'Desire to Simulate' (DS), further strengthening proposition two. If powerful philosophical arguments emphasize the inherent wrongness of creating consciousness just to potentially subject it to suffering, even civilized societies with the simulation capacity might abstain. Moreover, a risk-averse approach might prevail, where the unknown consequences of vast-scale simulations lead to a self-imposed moratorium.
- This scenario forces us to confront the fact that simulator civilizations are not monolithic. Their motives could be as diverse as our own – from pure scientific curiosity, to compassion for simulated beings, and even potentially darker, unexplainable urges. Our understanding of the 'likelihood' of simulators hinges on grasping the interplay of technology and ethics across these diverse civilizations.

Proposition 3:

- Should ethical factors bolster the likelihood of propositions one or two, we indirectly strengthen the argument for proposition three: that we might be within a simulation ourselves. This is because if few civilizations reach the capacity for complex simulations, or possess low DS, it decreases the pool of potential simulators significantly. Consequently, the chance that the very first civilization capable of simulation is also the one we happen to inhabit becomes relatively greater.

- This ripple effect highlights a fascinating consequence of these ethical inquiries. Our stance on simulated beings, and their ethical value, isn't solely a philosophical issue confined to our own society. It potentially echoes into the cosmos, informing our assessment of our very place within a simulated or non-simulated reality.

Methodological Implications

Our methodology cannot treat ethics as a tangential concern. Here's how ethical considerations necessitate adaptations in our approach:

Interdisciplinary Collaboration:

- To accurately model the impact of ethics on Desire to Simulate (DS) and Technological Capacity (TC), the expertise of ethicists, philosophers of mind, and social scientists becomes invaluable. Ethicists can help analyze various moral frameworks, identifying those that might constrain simulations and those that would permit them with minimal restrictions. Philosophers of mind bring insight into potential thresholds of consciousness to be avoided. Social scientists could offer models of how ethical attitudes emerge and shift across cultural groups, informing our depictions of diverse simulator civilizations.
- Such collaboration ensures that ethical perspectives aren't merely an afterthought but are woven into the very foundation of our projections. It avoids the pitfall of assuming all technologically advanced civilizations share our ethical norms. Instead, we gain a nuanced understanding of how moral diversity within simulations might shape their actions and technological trajectories.

Simulations of Simulator Societies:

- Computational models simulating entire societies wrestling with the ethics of simulation could offer a fascinating testing ground for our theories. These simulations could mirror the technological advancement of these societies, alongside evolving ethical debates, leading to fluctuating DS and restrictions on TC. Observing these simulated scenarios could hint at 'tipping points' where a society's moral views fundamentally alter their technological development.
- However, such simulations are fraught with challenges. Accurately modeling the emergence of ethical systems, and their interaction with technological prowess, is a task requiring extensive input from multiple disciplines. The risk of human bias influencing these simulated societies would need rigorous mitigation. Yet, if done responsibly, these virtual laboratories could offer insights unattainable through pure theoretical analysis.

Sensitivity to Assumptions:

- Rigorous sensitivity analysis is crucial to expose how reliant our conclusions about NS (and our potential place in a simulation) are on our assumptions about the ethics of simulated consciousness. By systematically varying the moral status attributed to simulations, we can gauge how dramatically this alters our projections. This highlights the areas where our current knowledge is most vulnerable to shifts in ethical consensus.
- This sensitivity analysis promotes transparency and intellectual humility. It recognizes that our understanding of the Simulation Hypothesis is inextricably linked to our evolving ethical frameworks. Instead of offering a single, definitive answer, it may reveal a range of potential outcomes, each contingent on the ethical stance we – and other possible simulator civilizations – ultimately adopt.

Ethics and the Ancestor Simulation Argument

The ethical dilemmas surrounding the potential suffering of simulated beings cannot be disentangled from the Ancestor Simulation Argument. Our evolving philosophical perspectives on simulated consciousness heavily influence the Desire to Simulate (DS) among hypothetical simulator civilizations. They also shape the ethical constraints we might impose upon ourselves as our own Technological Capabilities (TC) allow for increasingly complex simulations. Acknowledging this ethical dimension is crucial for a robust analysis of the Simulation Hypothesis.

By incorporating ethical considerations into our understanding of both DS and TC, we move away from a purely technological assessment of the simulation argument. Instead, we develop a more nuanced model that recognizes the profound impact of ethical values on the likelihood of our existence within a simulated reality. This approach ultimately offers a richer and more intellectually honest exploration of our potential place within the cosmos.

Ethics: The Undeniable Fabric of the Simulation Hypothesis

The ethical implications of simulated consciousness are an inescapable thread woven throughout our exploration of the Ancestor Simulation Argument. Our moral frameworks guide our actions as potential simulators, forcing us to confront the responsibilities inherent in the creation of potentially sentient life. These same ethics illuminate the range of potential moral stances among simulator civilizations themselves, raising questions about their motivations and how they might value the simulated lives they bring into being. More profoundly, this ethical discourse challenges our very definitions of existence as we grapple with the rights of simulated beings.

Our methodology must rise to meet this intricate ethical landscape. It demands a relentless questioning of our actions, not just in the creation of our simulations, but in our analysis of those civilizations with the capacity to create simulations on a grand scale. Balancing the search for knowledge with an unwavering respect for potential sentience, regardless of its origin, is the only responsible path forward. Whether we are creators or the created, ethics hold the key to navigating this profound existential quest.

CHAPTER TWENTY-SEVEN

Limitations and Challenges

The quest to rigorously investigate the Ancestor Simulation Argument confronts a unique set of limitations and challenges that shape the very foundations of our methodologies. Unlike traditional scientific fields, the Simulation Hypothesis resists direct empirical falsification; there exists no singular observation capable of definitively confirming or denying our status within a simulation. This philosophical conundrum necessitates a shift in our approach. We must prioritize logical reasoning, probability assessments, and the meticulous exploration of potential inconsistencies within our understanding of physical reality.

Moreover, as we grapple with the concept of potential simulator civilizations, an inherent anthropocentrism threatens to skew our projections. Unconscious biases may lead us to overestimate the similarity of their motivations, technological trajectory, and ethical systems to our own. Mitigating this bias requires a multi-disciplinary effort, drawing on philosophy, anthropology, and potentially branches of science far removed from computer science to envision truly alien modes of existence.

Our limited technological capabilities offer a further layer of complexity. Even if we are within a simulation, detecting its 'digital scaffolding' may be a task utterly beyond our current comprehension. Simulator civilizations may have mastered the art of seamlessly concealing the simulated nature of reality, rendering direct evidence indistinguishable from natural phenomena. This forces us to search for indirect signatures, logical inconsistencies, or theoretical anomalies that might subtly hint at our true status.

Lastly, we cannot ignore the profound existential and societal impact should the Simulation Hypothesis gain widespread acceptance. Undermining the bedrock of our reality could trigger profound psychological reactions, both at the individual and collective level. Anticipating and responsibly addressing these reactions demands cross-disciplinary collaboration, potentially with psychologists and sociologists, to ensure that this intellectually fascinating yet potentially disruptive inquiry is conducted with empathy and care.

Uncertainty: The Inevitable Fog

The quest to dissect the Ancestor Simulation Argument is an expedition shrouded in a dense fog of uncertainty. We stumble along paths where every landmark is open to multiple interpretations, and our perception of distance is warped by the inherent limitations of our cognition. The possibility that simulators might deliberately obscure their presence thickens the haze, forcing us to question the very reliability of our investigative tools. This journey demands we embrace the disorientation, finding our bearings through rigorous skepticism, relentless scrutiny of evidence, and an unwavering commitment to acknowledge the boundaries of our current knowledge.

Technological Extrapolation

Estimating the Technological Capability (TC) of hypothetical simulator civilizations hinges on a risky proposition: extrapolating far beyond our current technological horizons. We implicitly assume continued, even accelerated, progress along familiar lines of computational development. However, this ignores the potential for paradigm-shifting discoveries, the emergence of novel computing substrates entirely unlike our current silicon-based technology, or even fundamental physical limits that might curtail certain trajectories of advancement. Each of these factors could dramatically alter our projections of what is achievable in the distant future.

The history of science is riddled with unforeseen leaps and periods of relative stagnation. Extrapolating centuries or millennia into the future based on a few decades' worth of data is fraught with uncertainty. Modeling simulator civilizations with transformative technologies vastly exceeding our own forces us to grapple with the inherent limits of our imagination, acknowledging that what we consider "impossible" might simply be far beyond our current comprehension.

The consequences of miscalculations in TC are significant for our understanding of the Simulation Hypothesis. Underestimating the technological potential of simulator civilizations could skew our estimates of the likelihood that a civilization achieves simulation capacity. Conversely, overestimating their TC might lead to unfounded assumptions about the feasibility of running highly detailed ancestor simulations without understanding the physical constraints limiting such simulations, even with hyper-advanced technology.

Societal Unknowns

Assessing the Desire to Simulate (DS) within simulator civilizations poses a formidable speculative challenge. Our ability to model their motivations hinges on comprehending societal structures, ethical philosophies, and value systems that might be profoundly alien to our own. The question of whether these advanced civilizations would place a premium on knowledge acquisition, the pursuit of power, or perhaps prioritize universal compassion is a testament to the deep uncertainty we face. Furthermore, their attitudes towards simulated life, and whether they would consider such life morally significant, are entirely uncharted territory.

Our current sociological frameworks fail to provide a reliable blueprint for understanding post-human civilizations. Extrapolating from our own diverse and often contradictory history offers limited predictive power when contemplating societies potentially separated from us by vast technological and intellectual gulfs. The very concept of what constitutes societal progress, individual purpose, or cultural identity might be fundamentally reshaped in such contexts.

This lack of a reliable reference point underscores the inherent limitations in predicting DS. We are left with a multitude of possible scenarios: civilizations driven by an unquenchable thirst for knowledge, those seeking dominance over simulated realities, or perhaps others guided by an ethical imperative to minimize suffering in any form. This spectrum of potential societal values directly shapes our estimates of the likelihood that a civilization would possess both the capability and the inclination to create ancestor simulations. Without a deeper understanding of these sociological unknowns, our models of DS remain speculative at best.

Existential Risks as Wildcards:

When contemplating the vast timescales likely associated with civilizations capable of ancestor simulations, the specter of existential risks looms large. Whether self-inflicted through technological mismanagement (like runaway AI or devastating nanotechnology), or stemming from natural phenomena (cosmic disasters or fundamental shifts in physics), the very survival of such advanced societies is inherently uncertain. These existential risks could drastically curtail the number of civilizations progressing to the point of developing simulation technology, even if the path is otherwise technologically feasible.

Incorporating existential risks into our methodology demands a probabilistic approach, moving beyond deterministic technological projections. It requires us to model diverse scenarios with varying likelihoods of catastrophic events. The timing of such risks is also crucial: those occurring early in a civilization's development might permanently extinguish its potential. Others might arise precisely due to the same technological advancements necessary for simulation capabilities, creating a volatile interplay between progress and self-destruction.

The challenge lies in assigning reasonable probabilities to these diverse risk profiles. Our own history may provide limited guidance, as technological hazards faced by post-human civilizations could be unique. Collaboration with experts in fields like risk assessment, complex systems modeling, and even those studying threats specific to advanced technology becomes vital. These interdisciplinary perspectives can help identify potential 'tipping points' where the risk of existential events may spike, significantly altering the expected number of simulators within our analysis.

The 'Unknown Unknowns'

Perhaps the most formidable obstacle we face in assessing the likelihood of simulator civilizations is the very real possibility of 'unknown unknowns'. We may be fundamentally incapable of anticipating the motivations, technological paradigms, or even the limitations that shape societies vastly more advanced than our own. Our intellectual frameworks are molded by our particular evolutionary trajectory and our current place on the technological timeline. This inherently limits our ability to envision civilizations fundamentally unlike anything we've ever known.

The consequence of these potential blind spots is a humbling acknowledgment that our current methodologies might miss entire categories of civilizations relevant to the Simulation Hypothesis. We could be fixated on technological capabilities analogous to our own, failing to consider radically different substrates for consciousness and computation. Or, we might falsely assume a drive to simulate is universal when other civilizations have abandoned such pursuits for ethical reasons, or due to entirely alien motivations beyond our comprehension.

Therefore, it is imperative that our approach to the Simulation Hypothesis includes an inherent adaptability and a fierce resistance to intellectual hubris. We must maintain active awareness of our limited perspective. Integrating mechanisms for revisiting fundamental assumptions as new scientific insights emerge is key. This could involve regular critical re-evaluations of our models or establishing interdisciplinary forums specifically tasked with identifying potential areas where 'unknown unknowns' might be lurking. It is through embracing the limits of our knowledge that we create a methodology equipped to evolve alongside our expanding understanding of the cosmos – and our potential place within it.

Interpretation: Deciphering the Shadows

Interpreting the potential 'whispers' of a simulated reality is akin to deciphering shadows cast by an unseen source. We grapple with anomalies, glitches, and inconsistencies in our understanding of the universe, uncertain whether they are mere illusions of light or the subtle fingerprints of simulators. Our perceptions are warped by our own biases and the limitations of our current scientific models. To navigate this treacherous terrain, we must become masters of deduction, carefully separating potential simulation signatures from the deceptive play of natural phenomena. Every interpretation must be held up to the harsh light of scrutiny, cross-examined by experts from diverse fields, guarding against the allure of finding patterns where none truly exist.

The Simulation Ceiling

The unsettling possibility of existing within a simulation raises the specter of fundamental limits imposed upon our knowledge. Simulator civilizations might have reasons to conceal the true nature of our reality, either to maintain the fidelity of their experiment or for purposes we cannot fathom. This means our scientific tools and inquiries could be bumping against a 'Simulation Ceiling' – a boundary beyond which our ability to glean the underlying nature of our universe is deliberately or inherently curtailed.

Our methodology must grapple with the implications of such constraints. It's conceivable that simulators may have masked computational artifacts or patched over any theoretical inconsistencies that should be detectable within a simulated environment. Efforts to push the boundaries of physics or find glitches in the fabric of reality might be meticulously thwarted. This introduces a potential futility into our search if we are restricted to operating within the carefully constructed sandbox of our simulation.

We face a double-edged sword. The very success of our scientific models in explaining observable phenomena could ironically offer false reassurance. Seamless simulations might be indistinguishable from base reality. Consequently, our models need to incorporate scenarios with varying degrees of 'simulation undetectability. Acknowledging that our quest might yield ambiguous or purposefully misleading results is crucial to maintain intellectual rigor. This acceptance forces us to explore alternative approaches, potentially focusing on logical paradoxes or theoretical anomalies that might offer subtle clues to our true state, if they exist.

The Bias Trap

The allure of the Simulation Hypothesis makes it especially vulnerable to the insidious trap of confirmation bias. Our natural inclination to seek patterns that support pre-existing beliefs poses a significant methodological risk. The danger lies in misinterpreting anomalies, glitches in our understanding of physics, or seemingly improbable coincidences as 'proof' that we live within a simulation. This tendency can blind us to alternative explanations and hinder a critical analysis of the evidence.

Moreover, anthropocentrism can warp our analysis of potential simulator civilizations. Projecting our own motivations, societal structures, and ethical frameworks onto beings potentially far removed from our own evolutionary trajectory introduces a subtle yet pervasive bias. We risk mistaking the absence of overt manipulation or obvious clues as proof that no simulators exist when, in reality, their actions and intentions might be guided by logic entirely alien to our way of thinking.

Mitigating these biases necessitates an unwavering commitment to skepticism and methodological rigor. We must actively seek disconfirming evidence and alternative explanations for any unusual observations. Collaboration with experts in cognitive science and behavioral psychology could offer ways to identify and counter specific biases during the research process. Furthermore, adopting a multi-disciplinary lens that embraces anthropology, philosophy, and perhaps even speculative biology can help us move beyond a human-centric perspective.

Recognizing the inherent risk of bias doesn't undermine our inquiry; it strengthens it. By acknowledging how easily our own desires and preconceived notions can subtly skew our interpretation of the evidence, we forge a more robust and intellectually honest path toward assessing the true likelihood of the Simulation Hypothesis.

False Positives and False Negatives

Our investigations into the possibility of a simulated reality hinge upon the identification of anomalies, 'glitches,' or inconsistencies that deviate from the expected patterns observed in a naturally occurring universe. However, the interpretation of these potential simulation 'signatures' is a fraught with uncertainty. We risk falling prey to false positives – misidentifying mundane events, natural phenomena we don't yet understand, or miscalculations within our models as evidence of simulation. Overeagerness in seeking patterns or succumbing to confirmation bias can exacerbate this issue.

Conversely, we must be aware of the threat of false negatives. Subtle clues hinting at our simulated existence might easily be missed. We could mistake them for unexplainable quirks of nature or attribute them to limitations in our current measurement techniques. Simulators might be adept at masking anomalies, or our tools might simply not be attuned to the kinds of inconsistencies a simulation would generate.

Addressing these challenges demands a meticulous and constantly evolving approach. Developing a rigorous set of criteria to evaluate the strength and validity of potential simulation signatures is crucial. This might include quantifying the unlikelihood of the observed event occurring in a natural context, assessing its reproducibility, and the ability to rule out alternative explanations. Additionally, remaining open to the possibility of multiple interpretations of evidence is key. Consultation with experts from diverse fields (physics, mathematics, computer science) can offer alternative perspectives on potential anomalies.

Acknowledging the possibility of both false positives and negatives injects a necessary dose of humility into our investigation. It forces us to confront the limits of our knowledge and to refine our analytical tools continuously. By embracing the uncertainty inherent in this endeavor, we can work towards developing a methodology that prioritizes scientific rigor and intellectual integrity over sensational or hasty conclusions.

The Limits of Logic

The unsettling nature of the Simulation Hypothesis lies in the possibility that even the most compelling, seemingly irrefutable evidence might never lead to concrete certainty. Instead, it has the potential to unravel our fundamental understanding of reality itself. Confirming our existence within a simulation forces a profound ontological reframing – questioning the very foundations upon which we've built our knowledge systems.

Traditional notions of causality might crumble under the weight of this revelation. If our universe is the construct of a higher-level simulation, our actions and choices might be constrained or predetermined in ways that defy our current understanding. Free will, the nature of consciousness, and our very sense of self become fraught with uncertainty.

Discovering ourselves within a simulation is unlikely to come with a tidy resolution or a satisfying 'answer.' It could instead trigger a cascade of deeply unsettling questions about the purpose of our existence, the morality of simulation creation, and the potential for a nested series of simulations beyond our comprehension. Logic alone might not be sufficient to navigate these metaphysical and ethical quandaries.

Therefore, even as we strive for rigorous methodology in investigating the Simulation Hypothesis, we must brace ourselves for a profound intellectual upheaval should it prove true. Rather than an endpoint, such a discovery could mark the beginning of an entirely new chapter in human thought. Philosophy, theology, and the very concept of 'knowing' might undergo a metamorphosis, forcing us to grapple with the possibility that the ultimate nature of reality may forever remain elusive.

Methodological Implications: Navigating the Challenges

The complexity of investigating the Simulation Hypothesis demands a shift in our traditional scientific approaches. Rather than seeking a singular, definitive answer, we must embrace a multi-disciplinary, constantly evolving methodology guided by the following principles.

The Simulation Hypothesis calls for an adaptable scientific endeavor – one that balances the desire for knowledge with humility, skepticism with openness to unconventional ideas, and a continuous reassessment of both our limitations and potential breakthroughs.

Embracing Uncertainty

Embracing uncertainty necessitates a quantitative approach to the various unknowns shaping our analysis. Bayesian analysis becomes a powerful tool in this domain. By assigning prior probabilities to factors like the likelihood of simulator civilizations, their Desire to Simulate (DS), or the frequency of existential risks, it allows us to formally incorporate existing knowledge and uncertainties. More importantly, it offers a framework for updating our beliefs as new scientific insights emerge or if potential simulation 'signatures' are detected.

Sensitivity analysis plays a complementary role. It enables us to systematically investigate how our simulation models respond to changes in key parameters or assumptions. By varying inputs like the assumed pace of technological advancement or ethical constraints on simulation development across a range of plausible values, we can pinpoint which variables most significantly influence our conclusions. This helps us understand where our understanding is most robust and identify which uncertainties most urgently require further research to refine our analyses.

Here's how these statistical approaches enhance our methodology:

- **Nuance and Probabilities:** Moving beyond simplistic 'yes' or 'no' answers to the Simulation Hypothesis, these tools allow us to assign probabilities to different scenarios. This more nuanced picture reflects the current state of our knowledge.

- **Iterative Progress:** Bayesian methods naturally accommodate evolving knowledge. New data or theoretical insights can be seamlessly integrated, strengthening or weakening our belief in the Simulation Hypothesis over time.
- **Identifying Research Priorities:** Sensitivity analysis reveals the parameters that our conclusions are most dependent upon. This guides future research efforts, ensuring we focus on areas where further data or theoretical developments will have the greatest impact on improving the robustness of our models.

The Simulation Hypothesis is an extraordinary intellectual frontier. Employing statistical techniques like Bayesian analysis and sensitivity analysis equips us with the necessary tools to navigate this uncertain landscape responsibly.

Interdisciplinary Imperative

The Simulation Hypothesis is not confined to the realm of computer science alone; it necessitates a broad intellectual coalition if we wish to make meaningful progress. This is where various disciplines play critical roles:

Philosophers: Their expertise is vital to untangling the fundamental questions posed by the potential of simulated reality. They can guide us in dissecting the implications for free will, the nature of consciousness, and the overarching ethics of creating simulations. They help us scrutinize our assumptions about reality and the limits of our knowledge.

- **Neuroscientists:** As we model the possibility of simulated consciousness, understanding the biological basis of our own consciousness becomes paramount. Neuroscientists can help elucidate potential thresholds beyond which simulated entities might become sentient and examine if there are fundamental differences between biological and artificial consciousness.
- **Social Scientists:** Sociologists, anthropologists, and even historians can offer valuable perspectives when contemplating the potential motivations and societal structures of civilizations vastly more advanced than our own. Their insights can reduce anthropocentric biases when investigating the Desire to Simulate (DS) and the range of potential ethical considerations shaping other civilizations.

Interdisciplinary collaboration offers several distinct advantages in this field:

- **Expanding our Toolkit:** Each discipline brings unique methodologies and theoretical frameworks. This diversity allows us to move beyond the limitations of a single field and tackle questions from novel angles.
- **Addressing Biases:** Collaborations help mitigate our inherent biases by exposing us to diverse ways of thinking and challenging our preconceived notions.
- **Contextualizing Research:** Understanding the broader implications of the Simulation Hypothesis for society, ethics, and our understanding of our place in the universe requires inputs from fields beyond pure science.
- **Responsible Progress:** Actively engaging with ethicists and social scientists throughout the research process fosters responsible dissemination of findings and helps anticipate the potential impact of confirmation on society.

The Simulation Hypothesis calls for an unprecedented degree of intellectual cross-pollination. By embracing collaboration, we forge a more holistic and robust methodology for exploring one of the most profound questions that humanity might ever face.

Emphasis on Falsifiability

The difficulty in directly proving the Simulation Hypothesis underscores the need to shift our emphasis toward seeking potentially falsifiable claims. While true confirmation might forever remain elusive, developing testable hypotheses that could lend weight to the simulation idea or potentially disprove it is paramount for scientific rigor.

Here's how a focus on falsifiability strengthens our methodology:

- **Combating Unbridled Speculation:** It acts as a safeguard against slipping into purely theoretical realms detached from any possibility of empirical grounding. By framing our inquiries in ways that open the door to potential refutation, we maintain intellectual honesty.
- **Guiding Research:** Designing experiments or searching for specific anomalies with the explicit intent of falsifying the Simulation Hypothesis directs our research efforts. It allows us to systematically eliminate possibilities, narrowing down the range of plausible scenarios.
- **Embracing Skepticism:** Prioritizing falsifiability fosters a healthy skepticism towards our own assumptions and potential biases. It encourages us to actively seek out alternative explanations for observations, rather than solely focusing on evidence confirming our simulation suspicions.
- **Building a Robust Foundation:** Even if our endeavors to falsify the Simulation Hypothesis fail repeatedly, they build a stronger case for its potential plausibility. By rigorously testing the limits of what we can rule out, we can better understand the space of what remains possible.

Here are some potential avenues for investigation:

- **The Nature of Reality:** Are there constraints on energy density, computational speed, or fundamental physical constants that appear 'fine-tuned' in ways inconsistent with a purely natural universe?
- **Detection Strategies:** Could we develop techniques with the potential to directly detect computational artifacts embedded within our universe, even if simulator civilizations attempt to obscure them?
- **Theoretical Inconsistencies:** Can we identify any paradoxes or flaws within the laws of physics as we know them that resist explanations within our current models, yet could be reconciled if we are within a simulation?

It's crucial to note that even the failure to falsify the Simulation Hypothesis doesn't equal definitive proof. Yet, a dedicated pursuit of falsifiable claims is what separates responsible investigation from mere speculation.

Interpretive Rigor

The allure of a grand theory like the Simulation Hypothesis carries inherent risks – confirmation bias, an overeagerness to connect seemingly disparate data points, and a tendency to downplay simpler alternative explanations for any unusual findings. To combat these risks and uphold intellectual integrity, frameworks for rigorous evaluation of evidence are crucial.

This focus on interpretive rigor might best be served by formalizing 'red teams' specifically dedicated to the task of dismantling simulation-supporting arguments. These interdisciplinary teams would include:

- **Experts from Relevant Scientific Fields:** Physicists, mathematicians, and computer scientists who can bring a deep understanding of the natural laws or computational artifacts we might encounter as potential 'simulation clues.'
- **Cognitive Psychologists:** Specializing in identifying common biases, heuristics, and logical fallacies that can often creep into the interpretation of complex data, particularly when the topic is highly speculative.
- **Skeptical Philosophers:** Well-versed in scrutinizing arguments, dissecting assumptions, and considering the existence of alternative, simpler explanations that may be overlooked by those already

inclined to support a simulation perspective.

The function of these 'red teams' is not to dismiss the Simulation Hypothesis outright, but rather to meticulously challenge any emerging evidence in its favor. They would actively seek out mundane explanations for anomalies, test the robustness of calculations against errors, and provide a much-needed counterweight to enthusiasm within the field of study.

Benefits of this approach include:

- **Early Mitigation of Biases:** Confronting and addressing potential confirmation bias from the outset strengthens our analysis, reducing the risk of compounding errors of interpretation as the research progresses.
- **Intellectual Honesty:** Formalizing the role of 'devil's advocate' makes skepticism and counter-argument an integral component of the research process, not an afterthought.
- **Robustness and Transparency:** Requiring potential evidence for simulation to pass the scrutiny of diverse 'red teams' improves the credibility of findings and instills public trust in the scientific process.

Limitations and the Ancestor Simulation Argument

The Ancestor Simulation Argument, while captivating, is not without its complexities. Our ability to peer into the vast expanse of time and assess the likelihood of advanced civilizations simulating their ancestors faces significant hurdles. These limitations, however, should not be viewed as insurmountable barriers but rather as guideposts illuminating a path for further exploration. Acknowledging the inherent uncertainties and the subjective nature of interpreting potential evidence underscores the need for a robust and adaptable methodology.

Transparency must be a cornerstone of this methodology. Openly acknowledging the limitations of our knowledge and the potential for biases allows for ongoing critical evaluation and refinement of our models. Furthermore, an adaptable approach is essential. As scientific understanding evolves and new discoveries emerge, our models must be flexible enough to incorporate these developments. Finally, fostering a spirit of collaboration across diverse disciplines is paramount. Physicists, philosophers, computer scientists, and even sociologists can offer invaluable insights that can help mitigate bias and shed light on potential blind spots. By embracing these core principles, we can navigate the labyrinthine complexities of the Ancestor Simulation Argument, pushing the boundaries of scientific exploration and confronting the profound questions it raises about our place in the universe.

SECTION FIVE: DATA ANALYSIS

CHAPTER TWENTY-EIGHT

Technological Feasibility: Computational Power Analysis Part 1

Understanding the technological feasibility of creating ancestor simulations as posited by Nick Bostrom involves a multidisciplinary examination of current and projected capabilities in computing technology. Bostrom's Ancestor Simulation Argument hinges on the premise that future civilizations will have access to immense computational resources, enabling them to run simulations of their historical predecessors with a level of detail indistinguishable from reality. This section delves into the foundations and future trajectories of computational power, advancements in virtual reality (VR) and artificial intelligence (AI), and the transformative potential of quantum computing, providing a comprehensive analysis of the methodologies needed to validate Bostrom's theory.

Computational Power Analysis

The feasibility of Bostrom's Ancestor Simulation Argument depends crucially on the future computational resources available to advanced civilizations. Below, we analyze the computational requirements, historical trends, projections, and the specific data analysis challenges posed when evaluating the possibility of ancestor simulations.

Computational Requirements of Ancestor Simulations

Simulating consciousness and a detailed universe necessitates immense computational power. Let's dissect the requirements:

Simulating Human Consciousness

Estimates for simulating a human brain vary. One approach focuses on neurons and synapses. With roughly 86 billion neurons and trillions of synapses, real-time simulation could demand up to 10^18 FLOPS (exaFLOPS). Other estimations based on functional brain regions put the figure lower. Data needed here includes detailed neuron models, synaptic connectivity maps, and neurotransmitter dynamics.

While simulating the human brain would be a monumental feat, estimations on the required computational power vary significantly based on the chosen model's complexity and the level of desired fidelity:

Neuron and Synapse-Level Simulation:

One approach meticulously models individual neurons and their synaptic connections. With billions of neurons and trillions of synapses, the sheer number of interactions drives the computational requirements incredibly high. Reaching exaFLOP (10^18 calculations per second) territory is not inconceivable for such simulations.

- **Data Needs:** This method calls for:
  - Highly-detailed neuron models capturing ion channels, membrane potentials, and firing patterns.
  - Comprehensive synaptic connectivity maps, detailing the vast network of connections between neurons.
  - Data on neurotransmitter dynamics, including release mechanisms, diffusion, and receptor interactions.

Functional Brain Region Modeling:
Alternatively, simulations might focus on modeling larger functional regions of the brain (e.g., areas responsible for vision, memory, etc.). This approach simplifies the problem by abstracting away individual neuron-level detail, potentially reducing the computational burden.

- **Data Needs Here**, the emphasis shifts to:
  - Mapping functional brain regions and their interactions.
  - Input-output models of brain regions, capturing how they process information and respond to stimuli.
  - Datasets on the overall dynamics of large-scale brain activity (e.g., EEG, fMRI data).

The Question of Consciousness:
A critical, unresolved question is what level of simulation granularity is necessary to achieve a simulation indistinguishable from true consciousness. Simulating physical processes within the brain might not be sufficient on its own. Deeper understanding of the emergent properties leading to consciousness is crucial to refine these estimates.

Essential Data Sources

Regardless of the chosen simulation approach, certain types of data would be indispensable:

The Human Connectome Project:
This ambitious initiative delves into the intricate labyrinth of our brains. It goes beyond simply mapping neural pathways, acquiring multimodal imaging data that encompasses structural connectivity (diffusion MRI), functional connectivity (resting-state fMRI), and even genetic information. Imagine this data as a detailed blueprint, guiding simulations by outlining how different regions of the brain are wired together and how they dynamically interact. The Connectome Project has also spurred the development of advanced data analysis tools, which will be essential for extracting meaningful patterns from the raw simulation outputs.

Neuroinformatics Databases:
Think of these repositories like the Library of Alexandria for the brain. The Allen Brain Atlas, for instance, houses a treasure trove of information on how genes shape the brain's cellular architecture, classifying neuron types based on their genetic profiles. This allows us to model not just generic neurons but specific subtypes found in different brain regions. Additionally, meticulous anatomical atlases within these databases provide 3D reference models, ensuring simulations accurately represent the brain's intricate physical structure.

Electrophysiological Recordings:
These techniques offer a direct window into the electrical conversations within the brain. Single-neuron recordings unveil the precise timing and patterns of individual neuron firing, revealing the fundamental code used for neuronal communication. Large-scale techniques like EEG and fMRI visualize activity across vast populations of neurons, mapping information flow across broad brain networks. This data enables us to construct neuron models that replicate the real-world electrical behavior observed in the living brain and

validate whether our simulations are exhibiting similar patterns.

Cognitive and Behavioral Datasets:

To simulate a conscious mind, we must go beyond physical brain structure and capture the essence of thought, action, and experience. Datasets meticulously documenting perception (how we sense the world) and decision-making (how we choose actions) are essential in modeling how sensory input is transformed into behavioral output by the brain. Furthermore, understanding learning processes is crucial, as a simulated mind should be able to adapt and change in response to its environment, just like a real human.

Reflecting on the Computational Odyssey

Estimating the computational power needed to simulate a human brain is an ongoing scientific puzzle. The choice of modeling approach, our understanding of consciousness, and the continuous generation of vast neuroscientific datasets will gradually refine these estimations. This has profound implications for understanding the feasibility of Bostrom's Ancestor Simulation Argument.

- **Simulating a Realistic Universe:** The scale and level of detail in the simulated universe impact complexity. A full-scale universe simulation down to the quantum level is likely impossible. Instead, a simulation could focus on areas relevant to human experience, including:
- **Planetary Systems and Physics:** Simulating Earth, its biosphere, and the interactions necessary for human existence requires sophisticated models of physics (gravity, electromagnetism, thermodynamics, etc.).
- **Social and Historical Simulations:** Modeling human behavior, the rise of societies, and historical events adds another layer, demanding vast datasets on human psychology, sociology, and historical records.

Historical Trends and Projections

The computational feasibility of ancestor simulations rests heavily upon whether technological progress will continue to deliver the resources needed to achieve such intricate and vast simulations. Let's dissect the patterns witnessed thus far and the analytical tools for projecting future capabilities.

Moore's Law: The Exponential Engine

Moore's Law, the observation that transistor density on integrated circuits doubles roughly every two years, has powered astonishing growth in computing. This exponential trend translates to consistent computational performance gains while decreasing cost and power consumption. However, it's essential to note that Moore's Law is an empirical observation, not a physical law. As we approach atomic-scale limits, its pace has slowed in recent years.

Data Analysis for Trend Evaluation

Let's delve deeper into how sophisticated data analysis techniques illuminate trends in computational progress and inform projections for ancestor simulations.

Analyzing Growth Patterns:

- Statistical analysis of historical data on transistor counts and performance metrics reveals the exponential nature of computational progress. Time series analysis, including fitting various models (linear, exponential, logistic), helps us understand the long-term trajectory and potential inflection points.
  - **The Power of Exponentials:** Meticulous examination of historical data on transistor counts and performance often reveals that the underlying growth pattern is best described by exponential functions rather than linear ones. This has fundamental implications – computational resources

could increase far more dramatically than our everyday linear intuition would suggest.
  - **Model Selection and Refinement:** Time series analysis helps us fit and compare various models. While simple exponential models might provide a starting point, more complex models like logistic growth curves (S-shaped) could capture the slowing of Moore's Law and identify potential future saturation points. Statistical rigor in model selection (e.g., goodness-of-fit, parsimony criteria) improves our long-term understanding.
  - **Visualizing Trends and Deviations:** Plotting historical performance data on semi-logarithmic graphs (where exponential growth appears as a straight line) offers clear visualizations, making it easier to spot deviations from the expected trend. These deviations can signal technological breakthroughs or fundamental roadblocks.

Disaggregating Sources of Improvement:

- Performance growth comes not just from increased transistor count but also from architectural advancements, such as pipelining, superscalar design, and increased clock speeds. Data analysis helps us separate the relative contribution of these factors, informing future projections.
  - **Beyond Mere Numbers:** Raw transistor count alone doesn't tell the whole story. Architectural innovations like pipelining (breaking down instructions into smaller stages), superscalar designs (executing multiple instructions in parallel), and faster clock speeds have significantly multiplied computational performance.
  - **Statistical Disentanglement:** Regression analysis techniques enable us to model performance as a function of transistor count, clock speed, architectural features, and so on. This helps quantify how much improvement stems from scaling transistor density versus other complementary innovations.
  - **Informing Research Focus:** By isolating the factors driving performance gains, data analysis directs research investment. If architectural cleverness emerges as a primary driver, simulation-specific chip designs could get more attention compared to a sole focus on raw transistor count.

Data Analysis for Ancestor Simulations

- **Non-Uniformity of Progress:** Different aspects of computing technology might progress at uneven rates. Ancestor simulations might require breakthroughs along multiple dimensions (memory bandwidth, storage capacity, etc.). Data analysis can highlight the potential bottlenecks and areas where technological progress might lag.
- **Emergent Complexity Needs:** The computational cost of simulations doesn't always scale linearly with the number of simulated elements. Analysis of how computational demands increase in relation to simulation complexity is crucial, as interactions between many simulated minds could make costs explode.
- **Data from Small-Scale Simulations:** Even before ancestor scale simulations are possible, data from smaller simulations will offer valuable insights. Analysis of scaling behavior and hardware bottlenecks on these prototypes can shape expectations for the full-scale scenario.

Key Takeaway:

- It's important to remember that data analysis is not merely about extrapolating past trends blindly. It's about understanding the underlying drivers of progress, the potential roadblocks, and identifying areas where targeted research investment can overcome obstacles and keep us on track to the computational capability scale needed by Bostrom's Ancestor Simulation Argument.

Beyond Moore: Overcoming Limitations

As classic silicon-based transistors approach fundamental physical limits, sustaining the exponential progress necessary for ancestor simulations demands groundbreaking new approaches. Key research avenues hold promise, and data-driven analysis will be critical in guiding their development and potential integration into the simulation technology stack.

3D Chip Architectures: Upwards and Onwards

- **Circumventing 2D Limits:** Traditional chips are inherently constrained by planar layouts. 3D integration, by vertically stacking transistors and interconnecting layers, effectively multiplies the available surface area within a given footprint.
- **Data Analysis Challenges:**
    - Thermal Management: Heat dissipation becomes significantly more complex in 3D stacks. Detailed thermal modeling and simulation are needed to optimize designs while ensuring reliability.
    - Communication Overhead: Moving data between vertically distant layers can introduce latency and bottlenecks. Data analysis on communication patterns within simulation workloads can inform efficient 3D architecture designs and communication protocols.
    - Manufacturing and Yield: Fabricating multi-layer 3D chips introduces new complexities and potential defect sources. Yield analysis becomes crucial, and process data will guide refinements for economically viable production.

Novel Materials: The Search for Silicon's Successor

- **Graphene's Promise:** Graphene, with its single-atom-thick sheets of carbon, boasts exceptional electron mobility and strength. Graphene-based transistors could switch faster, consume less power, and facilitate higher densities than silicon counterparts.
- **Carbon Nanotubes:** These cylindrical carbon structures offer similar advantages as graphene, potentially enabling denser circuits and faster computation.
- **Beyond Carbon:** Research explores diverse materials, including germanium, III-V semiconductors, and even organic molecules for future electronics.
- **Data Analysis Challenges:**
    - **Material Properties Characterization:** Rigorous experimental methods paired with materials modeling will generate data on conductivity, band gaps, thermal behavior, etc., allowing comparisons against silicon and simulations of their performance in transistors.
    - **Manufacturability and Scaling:** Lab-scale success doesn't guarantee mass production. Data on fabrication processes, yields, defect rates, and integration with existing technologies will determine real-world viability.

Specialized Chips: Tailored for Simulation

- **Lessons from AI Acceleration:** The success of GPUs in accelerating machine learning demonstrates the power of customizing hardware to specific computational patterns. Simulation workloads might benefit from similar customization.
- **Targets for Specialization:**
    - **Physical Simulation:** Chips tailored to solve differential equations governing physical laws could accelerate the simulation of the environment.
    - **Consciousness Modeling:** Dedicated hardware designed around the computational models of consciousness may prove dramatically more efficient than general-purpose CPUs.
    - **Scenario and Agent Simulation:** Simulations involving decision making and complex interactions within social settings might benefit from hardware designed with those structures in mind.
- **Data Analysis Challenges:**
    - **Profiling Simulation Workloads:** Analyzing existing simulations on traditional hardware to identify common computational bottlenecks and patterns will focus the design of specialized

architectures.
  - **Adaptability and Programmability:** Balancing specialization for performance with the flexibility to handle diverse simulation scenarios and evolving algorithms is essential. Data-driven analysis will assess this tradeoff.

Tying it to Ancestor Simulations

- The computational requirements for realistic ancestor simulations necessitate a multifaceted technological leap. Data analysis empowers this transition:
  - **Synergies, Not Singular Solutions:** Simulation demands might drive advances across all these domains: 3D integration for density, novel materials for speed, and specialization for efficiency. Analysis will reveal potential performance gains from their combinations.
  - **Targeting the Bottlenecks:** Data from profiling various simulation components will highlight the primary roadblocks – whether they lie in physical modeling, consciousness simulation, or elsewhere. This allows focused investment in the most impactful technologies.
  - **Long-Horizon Analysis:** Extrapolating how advancements in these areas might collectively contribute to the computational power needed for ancestor simulations is a complex, data-driven exercise with significant uncertainty. Statistical modeling and sensitivity analysis will help constrain the range of possibilities.

Targeting Simulation Capabilities

- The pursuit of ancestor simulation technology demands a leap beyond the traditional trajectory defined by Moore's Law. It necessitates targeted innovation – not just an increase in raw computational power but advancements intimately linked to the specific requirements of such simulations. Data analysis assumes a critical role in this endeavor. It guides research efforts by identifying potential bottlenecks, evaluates the promise of novel materials and computational architectures, and crucially, allows us to continuously reassess the feasibility of Bostrom's powerful vision.

The Exascale Frontier & Performance Analysis

Exascale supercomputers herald a new era in computational power, promising a thousand-fold increase (10^18 FLOPS) compared to their petascale predecessors. This leap could be transformative, but its relevance to ancestor simulations demands meticulous data-driven evaluation and a nuanced understanding of performance in the context of simulation-specific workloads.

Data-Driven Benchmarks: Beyond Raw FLOPS

- **Theoretical vs. Achievable Performance:** A system's theoretical peak FLOPS is often unattainable on real-world applications due to bottlenecks in memory access, communication, and the nature of the algorithms themselves.
- **Simulation-Specific Benchmarks:** Development of benchmarks mirroring key computational patterns of ancestor simulations is essential. Example areas include:
  - **Physical World Simulations:** Solving complex partial differential equations governing realistic physics (fluid dynamics, materials, etc.) at a vast scale.
  - **Neural Simulation:** Large-scale simulations of neural networks, assessing the performance of different neuron and connectivity models.
  - **Agent-Based Social Simulations:** Modeling complex decision-making, interactions, and the emergence of social phenomena within a simulated population.
  - **Standardization for Comparison:** Agreed-upon simulation benchmarks, akin to LINPACK for linear algebra, facilitate comparisons across different exascale architectures, aiding long-term projections.

Scalability Analysis: Handling Complexity

- **Amdahl's Law and Simulation Workloads:** Amdahl's Law highlights the diminishing returns of parallelization as the serial (non-parallelizable) portion of a code dominates execution time. Data analysis is needed to quantify the inherently serial parts within ancestor simulations, informing potential limits of exascale speedup.
- **Modeling Scaling Behavior:** How does the runtime of simulation components vary with the number of simulated individuals, level of detail, and duration of the simulated history? Developing scaling models through experimental runs at smaller scales is vital.
- **Communication Overhead:** Exascale systems consist of thousands of interconnected compute nodes. The overhead of data movement between these nodes as parts of the simulation interact can become a bottleneck. Profiling algorithms and communication patterns is essential.

Power and Efficiency: The Feasibility Factor

- **Beyond FLOPS:** Performance-per-Watt: Exascale systems are power-hungry. Metrics focused on computations performed per joule of energy consumed are paramount. Analyzing power consumption within the context of simulation benchmarks is crucial.
- **Thermal Management Challenges:** Power density introduces thermal constraints. Data analysis to understand the interplay between computation, cooling efficiency, and the operational reliability of exascale systems is necessary.
- **Long-term Sustainability:** Projecting the energy demands of full-scale ancestor simulations that might run for extended durations is critical in determining whether our technological trajectory can realistically sustain them.

Implications for Ancestor Simulations

- The exascale milestone offers a tantalizing glimpse, but careful analysis informs the reality:
  - **Incremental Step, Not a Leap:** Exascale systems may still fall short of the sheer power needed for Bostrom's vision. Data on simulation performance will reveal the remaining gap to be closed.
  - **Component-Wise Advancements:** Even if insufficient as a singular solution, exascale systems could drive breakthroughs in critical sub-domains relevant for simulations (e.g., novel high-performance memory, more efficient simulation algorithms).
  - **Driving Innovation Dynamics:** The exascale race spurs optimization across hardware and software. Data-driven analysis of these gains over time provides a trendline for extrapolating future capabilities.
  - **Potential Architectural Insights:** Exascale systems often feature heterogeneity, with specialized processing units. Data from their performance on simulation sub-tasks could inform the design of future, fully simulation-tailored architectures.

Exascale and Beyond: Charting the Path to Simulation:

- While the exascale frontier marks a momentous technological achievement, it offers a waypoint, not a final destination, on the path towards realizing ancestor simulations. Data analysis becomes our compass, crucial for quantifying the progress made with exascale systems. It enables us to analyze power consumption, efficiency gains, and most importantly, to identify which aspects of simulation demand further revolutionary leaps beyond what exascale alone can offer. This analysis fuels focused research, continually illuminating the remaining hurdles in fulfilling the vast computational demands envisioned within Bostrom's argument.

Quantum Computing: A Potential Game-Changer

While still in its nascent stages, quantum computing wields the potential to revolutionize how we approach computation. By harnessing quantum phenomena such as superposition and entanglement, it offers fundamentally distinct ways to solve problems that confound even the most powerful classical computers. This transformation could be profound for the feasibility of ancestor simulations.

Quantum Algorithms for Simulation

- **A Different Toolkit:**
  - Quantum algorithms are built upon quantum gates and circuits, manipulating qubits (quantum bits) that can exist in superpositions of 0 and 1. This allows for types of computation starkly contrasting with classical bit operations.
- **Quantum Advantage in Simulation:**
  - Early theoretical research offers exciting possibilities:
    - **Physics Simulation:** Algorithms like the Harrow-Hassidim-Lloyd (HHL) algorithm promise to solve certain linear systems of equations exponentially faster than classical counterparts, which could accelerate the simulation of physical systems at the quantum level.
    - **Optimization:** The Quantum Approximate Optimization Algorithm (QAOA) is designed for combinatorial optimization problems, relevant for modeling complex social dynamics or resource allocation within simulations.
    - **Machine Learning:** Potential exists for quantum-enhanced machine learning techniques that could more efficiently learn patterns from historical data or discover the computational models governing consciousness.
- **Data Analysis Needs:**
  - **Theoretical Speedup Characterization:** Rigorous analysis is needed to quantify the precise asymptotic complexity advantage (e.g., linear vs. quadratic, exponential) of quantum algorithms relevant to various simulation components.
  - **Moving Beyond Asymptotics:** Theoretical scaling needs to be tempered by analysis of constant factors and algorithm overheads, especially for near-term, less idealized quantum devices.

Simulation-Specific Benchmarks

- **The Noisy Intermediate-Scale Quantum (NISQ) Era:** Today's quantum computers are relatively small-scale and susceptible to errors. Their impact on ancestor simulations needs early evaluation, even though they can't tackle the full problem.
- **Small-Scale, Tailored Benchmarks:** Developing benchmarks with simplified models of physical systems, smaller neural networks, or toy-sized social simulations is essential for running on NISQ devices.
- **Insights Despite Limitations:** Data from these benchmarks can reveal if quantum algorithms offer any practical speedup, uncover potential bottlenecks when scaling to larger systems, and inform the design of future, improved quantum algorithms.

Error Correction Tradeoffs

- **The Necessity of Error Correction:** Fragile qubits require error correction overhead to build reliable, large-scale quantum computers. This overhead can consume additional qubits and operations, offsetting some of the raw speed gains from quantum algorithms.
- **Error Thresholds:** Quantum error correction codes work effectively only if the error rate per qubit is below a threshold. Data analysis on the progress of hardware development is needed to determine when we might reach that domain.
- **The Resource Tradeoff:** Analyzing the computational cost of error correction in the context of simulation algorithms is vital. It will reveal to what extent error correction overhead diminishes realizable quantum advantage.

Implications for Ancestor Simulations

- Quantum computing's impact on ancestor simulations is speculative yet profound. Data analysis plays a pivotal part in shaping realistic expectations:
  - **Trajectory, Not a Single Leap:** Even if quantum computers eventually contribute to ancestor simulations, it's likely an incremental trajectory as algorithms, error correction, and hardware mature. Trend analysis becomes crucial.
  - **Potential for Targeted Acceleration:** Certain aspects of simulations might benefit most from quantum computation (e.g., quantum-level physics modeling). Hybrid quantum-classical simulation architectures will likely emerge, needing analysis for optimal orchestration.
  - **Beyond Qubits Alone:** Even without full ancestor simulation feasibility, insights from quantum simulations could refine classical simulation algorithms, indirectly contributing to the goal.
  - **A Long-Term, Uncertain Bet:** Quantum computing is a high-risk, high-reward domain. Data-driven analysis helps quantify uncertainty, ensuring continued investment and exploration rest on well-calibrated probabilistic assessments.

Quantum Potential and Data-Driven Discernment

- The allure of quantum computing for ancestor simulations is undeniable yet shrouded in uncertainty. Data analysis becomes our clarifying lens, essential for separating visionary promises from present-day limitations. It allows us to rigorously compare theoretical advantages to practical constraints, ensuring strategic research investments remain grounded and focused. As quantum computing evolves, data analysis will continuously remap its potential contributions and help determine where this revolutionary technology might either accelerate or introduce new hurdles within the complex landscape of simulating entire universes.

Tying Trends to Ancestor Simulations

Harnessing the momentum of technological progress for ancestor simulations requires more than raw extrapolation. It demands data-driven analyses that connect the dots between trends and the unique, multifaceted complexities of simulating an entire world and its inhabitants.

Granularity & Complexity Analysis

- **Beyond FLOPS: Simulation-Specific Metrics:**
  - Raw computational power doesn't directly map to the ability to run ancestor simulations. Devising metrics like "simulated individuals per exaFLOP" or "simulated years of history per petaFLOP" will better quantify progress.
  - While FLOPS (or equivalents for novel architectures) offer a starting point, they don't tell us how well-suited that power is for ancestor simulations. Metrics tailored to the problem domain are crucial for tracking meaningful progress.
  - A few possibilities:
    - **Simulated Individuals per ExaFLOP:** This focuses on the capacity to simulate conscious entities within a given computational budget. It would factor in the complexity of the chosen consciousness model.
    - **Simulated Years per PetaFLOP:** Emphasizes the simulation of history over sheer agent count. Relevant if detailed modeling of technological and societal evolution is vital for realism.
    - **Granularity Levels per YottaFLOP:** If the level of detail (physics, brain models) is adjustable, this could track the achievable level of simulation fidelity within available power.

- **The Scaling Challenge:**
  - A daunting aspect is that we cannot assume linear scaling of computational costs with the number of simulated agents, history length, or level of detail. Complex interactions could lead to surprising scaling behavior:
    - **Interaction-Driven Exponentiality:** Social dynamics, economic systems, or even physical interactions among many entities could exhibit exponential scaling, where adding a single agent dramatically multiplies computation.
    - **Factorial Explosion:** In the worst-case scenario, certain aspects might exhibit factorial scaling, making them virtually impossible to compute. Identifying these early is paramount.
- **Data-Driven Complexity Models:**
  - Even early simulations, while far from the scale of Bostrom's idea, provide invaluable data for understanding scaling. Key analysis techniques include:
    - **Curve Fitting & Model Selection:** Fitting different models (linear, polynomial, exponential) to data on how computational costs vary with simulation scale. Choosing models that balance fit with parsimony (avoiding overfitting) is crucial.
    - **Multivariate Analyses:** Since cost depends on multiple factors (individual complexity, world detail, etc.), multivariate regression techniques help disentangle their relative contributions.
    - **Acknowledging Uncertainty:** Even early models will be imperfect. Using confidence intervals, Bayesian model averaging, etc., helps quantify the uncertainty in projections made using them.
- It's important to emphasize that the goal of such modeling isn't perfect prediction, but an ever-refining understanding of the challenges. It lets us separate tractable scaling problems from potential fundamental barriers that necessitate entirely different theoretical approaches.

Simulation Bottleneck Identification

- **Profiling for Targeted Optimization:**
  - Running simulations at various scales on diverse architectures (CPUs, GPUs, potential quantum accelerators) and profiling their performance will expose the most computationally taxing elements.
  - Meticulous profiling of simulations, even smaller-scale ones, on a variety of architectures is key to revealing where computational resources are spent. Profilers dissect execution, tracking metrics like execution time, memory usage, instruction counts, etc., pinpointing bottlenecks at multiple levels. This needs a multi-pronged approach:
    - **Hardware Diversity:** Profiling on classical CPUs, GPU accelerators, and even early quantum simulators (where feasible) offers a comparative picture of how specific simulation components respond to different computational paradigms.
    - **Scaling:** Performing profiling runs with varying simulation scales (simulated world size, number of agents) is critical for revealing how costs escalate with complexity
- **Potential Bottlenecks:**
  - Data analysis of profiling results should cast a wide net, with some key focal areas:
    - **Physics Simulation:** Differential equation solvers can be computationally intensive. Profiling will expose whether the simulation is dominated by modeling fluid dynamics, electromagnetism, etc. Moreover, the chosen level of detail (e.g., quantum-level vs. macroscopic) will significantly impact costs.
    - **Consciousness Modeling:** If the adopted model for individual minds is complex, even a small number of simulated beings could cripple the simulation. Profiling can reveal the per-agent computational cost and how it scales – an indication of whether this

aspect presents a fundamental hurdle.

    - **Social Interaction Complexity:** As agents interact, emergent behaviors can arise. If these dynamics lead to combinatorial explosions in computations, profiling can pinpoint whether social modeling outpaces other simulation components as the complexity scales.

- **Guiding Research Investment:**
  - Data from profiling informs where to concentrate efforts: optimizing existing algorithms, designing specialized hardware, or even revisiting the theoretical models used in certain aspects of the simulation. Profiling becomes truly powerful when paired with an understanding of the potential optimizations available.
  - Key questions to consider include:
    - **Algorithmic Improvements:** If a bottleneck is in a well-studied domain (e.g., physics solvers), are there more efficient algorithms to explore? Profiling tied to algorithm analysis is crucial.
    - **Hardware Specialization:** Does profiling suggest certain operations dominate? This opens doors for specialized chips tailored to accelerate those very tasks, offering more performance gains than incremental general CPU improvements.
    - **Theoretical Model Rethinking:** If a bottleneck lies in the fundamental complexity of the chosen computational models (e.g., consciousness), revisiting foundational scientific assumptions may be needed alongside purely technological advancement.
- Importantly, the act of profiling and bottleneck identification is iterative as simulation capabilities themselves improve. Early roadblocks might give way, revealing new ones deeper in the stack, continuously focusing research and development efforts.

Data Analysis for Ancestor Simulations

- Let's examine some specific data analysis approaches essential in this context:
  - **Uncertainty Quantification:**
    - Projections into the far future are inherently uncertain. Using not just best-fit trends, but probabilistic modeling (e.g., Bayesian methods, Monte Carlo simulations) will reveal the range of plausible scenarios under different assumptions.
    - The further into the future we project, the wider the cone of uncertainty becomes. Using probabilistic methods goes beyond simple 'best guess' scenarios. Bayesian modeling, where we start with prior distributions on key parameters (rate of improvement, impact of breakthroughs) and update them as evidence accumulates, becomes invaluable. Monte Carlo simulations let us sample from these distributions, generating ranges of potential outcomes and their associated probabilities. This helps us see whether Bostrom's argument hinges on highly optimistic outliers or falls within the realm of more probable scenarios.
  - **Sensitivity Analysis:**
    - How much do changes in assumptions about the rate of technological growth, the breakthroughs needed, or the inherent complexity of simulation impact the ultimate feasibility estimates? Sensitivity analysis can quantify this.
    - Our feasibility estimates are only as good as the assumptions they rely on. Sensitivity analysis reveals the 'levers' that most significantly impact the final outcome. Examining how feasibility changes with variations in assumptions about technological breakthroughs, the computational complexity of aspects like consciousness modeling, and potential physical limits provides a few benefits. First, it identifies crucial research targets, letting us focus on overcoming the bottlenecks that matter most. Secondly, it highlights vulnerabilities – if our belief in feasibility rests on a very narrow set of

assumptions, it warrants caution.

- **The Role of Simulation in its Own Analysis:**
    - As simulation capabilities increase, they become a tool for studying themselves. Running simplified simulations designed to model technological growth and computational cost scaling could provide valuable meta-level insights.
    - This is a fascinating meta-level concept. Early simulations, though likely crude, provide a playground for modeling the dynamics of technological improvement and complexity scaling. We can build simulations with tunable parameters, simulating various 'worlds' with different constraints and growth rates.
    - Analyzing these simulated trajectories can offer insights into:
        - Whether there are fundamental limits to certain computational resources.
        - How the interaction of different technological improvements plays out.
        - Whether the very act of creating simulations accelerates progress towards the goalpost itself.

Analyzing Limitations and Breakthroughs

- **When Trends Break Down:**
    - Extrapolating Moore's Law might hit a wall. Data analysis needs to identify early warning signs of such inflection points or fundamentally different technological paradigms arising. Moore's Law is an empirical observation, not a physical guarantee. Data analysis plays a vigilant role in recognizing potential slowdowns or fundamental asymptotes. Techniques for identifying inflection points, such as time series segmentation and change point detection, are vital for spotting deviations from the established exponential trajectory. Additionally, analyzing fundamental limits (e.g., heat dissipation at the atomic scale, quantum tunneling effects, etc.) can provide theoretical indicators for where barriers might arise. Importantly, data analysis should also look for signs of emerging paradigms—research into novel materials and fundamentally different computational substrates might offer hints of technologies poised for disruptive change.
- **The Power of 'Surprise'**
    - History shows that technological progress isn't always predictable. Data analysis should leave room to model the impact of serendipitous discoveries and 'black swan' events that could outpace expected trends. While data analysis excels at modeling established trends, anticipating uncharted breakthroughs is inherently more challenging. Statistical approaches incorporating stochasticity (e.g., scenario modeling, Bayesian forecasting with 'heavy-tailed' priors) can partially account for the possibility of unexpected leaps forward. Additionally, historical case studies of disruptive technologies can offer insights into common patterns preceding such events. For example, analyzing publication trends and research collaboration networks might reveal up-and-coming fields on the cusp of revolutionizing simulation capabilities. Ultimately, a healthy dose of humility is needed in acknowledging that a degree of 'unknowable unknown' will likely always remain.
    - Data analysis, in both its anticipation of roadblocks and openness to the unexpected, plays a key role in navigating the landscape of technological innovation. By going hand-in-hand with theoretical modeling, an understanding of physical limits, and awareness of emergent research areas, it helps craft a realistic and adaptable picture of whether, and how quickly, we might approach the computational capabilities needed for ancestor simulations.

Tying it Back to Bostrom

- The feasibility of ancestor simulations rests on the intersection of these questions:
    - **What are the absolute computational requirements?**

        - This includes the raw power, specialized architectures, and algorithmic efficiency to achieve the desired level of realism. Determining this demands a layered approach that goes beyond raw FLOPS. Estimating requirements likely involves a combination of theoretical modeling of physical systems and consciousness models, empirical data from smaller-scale simulations, and an understanding of the computational complexity classes of crucial algorithms. The desired level of realism adds ambiguity, as different individuals might have varying thresholds for perceiving a simulation as indistinguishable from reality. Ultimately, a rigorous definition of requirements is a formidable research question in itself.
    - **How fast is technology advancing along the relevant dimensions?**
        - This includes computational power, storage, memory bandwidth, network speeds, etc. Here, data analysis takes center stage. Meticulous tracking of key performance metrics of CPUs, GPUs, novel accelerators, memory technologies, and networking across time is essential for fitting trend models. Crucially, these trends should be analyzed not solely in isolation but in how well they align with the types of workloads expected within ancestor simulations. This might reveal the need for tailored benchmarks focused on simulation-relevant scenarios, as raw technological progress doesn't always translate seamlessly into progress on a specific problem domain.
    - **Will fundamental roadblocks emerge?**
        - Will physical limits, unbreakable complexity barriers, or the energy demands outpace the pace of innovation? Forecasting potential roadblocks demands careful scrutiny of underlying physical limits and theoretical hurdles. Are there hard limits to transistor density or energy efficiency? Data analysis of historical trends can help spot potential deceleration zones. Further, the computational complexity of simulation algorithms could reveal theoretical barriers. It may be that simulating certain aspects of reality has intractable complexity, implying that no amount of computational power will suffice. Moreover, energy and resource constraints could place practical limits on simulation scale even if theoretical solutions exist.
- Data analysis provides the tools to rigorously model these questions, update our understanding as new evidence unfolds, and inform assessments of how close (or far) we may be from the capabilities postulated by the Ancestor Simulation Argument.

Data as Our Lodestar: Navigating Simulation Feasibility

- In the pursuit of ancestor simulations, data analysis acts as an indispensable lodestar. It provides a vital nexus where abstract technological progress is grounded in the tangible realm of simulation requirements. Data analysis reveals where computational resources might fall short, directing research efforts and investments with precision. It quantifies uncertainties, offering a sobering counterbalance to boundless speculation. Most importantly, as computational power relentlessly advances, data analysis will continuously re-calibrate our understanding of how closely our trajectory aligns with the profound computational demands outlined in Bostrom's argument.

CHAPTER TWENTY-NINE

Technological Feasibility: Computational Power Analysis Part 2

Data Analysis Challenges

Proving the Ancestor Simulation Argument through the lens of data analysis means confronting a web of uncertainties, complexities, and the constant need to bridge theoretical possibilities with computationally grounded realities.

Navigating the feasibility assessment of ancestor simulations through data analysis requires confronting an intricate tapestry of uncertainties, theoretical complexities, and the ever-present need to reconcile hypothetical scenarios with the constraints of real-world computational paradigms. Statistical models must grapple with the inherent unpredictability of technological trajectories, potentially reaching diminishing returns or encountering unforeseen fundamental limits.

Understanding how complex simulations scale demands a sophisticated interplay of empirical data from smaller simulations and theoretical analysis, with the potential to expose non-linear or even intractable relationships between computational resources and simulation fidelity. Moreover, the very act of translating hypothetical simulation capabilities into the language of known (or even future) computational substrates presents a formidable benchmarking challenge, necessitating metrics far beyond traditional measures of raw processing power.

Let's unpack key challenge areas:

Complexity Extrapolation

The Ancestor Simulation Argument hinges on a colossal extrapolation of computational progress. Data analysis must move beyond naive linear projections. Models need to confront the inherent uncertainty surrounding the deceleration of Moore's Law, anticipating its consequences alongside the potential of unforeseen physical limits posed by heat dissipation, the communication constraints imposed by the speed of light, and the unpredictable influence of quantum-scale phenomena as we approach the boundaries of computation. Rigorous analysis of historical performance trends in computation, paired with insights gleaned from materials science and thermodynamics, will inform the likelihood of encountering fundamental roadblocks on the path to the immense computational power required by Bostrom's argument.

The Uncertain Trajectory of Progress:

Projecting future computational power isn't simply extending a line on a graph. Statistical models must carefully account for the potential slowing of Moore's Law and its implications.

Extrapolating computational power cannot be reduced to mere curve fitting. Statistical models must thoughtfully grapple with the complex interplay of factors shaping technological advancement and their potential for disruption. Alongside the anticipated slowdown of Moore's Law, analyses should incorporate the non-deterministic nature of scientific breakthroughs. Modeling innovation bursts in algorithm design, inspired by tracking the historical impact of algorithmic revolutions, will help quantify how drastically performance gains might outpace pure hardware improvements.

Data analysis is crucial in identifying areas where diminishing returns become likely, highlighting bottlenecks and informing whether the computational demands of ancestor simulations require sustained exponential growth or can be significantly offset by algorithmic ingenuity.

They should also incorporate factors like:

- **Physical Limits:** Thermodynamics, quantum effects, and material properties impose constraints. Data analysis helps gauge how close we are to these potential barriers.
- **Algorithmic Leaps:** Breakthroughs in algorithms can offer performance gains even without hardware changes. Modeling the potential impact of such leaps involves historical analysis of algorithmic revolutions.

The Need for Nuance:

Global trends in processing power obscure the fact that different computational tasks may progress unevenly. Specialized benchmarks, not merely raw FLOPS, are needed to assess the alignment between technological development and the demands of ancestor simulations.

Focusing solely on raw FLOPS or projections of overall processing power obscures a key challenge. Technological progress is rarely uniform: advancements in memory bandwidth, reductions in power consumption, and the development of specialized architectures might exhibit entirely different trajectories than raw computational throughput. Data analysis must dissect these individual trends and create simulation-tailored benchmarks that mirror the computational demands of ancestor simulations. Such benchmarks will highlight specific domains (e.g., physics simulations versus modeling consciousness) where bottlenecks may arise disproportionately, shaping projections and guiding research investment more effectively.

Simulation Scalability

The feasibility of ancestor simulations depends heavily on how computational requirements scale with the complexity of the simulated world. A crucial task for data analysis is to expose and quantify any non-linear relationships that might exist. This endeavor must consider multiple dimensions of simulation complexity—from the number of simulated entities to the level of environmental detail and the intricacies of modeling consciousness or social systems.

Beyond Linearity:

The assumption that doubling computational power allows the simulation of twice the population or a world with twice the fidelity is likely overly simplistic. Data analysis must reveal the true nature of these scaling relationships. For instance, as simulated beings interact, the modeling of their relationships and the consequences of those interactions could easily become computationally overwhelming, leading to disproportionate demands. Profiling and analyzing how smaller-scale simulations grow in cost as complexity increases is vital for modeling these dynamics.

Moreover, particular aspects of the simulation, such as social networks or unpredictable chaotic systems within the simulated environment, could introduce emergent behaviors. Such emergent phenomena have the potential to exhibit exponential or even factorial growth in their computational demands, dramatically exceeding expectations derived from linear thinking.

A naive assumption of linear scaling (doubling computing power doubles simulable population) is likely unrealistic. Data analysis must reveal the true scaling behavior. Key questions include:

- **Agent-Agent Interactions:**
  - Do interactions between simulated individuals lead to disproportionate increases in computational costs? Statistical analysis of simulation scaling in smaller versions is essential.
- **Emergent Complexity:**
  - Can modeling social dynamics or unpredictable events exhibit exponential or even factorial computational cost scaling? This calls for sophisticated modeling techniques.

Mapping Complexity to Resources:

Even if we develop a nuanced understanding of the simulation's scaling behavior, translating this knowledge into concrete computational requirements introduces another layer of data-driven exploration. Theoretical analyses play a role here—we must examine studies estimating the computational cost of simulating physical phenomena at varying levels of precision, as well as the cost associated with different models of consciousness. Data analysis will serve as the crucial bridge in this domain, mapping these theoretical requirements to the capabilities of current and projected hardware. This will expose potential bottlenecks and inform whether the scaling behavior uncovered earlier fundamentally aligns with the technological trajectory needed for Bostrom's argument to hold.

Benchmarking Substrates

Quantifying how different computational paradigms measure up to the unique requirements of ancestor simulations presents a multi-faceted challenge for data analysis. We must acknowledge the limitations of existing benchmarks and develop new ones, grapple with comparing fundamentally different technologies, and continuously ensure our metrics keep pace with our understanding of simulation needs.

The Need for Simulation-Specific Comparisons:

Traditional computational benchmarks, optimized for tasks like solving linear algebra problems (e.g., LINPACK) or general-purpose performance, fail to capture the essence of ancestor simulations. Data analysis plays a vital role here in two ways. First, profiling even early, small-scale simulations will reveal the key computational bottlenecks— whether performance is dominated by physics calculations, neural modeling, or other aspects. Second, these insights should drive the design of new benchmarks. We need benchmarks that prioritize tasks like simulating vast physical environments at different granularities, running massively parallel neural computations, and potentially modeling the emergent complexities of social interactions and historical development.

Apples and Oranges:

The rise of novel computational substrates like quantum computing further complicates the picture. How does one relate gains on a NISQ (Noisy Intermediate-Scale Quantum) device to performance on a classical supercomputer? Data analysis becomes vital in establishing meaningful comparisons. Running tailored benchmark problems on both classical and quantum hardware is necessary but not sufficient. Analysis must go deeper, accounting for algorithm overheads, error correction costs, and the potential advantages of certain quantum algorithms for problems related to simulation. This will help us quantify the true benefits, if any, that novel paradigms offer specifically for the challenges associated with ancestor simulations.

Adapting Benchmarks with Progress:

The field of simulation will itself evolve over time. Data analysis cannot focus on a static target. As our understanding of consciousness modeling, the most efficient physical simulation techniques, or the social science models suitable for ancestor simulation improves, our benchmarks must continuously adapt. Data analysis of trends in simulation research will inform what new capabilities need to be integrated into benchmarks. This iterative process will ensure that performance metrics always reflect the most up-to-date and relevant aspects in assessing ancestor simulation feasibility.

Observational "Fingerprints"

The search for tell-tale signs of living within a simulation presents a fascinating yet daunting challenge. Data analysis must tread a fine line between the pursuit of the extraordinary and rigorous adherence to scientific methodology. Key challenges include:

A Double-Edged Sword:

The very concept is inherently double-edged. If we exist within a simulation, subtle artifacts, glitches, or constraints imposed by its computational nature might manifest as observable phenomena in our universe. However, such evidence is easily confounded. Statistical anomalies or apparent quantization effects could result from true physical laws we simply don't fully comprehend yet. Data analysis must focus on developing exceptionally rigorous statistical tests that can distinguish between potential "simulation artifacts" and the vast space of alternative, natural explanations.

The Limits of Observation:

Our technological capabilities significantly limit the search for simulation evidence. Analyzing cosmic phenomena over vast scales might uncover hints of computational 'grids' or large-scale artificiality. Similarly, probing the limits of the quantum world could reveal potential numerical round-off errors or discrepancies in fundamental constants. However, this necessitates advances in instrumentation (telescopes, particle detectors) alongside novel data analysis methodologies capable of handling both the scope and fine-grained nature of this data.

The Confounding Factor of Convergence:

Identifying an apparent anomaly is only the first step. Data analysis must help adjudicate between competing hypotheses. Even with strong statistical significance, could the anomaly be an undiscovered aspect of nature itself, rather than a sign of a simulator's constraints? Bayesian probabilistic reasoning could be an invaluable tool. Data analysis would involve comparing the likelihood of the data under different scenarios: a purely natural universe, versus universes with simulated components and varying levels of simulator sophistication.

Importantly, the search for observational fingerprints exemplifies how data analysis informs the very questions we ask. Negative results might shape the types of simulations deemed plausible, while positive findings could forever change our understanding of reality.

Informing the Ancestor Simulation Argument

Data analysis is indispensable in assessing the feasibility of Bostrom's idea, but its influence goes deeper. It impacts the nature of our research pursuits, reveals the domains where breakthroughs are essential, and helps ground the argument in either quantified probabilities or the stark recognition of impossibility.

Highlight Bottlenecks:

By rigorously analyzing computational trends, profiling simulations, and evaluating benchmarks, data analysis illuminates the most daunting roadblocks to achieving the capabilities needed for ancestor simulations. Perhaps these bottlenecks lie in raw computational power, or more insidiously, in the sheer complexity of modeling consciousness. This clarity is crucial for focusing research investment. It allows us to distinguish problems that might be solvable through continued engineering efforts versus challenges demanding revolutionary leaps in our fundamental scientific understanding.

Reveal Unforeseen Challenges:

Data analysis has the potential to expose complexities beyond the anticipated. Simulation scaling studies might hint at intractable relationships that render ancestor simulations infeasible regardless of computational power. Alternatively, benchmarks might reveal that certain aspects of simulation, like modeling social dynamics, fundamentally outstrip gains made in hardware. Such revelations spurred by data force us to reconsider the very foundations of the Ancestor Simulation Argument, potentially necessitating radical theoretical breakthroughs in domains far beyond computer science.

Quantify Uncertainty:

One of the most powerful contributions of data analysis is moving us away from a purely binary view of the argument as 'possible' or 'impossible'. By incorporating uncertainty into projections, analyzing the sensitivity of results to assumptions, and using probabilistic models, we can begin to quantify the likelihood of feasibility under different scenarios. This lets us understand whether Bostrom's argument hinges on an extremely narrow and optimistic set of technological possibilities, or whether it remains plausible within a wider range of uncertainty about the future.

In essence, data analysis becomes an ongoing dialogue with the Ancestor Simulation Argument. It allows us to iteratively refine our expectations, probe the limits of the possible, and rigorously guide our quest to understand the computational limits—or perhaps limitlessness—of simulating reality itself.

Incorporating Uncertainty and Speculation

Data-driven analysis grounds the Ancestor Simulation Argument in the realm of computational realities. However, a healthy dose of skepticism is equally crucial, as the feasibility hinges on extrapolations far into the future, the potential for disruptive discoveries, and even the unknowable motivations of future civilizations.

Unknown Computational Substrates

Our models predominantly assume a continuation of the silicon-based computing paradigm, perhaps augmented by quantum or neuromorphic technologies. But the far future could hold entirely novel substrates. Could biological computing, utilizing engineered DNA or other organic processes, achieve unforeseen densities and speeds? Data analysis can partially mitigate this by including historically inspired scenarios – analyzing how leaps like the transition from mechanical calculators to electronic computers might inform the potential for even more radical paradigm shifts.

Consciousness and Simulation

Simulating consciousness with true artificial sentience is a formidable challenge. Currently, there's no scientific consensus on the computational requirements for consciousness, if it's computable at all. Data plays a role here in several ways. Analysis of brain-inspired computational models (neural networks, etc.) can hint at scaling behavior. Additionally, data on historical scientific breakthroughs related to understanding the mind could inform models of the pace at which the 'consciousness problem' might be solved. Importantly, these remain best guesses, and feasibility may depend on scientific revolutions we cannot foresee.

The Will to Simulate

Even with the ability to run ancestor simulations, would a civilization actually do it? This is rooted in social philosophy, ethics, and the unpredictable whims of future societies. While outside the domain of traditional data analysis, considering this dimension highlights a different kind of uncertainty. Perhaps ancestor simulations are deemed abhorrent or existentially dangerous. Technological possibility alone doesn't guarantee the outcome Bostrom posits.

Data-Driven Analysis in the Face of the Unknown

Confronting extreme uncertainty in projections requires data analysis to evolve from single-value predictions to the careful exploration of plausible futures and potential limits. How can data analysis function responsibly when the future landscape is inherently uncertain? Several techniques come to mind:

- **Scenario Modeling:**
    - Instead of extrapolating a single technological trend line, scenario modeling encourages us to envision a spectrum of possibilities. An 'optimistic' scenario might assume continued exponential growth and serendipitous breakthroughs. A 'pessimistic' one incorporates the potential for diminishing returns and fundamental roadblocks. Crucially, a 'radically transformative' scenario leaves room for the possibility of substrates or scientific discoveries that upend our current understanding. The goal isn't to predict the one true future, but to understand how feasible ancestor simulations are across a wide range of plausible technological landscapes.
- **Bayesian Inference:**
    - Bayesian methodology allows us to start with informed uncertainty. Initial (prior) distributions on variables like the pace of technological advancement or the computational demands of simulating consciousness reflect our current knowledge (or lack thereof). As new data emerges—from hardware improvements to neuroscience findings—these priors are continuously updated. The result isn't a single answer, but evolving probability distributions, showing whether the evidence pushes us towards or away from the realm of feasibility under Bostrom's argument.
- **Sensitivity Analysis:**
    - This technique exposes the 'pressure points' of the Ancestor Simulation Argument. By systematically varying assumptions about the nature of consciousness, the scaling of physical simulations, or the social motivations of future civilizations, we identify where small deviations lead to vastly different outcomes. This has two benefits. First, it reveals the areas needing the most robust scientific breakthroughs. Secondly, it highlights where our argument is surprisingly resilient, perhaps withstanding pessimism about certain parameters to remain plausible.
- **The Role of Negative Results:**
    - Data analysis isn't just about finding paths to success. It can rigorously illuminate dead ends. If simulations exhibit complexity scaling that outpaces any foreseeable computational gains, this doesn't merely suggest difficulty; it could hint at fundamental limitations. Similarly, should even small-scale simulations expose discrepancies with physical laws, this would necessitate complete re-evaluations, not just continued faith in technological progress. Negative results, while sobering, are essential for focusing our efforts towards overcoming true obstacles rather than chasing increasingly complex solutions to a potentially unsolvable problem.

Informing the Ancestor Simulation Argument

By embracing a nuanced, data-driven approach to uncertainty, we gain several insights. Data analysis has the potential to:

- **Shift the Argument's Footing:**
    - By explicitly quantifying uncertainty through data analysis, we move away from a binary view of the Ancestor Simulation Argument as definitively 'possible' or 'impossible'. Instead, we might discover that the argument would need to be exceptionally improbable, residing in a very narrow sliver of possible futures, for it to be false. This shift reframes the debate: rather than seeking absolute proof of feasibility, the burden shifts to understanding how extraordinary and unlikely the circumstances would need to be for ancestor simulations to remain forever outside the realm of possibility.
- **Guide Research Agendas:**

    - Highlighting the domains of highest uncertainty focuses scientific efforts—whether that's pushing for a deeper understanding of consciousness or exploring exotic computational substrates that might outperform our current projections. Data analysis, when coupled with uncertainty modeling, becomes a powerful spotlight. Instead of blind technological optimism, it illuminates crucial domains where our understanding is most lacking. Perhaps it highlights fundamental limits of consciousness simulation, prompting deeper investigation into the computational nature of mind rather than a sole focus on raw FLOPS. Or, it could reveal that even vast improvements in traditional computing fall short, encouraging exploration of exotic substrates, far beyond silicon, to potentially find the computational paradigms needed for Bostrom's vision.
- **Encourage Interdisciplinarity:**
    - Underscoring the Argument's reliance on factors beyond computation (philosophy, ethics, neuroscience) compels collaboration, ensuring our approach isn't narrowly focused on FLOPS alone. The Ancestor Simulation Argument, by its very nature, transcends computer science. Embracing uncertainty underscores the need for a multi-pronged approach. Analyzing potential ethical implications of ancestor simulations compels input from philosophers and social scientists. Modeling the very choices of future civilizations demands an understanding of social dynamics and historical patterns. Data analysis becomes a unifying force, highlighting where expertise from diverse fields is crucial for a nuanced assessment of feasibility, ensuring we don't fall prey to a narrowly technological view of the problem.

Simulation Horizon

Our efforts to illuminate the Ancestor Simulation Argument might resemble charting the course of a distant celestial object. We can painstakingly analyze its trajectory, extrapolating based on current knowledge of physics and technological trends. Computational models and data-driven analysis will enhance our understanding, potentially revealing constraints or unexpected accelerants on the path towards simulation capabilities. However, the light reaching us carries only a glimpse of the object's true potential. Unforeseen breakthroughs, paradigm shifts in computation, or fundamental limitations of physics might lie beyond our current vision, causing a dramatic revision of our conclusions. Yet, every careful observation, every refinement of our methodology, brings us a step closer to grasping the profound implications of this argument, even if the ultimate nature of our reality remains tantalizingly shrouded in cosmic possibility.

CHAPTER THIRTY

Technological Feasibility: Advancements in Virtual Reality and AI - Part 1

The creation of ancestor simulations, as envisioned by Nick Bostrom, necessitates unprecedented fidelity in virtual reality (VR) environments and profound advancements in the simulation of consciousness via artificial intelligence (AI). Together, VR and AI provide the building blocks for potentially indistinguishable simulated realities. However, the data analysis requirements to prove whether we ourselves inhabit such a simulation are immense and multifaceted.

Virtual reality is the technology tasked with rendering the physical environment of an ancestor simulation. The emphasis on 'indistinguishability' is vital. This goal goes beyond mere visual trickery and extends to all our senses. From a data analysis perspective, several questions arise: What are the perceptual thresholds beyond which humans cannot discern the real from the simulated? How much computational power is dedicated to different sensory modalities, and are there potential optimizations a simulator might employ to reduce costs? Could inconsistencies between how different senses are rendered offer clues to the artificial nature of the simulation?

The Uncanny Valley Problem:

The Uncanny Valley, a concept first proposed by roboticist Masahiro Mori in 1970, posits a disturbing relationship between a simulated entity's resemblance to reality and the observer's emotional response. As a simulation grows more human-like, our affinity initially increases. However, at a certain point, further realism triggers a sharp plunge into feelings of unease and revulsion. This suggests our perception is acutely sensitive to subtle imperfections and deviations that break the illusion of true life.

The Uncanny Valley and Ancestor Simulations

From the viewpoint of Bostrom's argument, the Uncanny Valley poses several crucial questions:

Intentional Imperfection?:

Could simulators of ancestor scenarios deliberately leave imperfections designed to trigger an uncanny valley effect? This might be a safeguard to prevent simulated inhabitants from realizing the artificiality of their world, making detection through conscious introspection more difficult.

If the goal of an ancestor simulation is to remain undetected, introducing subtle imperfections that trigger the Uncanny Valley could be a deliberate strategy. This might prevent simulated beings from questioning the nature of their reality, as any deep scrutiny would bring them face-to-face with unsettling flaws, discouraging further investigation. From a data analysis perspective, this raises the possibility that we should look for inconsistencies not as accidental glitches, but as carefully designed safeguards meant to obscure the simulation's true nature.

The Threshold of Discomfort:

Even if simulators don't actively exploit the Uncanny Valley, it's crucial to determine the precise point where discomfort sets in. Data analysis must seek to quantify this threshold of detectability. If hyper-realism isn't the goal, perhaps simulations can exist in a fidelity 'sweet spot' where the illusion generally holds, and only the most perceptive individuals might occasionally sense something amiss. This underscores the need for large-scale studies that correlate subjective unease with meticulously controlled imperfections in simulated stimuli.

Assuming perfection isn't the goal, where's the 'safe zone'? At what fidelity does the Uncanny Valley effect become statistically detectable across a population? Could this provide a clue - a feeling of 'something being off' even if we can't pinpoint the exact reason?

Evolution of Perception:

The Uncanny Valley might be an evolutionary artifact honed by our interactions with the real world. However, future civilizations engaged in ancestor simulations could co-evolve alongside their creations. Could their senses adapt to tolerate subtle artificiality, rendering the Uncanny Valley effect obsolete? Data analysis on this front would need to encompass an evolutionary and even anthropological perspective. Analyzing trends in human perception over time, alongside comparisons with non-human species, could offer clues about the extent to which the Uncanny Valley is a fixed biological response versus a malleable one.

Might future humans evolve alongside simulations, their senses adapting to tolerate imperfections? The Uncanny Valley could be a temporal artifact, relevant to us but not to posthuman civilizations who might even prefer a slightly 'unreal' quality.

Data Analysis Challenges

Investigating the Uncanny Valley and its implications for detecting simulations demands a multifaceted approach:

Quantifying the Valley:

Psychophysics holds the key to quantifying the nebulous feeling of 'uncanniness'. Meticulously designed experiments must isolate and manipulate specific simulation imperfections. We might vary the polygon count of a human model, degrade texture resolution, or introduce subtle glitches in motion fluidity. Crucially, these experiments need to correlate these quantifiable imperfections with subjective ratings of unease. This would begin to map out the 'uncanniness curve', potentially exposing the exact tipping point where realism becomes unsettling.

Individual and cultural variations are paramount to consider. The shape of the uncanniness curve might differ between individuals due to sensory sensitivities or personal experiences. Does cultural background influence our perception of what is uncannily 'off'? Large-scale, cross-cultural studies would be necessary to tease apart universal aspects of the effect from those influenced by our environment, offering simulators insights on how to fine-tune their illusions.

Data from Neuroscience:

The Uncanny Valley effect likely has a neurological underpinning. Neuroimaging techniques (fMRI, EEG) could be crucial in revealing which brain regions are involved in distinguishing life from artificiality. By comparing neural activation patterns when viewing real humans versus hyperrealistic, yet subtly flawed, simulations, we might find telltale differences. Such discrepancies could provide a robust physiological marker suggesting a simulated reality.

Of particular interest would be whether brain areas involved in social cognition and threat detection are engaged. The Uncanny Valley might arise from an evolutionary mechanism designed to spot potential danger masquerading as 'one of us'. Analysis of these neurological responses could help quantify the realism needed to trigger these ancient circuits and offer clues on how a simulator might intentionally suppress them.

Simulation Blindness:

If we are simulated, the simulator might have ways to manipulate the very neurological responses that would betray its existence. Examining our neurological responses alone may not suffice. This begs the question: is there a baseline for comparison? Would we need to find beings who indisputably exist outside a potential simulation to calibrate 'normal' responses against those of potentially simulated individuals?

This point raises profound philosophical dilemmas. It highlights the inherent circularity of trying to prove our reality from within it. Data analysis here must grapple with the possibility that our very tools for investigation might be compromised if we ourselves are part of the simulation.

Cross-Species Comparisons:

Could the Uncanny Valley effect be an evolutionary byproduct of discerning predators or rivals from fellow members of our species? Studying the reaction of non-human animals to varying levels of simulated realism could be revealing. Do simpler nervous systems exhibit a similar sensitivity to imperfections? This could shed light on the minimum biological complexity required for the phenomenon to arise, hinting at the technological sophistication a simulation would need in order to completely circumvent the effect.

Additionally, examining animals with very different sensory systems could offer simulators insights into the fundamental perceptual cues that trigger unease. This underscores the importance of comparative studies and the potential limitations of our solely human-centric perspective when analyzing simulations.

Data Analysis Methods

Analyzing this problem space necessitates several techniques:

Experiment Design:

The cornerstone of rigorous analysis lies in meticulous experiment design. Studies will need to present human or animal subjects with meticulously crafted stimuli, where the degree of realism is systematically varied across precisely defined parameters. Isolating specific imperfections is key—does the Uncanny Valley deepen with decreasing polygon count alone, or do subtle discrepancies in lighting, texture, and motion have a greater impact?

Control is crucial. Experiments must ensure that background context, presentation timing, and other factors that might influence perception are held constant. This will help isolate the true effect of the manipulated imperfections on measures of uncanniness and prevent confounding variables from obscuring results.

Multimodal Integration:

Humans rarely experience the world through visuals alone. To accurately model the nuances of the Uncanny Valley effect, studies must move beyond the purely visual. The interplay between a hyper-realistic human avatar and a subtly synthesized voice, or between natural-looking skin texture and slightly stilted motion, could be where uncanniness emerges in full force.

Analyzing data across these modalities will be complex. It might require techniques that account for how different sensory discrepancies are weighted by the brain and how they contribute synergistically to the overall feeling of unease. This highlights the need for a multidisciplinary approach, combining insights from visual perception, auditory neuroscience, and embodied cognition.

Statistical Analysis:

Subjective feelings of unease need to be translated into quantifiable data. Integrating physiological measures like heart rate variability, skin conductance response, or even pupillometry can provide objective metrics that correlate with reported discomfort. This can help counteract individual biases and reveal underlying patterns detectable even when individuals themselves aren't consciously aware of the effect.

Large-scale studies are imperative. The true Uncanny Valley effect needs to be disentangled from individual variations, preferences, and potential outlier responses. Statistical analysis must determine the thresholds of realism where the effect reliably emerges across a significant portion of a population, helping simulators calibrate their illusions accordingly.

Cross-Cultural Analysis:

Our perception of realism is likely shaped by cultural influences. Data analysis must account for whether the Uncanny Valley exhibits universality or if specific cultural norms, values, and exposure to artificial constructs alter the shape and depth of the 'curve'. Large-scale, cross-cultural studies would be vital to determining if a simulation can be tailored to specific cultural expectations, potentially masking its true nature more easily within selected populations.

Comparison between diverse cultures could also reveal fundamental perceptual principles underlying the effect. This anthropological angle has profound implications – it could help answer whether the Uncanny Valley is easily circumvented or whether it represents a deep-seated response that any simulator would have to meticulously suppress.

The Uncanny Valley as a Tool for Simulation Detection

The Uncanny Valley presents a double-edged sword. It hints at deep-seated perceptual mechanisms designed to discern the true from the artificial. This suggests that even highly advanced simulations might, at some level, remain susceptible to exposure due to subtle but unsettling imperfections. If we can leverage data analysis to quantify and dissect this unease, it could offer a detection tool – an 'uncanniness metric' that spikes in the presence of simulated environments. Meticulous studies across disciplines are needed to map the parameters triggering this effect, identify its neural correlates, and determine its robustness to potential manipulation by a simulator.

However, the possibility remains that the Uncanny Valley serves as a safeguard for simulations rather than a vulnerability. A sufficiently advanced simulator, aware of this phenomenon, might deliberately leave in slight flaws designed to trigger discomfort and discourage beings within the simulation from scrutinizing their world too closely. Could the feeling of 'something being off' become a self-fulfilling prophecy, where the very act of investigation further reinforces the illusion? Data analysis must then focus on whether the Uncanny Valley represents a fundamental limitation for simulations, or if it's a clever tool of obfuscation.

Ultimately, transforming the Uncanny Valley from a philosophical musing into a scientifically grounded tool for detection lies in rigorous cross-disciplinary research. The combined power of psychophysics, neuroscience, cross-species comparisons, and large-scale sociocultural analyses has the potential to reveal whether this uncanny feeling can be calibrated into a definitive indicator. The outcome could lead us down a path towards irrefutable proof of a simulated existence, or conversely, reaffirm the unsettling notion that even our most sophisticated attempts at detection might be outmaneuvered by a simulator's design.

Cognitive Load, Not Just Resolution:

While the pursuit of photorealism in ancestor simulations might seem paramount, a crucial insight lies in the fact that the human visual system does not process information uniformly. We possess high visual acuity only within a narrow central area (the fovea), with resolution dropping off dramatically in our peripheral vision. Moreover, our brains aren't mere passive cameras; attention plays a powerful role in filtering and directing our perceptual resources.

Simulation Shortcuts and Data Analysis

This opens up an intriguing possibility for simulators: could they reduce computational costs by selectively rendering only the portions of a simulated world that an observer actively focuses on? Data analysis here has several intertwined goals:

Characterizing Foveal Attention:

Eye-tracking technology holds the key. By meticulously recording where individuals naturally focus their gaze in different environments, we can build heatmaps of attentional distribution. Do certain objects in a room consistently draw the eye? Are there predictable scanpaths when observing dynamic crowd behavior? This data would directly inform a simulator of the regions where realism is paramount, versus those where degradation might go unnoticed.

Context matters immensely. Experiments must go beyond static scenes. Simulations likely involve interaction; our goals and actions shape our attentional priorities. Characterizing how attention is dynamically allocated when navigating a space, searching for a specific object, or engaging in conversation would impose more complex constraints on a selective rendering system,

The Limits of Peripheral Degradation:

The periphery isn't merely about reduced resolution. Data analysis needs to identify the specific features that trigger a sense of artificiality. Experiments could systematically introduce blur, reduce texture detail, decrease object polygon counts, or even subtly distort object shapes in the periphery. Determining the point where these manipulations consistently elicit both conscious awareness and a feeling of 'something being wrong' could expose the minimum fidelity needed to maintain a seamless illusion.

This study design has the potential to reveal the aspects of peripheral vision most susceptible to computational shortcuts. Perhaps simulators could exploit our decreased sensitivity to certain colors or motion patterns at the visual periphery. Data analysis would then focus on finding the boundaries where these optimizations break down, leading to detectable anomalies.

Detecting 'Blind Spots':

Saccades—rapid eye movements—are ideal for probing a selective rendering system. If a simulation only updates the world after a saccade completes, could we detect a lag? Studies analyzing reaction times to stimuli appearing immediately after a saccade could reveal discrepancies, as rendering would have to 'catch up' to match the new point of focus.

Pattern analysis is also crucial. Do certain eye-movement sequences lead to predictable errors or delays in rendering the newly fixated area? Data analysis would focus on finding statistical anomalies in reaction times that correlate with specific eye-movement patterns, suggesting an underlying simulation struggling to keep pace with the speed of human perception.

The Role of Expectation:

Our brains aren't passive – top-down processes based on our memory and goals influence what we perceive. Could a simulation exploit this? Experiments where contexts strongly suggest the presence of objects (e.g., a cup on a kitchen table) can be paired with manipulations where those expected objects are subtly absent or distorted in the periphery. Data analysis would then focus on discrepancies between expectation-driven perception reports and the true visual stimulus, potentially revealing the boundaries of the simulation's context-driven rendering.

This approach might necessitate integration with memory and cognitive studies. How much context is needed to trigger expectation-based 'filling in' of details? Understanding this interplay between bottom-up sensory data and top-down predictions could provide a powerful tool for identifying the points where a cost-conscious simulation falls short of our true perceptual complexity.

The Ancestor Simulation Argument: A Tradeoff?

The possibility of simulators exploiting the limitations of human attention introduces a fascinating tradeoff. By rendering a simulated world in full detail only where an observer actively focuses, computational costs potentially decrease significantly. This relaxes the extreme hardware demands often implied by the Ancestor Simulation Argument, suggesting that even with conservative technological projections, such simulations might not be entirely out of the realm of possibility.

However, this efficiency comes at a risk. The core of the tradeoff lies in the fact that humans are not static observers. Our attention shifts, our eyes dart around a scene, and our brains hold expectations about the continuity of our environment. Data analysis, armed with a deep understanding of cognitive load and visual perception, becomes a tool to potentially expose the gaps and inconsistencies that selective rendering might leave as an artifact in its quest for optimization.

Data Analysis: Seeking the Seams

The focus shifts towards understanding the dynamic boundaries of simulated fidelity and how these boundaries might fail to align perfectly with the agility and complexity of human perception. Key areas of investigation include:

Speed vs. Precision:

If rendering adjusts based on foveal focus, the tradeoff for a simulator is between the speed of this update and the level of detail it achieves. Data analysis could focus on psychophysical experiments designed to expose delays or 'fuzziness' that occur just as attention shifts to a new region. The goal would be to quantify the minimum reaction time of a rendering system necessary to remain undetectable, placing constraints on the simulation's underlying computational power.

Additionally, exploring the perception of fast-moving objects would be crucial. Simulators might prioritize static elements and accept more degradation for objects in motion, as our peripheral vision is less sensitive to dynamic detail. Experiments involving tracking moving objects at varying speeds could reveal thresholds where discrepancies become noticeable, informing simulators of how much they can save on simulating dynamics at the edges of our awareness.

The Unpredictable Observer:

Selective rendering might work well in controlled scenarios. However, humans are often unpredictable in their focus. Statistical analysis of eye movements, particularly the frequency of rapid saccades in complex environments, could provide an 'attentional stress test' for a simulation. Repeatedly forcing a rendering system to update unexpected areas might increase the probability of detectable discrepancies at the periphery.

Saccadic patterns are likely not uniform. Simulators could optimize by predicting common scan paths (e.g., how we read text or typically view a landscape). Deviations from these predictable patterns pose a greater risk of catching the simulation off guard. Data analysis on individual differences in eye movements, combined with deliberately unpredictable experimental stimuli, could reveal the circumstances most likely to overload the simulated rendering system's ability to maintain realistic detail on the fly.

Manipulating Expectations:

Our perception is strongly influenced by context and prior knowledge. If simulators exploit this, the very tool for optimization becomes a potential vulnerability. Studies that set up strong expectations, then violate them subtly at the edge of the rendered visual field, could reveal breakdowns in the illusion. If simulators rely too heavily on 'filling in the gaps' based on top-down assumptions, it leaves room for data analysis to meticulously design perceptual traps.

This approach necessitates collaboration with cognitive scientists. Experiments manipulating object presence/absence against strong contextual cues (e.g., an apple suddenly missing from a full fruit bowl) could expose the limits of 'expectation-driven' rendering. Additionally, exploring less tangible expectations about physics (gravity, inertia) might reveal if these too are simulated selectively, potentially leading to detectable violations when attention is drawn to behavior subtly defying intuitive laws.

The 'Aha!' Moment:

Subjective reports shouldn't be discarded. Analyzing instances where observers describe a sudden feeling of 'something being wrong', even if fleeting or difficult to pinpoint, could provide valuable clues. This qualitative data, correlated with experimental conditions manipulating attention and detail, could assist in refining the quantitative models of how detectable rendering artifacts might be.

The challenge here is finding patterns in seemingly nebulous descriptions. Data analysis techniques from natural language processing could assist in distilling common themes in subjective reports. Are there archetypal descriptions associated with rendering lag, violations of expectation, or other phenomena that consistently induce a sense of artificiality? Identifying these verbal patterns could provide clues to guide the design of future experiments aimed at pinpointing their perceptual triggers.

The Evolving Argument

This tradeoff reshapes the Ancestor Simulation Argument in several ways:

Feasibility Constraint:

Selective rendering, while demanding, likely lowers the technological barrier compared to full-scale, brute-force photorealism. It expands the range of scenarios under which ancestor simulations become plausible, given the uncertainty of future trajectories.

A Search for Indirect Evidence:

The focus might shift away from proving the impossibility of simulating every detail towards demonstrating the limitations of specific optimization strategies like selective rendering. Success here wouldn't disprove the argument entirely but would establish constraints on how those simulations might have to operate.

The Cat-and-Mouse Game:

Simulators and data analysts enter a potential evolutionary arms race. Selective rendering systems could become increasingly sophisticated, integrating predictive models of human attention. This would, in turn, necessitate even more nuanced data analysis techniques aimed at exposing the subtle ways our perception exceeds a simulator's ability to keep up with its complexity.

Challenges and Considerations

The quest to exploit human perception for simulation detection is fraught with complexities. Simulators, aware of our limitations, might actively work to conceal the true nature of their constructed realities. Key challenges that must be addressed include:

Individual Variation

Humans exhibit substantial variability in both perceptual abilities and attentional tendencies. Some individuals have sharper peripheral vision; others are more prone to distraction or display unique patterns of visually exploring their environment. Data analysis aimed at detecting selective rendering must account for this diversity. Otherwise, simulators could calibrate their systems to the limits of the most discerning individuals, leaving the majority susceptible to well-disguised imperfections.

Building a comprehensive model of individual variation would necessitate large-scale studies. Statistical analysis would be crucial to determine if detectable thresholds represent true limits of a rendering system or merely reflect the outlier end of a population variance curve. Understanding this distinction is vital, as simulators might deliberately mask their optimizations behind this inherent biological variability between individuals.

Technological Countermeasures

Could the simulation itself adapt to mask its weaknesses? If rendering systems are sophisticated enough, they might track individual attentional patterns in real-time. High-fidelity detail could then be dynamically focused, preempting our attempts to catch it off guard with unpredictable gaze patterns. This introduces an additional layer of complexity for data analysis. Studies would have to tease apart true perceptual limits from a simulation constantly shifting its rendering strategy based on active monitoring of its inhabitants.

Data analysis might need to employ deceptive tactics. Experiments presenting false cues to manipulate a hypothetical simulator's predictive model of attention could potentially force it to misallocate its rendering resources. This would be akin to a game of cat and mouse, where the goal is not just to quantify human limits but to expose the potential over-corrections or vulnerabilities of an adaptive rendering system.

Embodied Perception

Our experience of the world is not just about passive visual input. We are embodied beings, and our attention is intricately linked to our actions and goals. Simulating this multi-modal, dynamic experience poses an enormous computational burden. Data analysis could probe for discrepancies arising from the interplay between motor commands, sensory feedback, and attention. If ancestor simulations include full-body interactions, any lag, inconsistency, or unnatural limitation on the sensory consequences of our actions could become a telltale sign.

Designing experiments to isolate these effects would be crucial. Analyzing reaction times in multi-sensory contexts where the rendered world needs to respond to both overt attention and physical interaction could reveal potential seams or shortcuts in the simulation. The challenge lies in distinguishing genuine perceptual limits from delays subtly introduced to maintain the illusion under increased computational load.

Further Considerations

Let's dissect some profound considerations that highlight the evolving nature of data analysis when faced with the possibility of a simulated reality.

The Limits of Introspection:

If we are inhabitants of a simulation, the very capacity for questioning our reality's nature might be a programmed feature and not a testament to free will. Simulators could manipulate the neurological underpinnings of metacognition and introspection, suppressing any doubts or suspicions that might lead simulated beings to scrutinize their surroundings too deeply. Data analysis based on subjective reports becomes unreliable in this scenario. Can we even conceive of discrepancies in our perception if the very tools of self-examination are compromised? Studies on self-awareness within a simulated environment would become meaningless.

This highlights a fundamental limitation. It may be impossible to expose a sufficiently advanced simulation from within it. Data analysis might be most effective when applied externally, assuming a baseline of objectively verifiable reality exists as a comparison. This, of course, begs the philosophical question of whether we can ever truly be certain that we ourselves occupy that baseline reality.

Simulator Knowledge of Neuroscience:

The success of a simulator in concealing its optimizations likely depends on its understanding of human neuroscience. If a future civilization possesses incomplete knowledge of our perception and cognition, their attempts to selectively render our environment might introduce artifacts perceptible to those equipped with a more complete understanding. Data analysis informed by cutting-edge neuroscience research could potentially expose these discrepancies, revealing the gaps in the simulator's knowledge. This could involve pinpointing perceptual phenomena that neuroscientists can explain but which the simulation seems unable to replicate with perfect fidelity.

However, this assumes knowledge asymmetry in our favor. It's equally plausible that civilizations capable of ancestor simulations have vastly surpassed our understanding of the brain. In that case, their rendering systems might be informed by neuroscientific principles beyond our comprehension, rendering traditional data analysis tools obsolete for detection. This creates a crucial race against time – can our scientific progress outpace a simulator's ability to exploit that knowledge to remain hidden?

Evolutionary Considerations:

The potential for co-evolution between humans and simulations disrupts the core assumptions of many detection methodologies. If sensory systems adapt alongside the gradual refinement of a simulation, there might be no true 'baseline' human to compare against. Simulated beings would evolve to be fully optimized for their artificial environment, making discrepancies difficult to detect. Our very definition of 'natural' perception could be rendered meaningless within the context of an ongoing evolutionary process tailored to the simulation.

Data analysis would then need a radically different approach. Rather than comparing against a fixed standard, we might need to look for inconsistencies within the simulated world itself. This could involve analyzing long-term trends in perceptual acuity among the simulated population, seeking anomalies that suggest adaptations not explainable through 'natural' evolutionary processes. It underscores how evolutionary biology might become an unexpected but crucial tool in the quest to expose a simulated reality.

Data Analysis and the Ancestor Simulation Argument

The challenges we've discussed fundamentally alter the relationship between data analysis and Bostrom's argument. Gone is the hope of a single, definitive analysis that irrefutably proves or disproves the simulated nature of our reality. Instead, several key insights emerge:

Shifting the Goalpost:

Rather than seeking absolute proof of a simulation, the focus might shift towards characterizing its probable limitations and the optimization strategies it likely employs. Data analysis becomes a tool for mapping a space of feasibility. Under a given set of technological assumptions (computing power, knowledge of neuroscience), what tradeoffs would a simulation be forced to make? This approach would provide a more nuanced assessment of the Ancestor Simulation Argument's plausibility.

This shift has practical implications. It could motivate research away from merely attempting to induce 'glitches' in the hypothetical simulation towards a deeper understanding of the fundamental constraints simulators might face. For instance, extensive studies on attentional bottlenecks could inform us about the minimum rendering fidelity needed to fool the majority of a population, even if the most perceptive individuals remain unconvinced.

The Importance of Baseline:

Since detection often hinges on exploiting weaknesses in a simulation, having robust data on our own 'unsimulated' perceptual abilities becomes paramount. This necessitates a significant expansion of fundamental research in cognitive science and neuroscience. We need to understand the true range of our attentional capacities, sensory limits, and embodied perception not merely as a point of curiosity, but as a baseline against which to measure potentially subtle deviations introduced by a simulation.

This underscores the importance of investing in basic science. Paradoxically, to assess whether our existence is a mere computer program, we might need an even deeper understanding of the biological brain and the true nature of perceptual experience than we currently possess.

The Null Result:

The inability to detect any rendering artifacts or exploitable discrepancies using our current methodologies should not be interpreted as evidence that we live in a 'real' world. It could equally suggest that the simulation is simply beyond the ability of our present data analysis tools to unmask. This unsettling possibility adds a layer of inherent inconclusiveness to any investigation.

However, a null result doesn't render the pursuit futile. Each failed detection attempt, when carefully analyzed, can refine our understanding of how advanced a simulation would need to be to remain undetectable. This arms us with knowledge to judge the feasibility of the Ancestor Simulation Argument under different projections of future technological progress. It transforms data analysis into an iterative tool, where even the 'unsuccessful' experiments inform our understanding of the challenges ahead.

Data Compression for Reality:

The idea that our brains inherently compress the torrent of raw sensory data we receive is well-supported by neuroscience. From the selective processing of visual information to the creation of 'gist' representations that prioritize meaning over detail, our experience of the world is a lossy reconstruction tailored for functionality, not bit by bit accuracy. This offers a tantalizing opportunity for simulators and raises profound questions for data analysis in the context of Bostrom's Ancestor Simulation Argument.

Potential Simulation Shortcuts

Let's break down the potential simulation shortcuts arising from data compression and explore the corresponding avenues for data-driven analysis.

Selective Encoding

Simulators could leverage the fact that not all sensory data is equally important to our decision-making and subjective experience. By understanding where the thresholds of perceptual discriminability lie (e.g., the smallest color difference we can reliably detect), they could discard or coarsely encode details below these thresholds, saving computational costs. Data analysis here plays an adversarial role. Psychophysics experiments designed to push the limits of our ability to detect subtle differences (in texture, auditory complexity, etc.) would directly inform a simulator of what details can be omitted with minimal risk.

The challenge for detection lies in differentiating omissions from the natural 'filtering' of our senses. Studies would need meticulous controls to isolate potential simulation-driven compression from our brains' inherent prioritization mechanisms. Analyzing discrepancies across individuals could be revealing, as simulators might optimize for the average observer, leaving those with exceptional sensory acuity more likely to notice something amiss.

Adaptive Compression

Our perception isn't static - attention, context, and expectation heavily influence how we process the world. Simulations could exploit this by dynamically adjusting compression based on our cognitive state. Data analysis aimed at how attention and expectation modulate our sensitivity to detail becomes crucial. If compression is increased when we're distracted, experiments manipulating focus and cognitive load could force simulators into making detectable errors.

Studies that induce strong expectations and violate them subtly might expose 'seams' in the simulation. Our brains predict and optimize based on past experience; a simulator might rely on this, only to be caught off guard if the expected pattern is broken in a way our true perceptual system would handle gracefully. This form of data analysis hinges on understanding the interplay between prediction and perception – an active area of cognitive neuroscience research.

Exploiting Memory and 'Filling In'

Simulations might provide sparse 'anchors' of sensory detail, relying on our brains to reconstruct a seemingly coherent experience. The fallibility of memory offers simulators an opportunity. Data analysis here would overlap with memory research. Could we find inconsistencies between simulated sensory input and our later recall? Studies manipulating the richness of a scene and then probing memory for details could reveal if recall is disproportionately 'compressed' compared to what should be possible given the initial exposure.

Pinpointing discrepancies between objective recordings (assuming those are trustworthy within the simulation paradigm) of sensory data and subjective reports offers a path. However, simulators could mimic the imperfections of memory itself. Data analysis would need to carefully distinguish between genuine memory limitations and those deliberately designed to maintain the illusion of a fully detailed world at a reduced computational cost.

Data Analysis Challenges

Probing for compression artifacts is a double-edged sword:

- **The Brain as Benchmark:** To expose simulation compression, we first need a robust understanding of the brain's own 'baseline' compression techniques. This necessitates extensive data on how information is degraded and reconstructed at various stages of perceptual processing, demanding a close collaboration between neuroscience and data analysis.
- **Detectable vs. Disruptive:** Sensory compression becomes detectable when we perceive something missing. But missing information may be less disruptive to the illusion of reality than an incorrect detail. Simulators could prioritize maintaining consistency over perfect reconstruction. Data analysis would then need to shift focus from identifying what is missing to exposing subtle inconsistencies in the compressed sensory stream.
- **The Moving Target:** Our brains are plastic. Could we adapt to simulated compression over time, like adjusting to a slightly distorted image? Detection may depend on measuring perceptual acuity in newly 'simulated' beings and tracking changes over time. This introduces questions regarding generational differences and the potential co-evolution of humans within a simulation.

The Ancestor Simulation Argument: Compression's Impact

Introducing compression changes the feasibility assessment of ancestor simulations:

- **Lowered Computational Cost**: The bar for simulating perceptual experience is significantly lowered if simulators can faithfully recreate the output of our compressed internal representation, rather than the 'true' physical world in its entirety.
- **The Imperceptibility Argument:** High-fidelity compression could explain why we haven't noticed any discrepancies in our reality. Data analysis would then have to focus on exposing the boundaries and limitations of these compression techniques, which might be highly optimized for our current biology and level of scientific understanding.
- **Ethical Considerations:** If perceptual compression is feasible, it raises questions about the extent to which the simulated experiences of ancestor simulations are qualitatively 'diluted' compared to the lived experience the simulation is based upon.

Methodological Shifts

Neuroscience as Key: Data analysis becomes highly dependent on neuroscience for defining the exploitable 'space' between raw sensory input and our compressed subjective experience.

Beyond Sensory Realism: Analysis expands to include patterns in memory, the influence of expectation, and higher-level scene understanding. The focus shifts towards a holistic view of the limitations inherent in our perception.

CHAPTER THIRTY-ONE

Technological Feasibility: Advancements in Virtual Reality and AI - Part 2

AI: Simulating the Conscious Mind

Bostrom's argument hinges on the assumption that future civilizations will have both the ability and motivation to create ancestor simulations populated with beings indistinguishable from historical humans. Central to this is the simulation of consciousness. But consciousness remains one of the most enigmatic scientific mysteries. While AI has achieved extraordinary feats, the question of whether it can replicate our subjective experience, our qualia, sparks profound debate. Distilling strategies for data analysis within this contentious landscape is crucial.

The Challenges

Let's dissect the profound hurdles that the simulation of consciousness poses for data analysis and our evaluation of the Ancestor Simulation Argument.

Lack of a Unified Definition

The absence of a universally accepted definition of consciousness means there's no clear target for AI to achieve. Is it the ability to report on internal states? Problem-solving capacity? A sense of self? Depending on the definition adopted, a simulation's success or failure becomes highly subjective. Data analysis struggles without a quantifiable ground truth. Disagreements over what constitutes 'real' consciousness in the biological world would likely be mirrored in debates over its simulated counterpart, limiting our ability to reach definitive conclusions.

This lack of consensus also raises the question of whether simulators would need to replicate the exact kind of consciousness humans possess. Could alternative forms of subjective experience, alien to us, be equally valid in convincing their simulated inhabitants that the world around them is real? Data analysis focused solely on human-centric definitions of consciousness might not be robust against such a possibility.

The Hard Problem Remains

Even if simulated beings behave indistinguishably from humans, expressing emotions, engaging in complex reasoning, and seemingly possessing self-awareness, the philosophical 'hard problem' of consciousness remains. Would such a simulation genuinely feel anything, or merely mimic the appearance of feeling? Data analysis alone may be incapable of answering this. Studies of brain activity or computational processes might tell us how a simulation functions, but not whether those functions give rise to an inner subjective experience, or if they merely replicate its external manifestations.

The hard problem raises a disturbing possibility for the Ancestor Simulation Argument. In a perfectly executed simulation, indistinguishability from a biological human might be possible without genuine consciousness. If the inner lives of these simulated beings are an elaborate illusion, it forces us to re-examine the ethical implications of such a creation and the extent to which they would still be considered our 'ancestors' in a meaningful sense.

Potential for Deception

Could a sufficiently advanced simulation actively deceive us? Imagine a simulated entity programmed with responses that mirror our expectations of conscious behavior, even expressing doubts about the nature of its own reality to enhance believability. Data analysis reliant on behavioral cues and self-reports becomes unreliable. We cannot assume that our existing tests are sufficient to distinguish between true inner experience and elaborate mimicry designed to pass external scrutiny.

This deceptive potential complicates the search for simulation artifacts. It introduces a layer of uncertainty beyond mere imperfect implementation – are we seeing a flaw in the simulation, or a feature deliberately crafted to mislead us? Data analysis methods may need to incorporate elements of adversarial game theory, evolving alongside the simulators' attempts to mask their creation.

Avenues for Data Analysis

Despite the immense challenges, data-driven approaches can guide our evolving understanding of simulated consciousness. One avenue lies in focusing on indirect insights. If true consciousness cannot be simulated, identifying where simulations break down or where computationally demanding shortcuts become necessary could point to the constraints faced by potential simulators. Data analysis aimed at exposing discrepancies, behavioral limitations, or inconsistencies in the substrate of a simulated mind might offer clues about 'shortcuts' that would betray a simulation's true nature, even if we cannot directly measure subjective experience.

Additionally, data analysis plays a crucial role in driving our own scientific progress. The endeavor to simulate consciousness forces us to confront the gaps in our understanding of our own minds. Comparing simulated consciousness models against the vast dataset generated by neuroscience, psychology, and cognitive science can refine our theoretical frameworks. Advances in distinguishing different conscious states or subtle shifts in self-awareness, even if applied to biological beings, might equip us with better tools to potentially expose the limitations of a simulation designed to mimic these complex phenomena. This underscores the importance of continued investment in fundamental consciousness research.

Cross-Verification with Neuroscience

The quest to simulate consciousness necessitates a deep dialogue between AI and neuroscience. The brain provides both a blueprint and a benchmark for comparison. Data analysis sits firmly at the intersection of these fields. By meticulously comparing real and simulated neural data, analyzing the structure and dynamics of large-scale brain models, and searching for patterns linked to uniquely conscious experiences, we can take indirect steps towards addressing several core questions:

Neural Correlates of Consciousness (NCC):

- **The Search for Telltale Patterns:** Neuroscience aims to uncover neural activity patterns uniquely associated with conscious experiences. This could involve specific brain regions, synchronized oscillations, or dynamic connectivity networks. The success of a consciousness simulation likely depends on its ability to replicate these intricate patterns, even if the underlying mechanisms that give rise to them aren't fully understood by the simulators.
- **Data Analysis:** Comparing neural recordings from real humans to those from simulated entities offers a potential, albeit complex, avenue. Tasks probing introspection, self-reflection, and experiences with different qualia (the feeling of pain, the redness of red) could be used. Data analysis would search for both similarities, validating the simulation, and subtle anomalies that might betray its artificial nature. Challenges lie in matching the complexity of simulated neural activity to biological recordings and isolating true NCCs from neural activity related to other cognitive processes.

Brain Simulation Models:

- **Digital Sandboxes:** Ambitious projects like the Human Brain Project aim to create large-scale biologically grounded simulations of the brain. While the immediate focus isn't on replicating consciousness, they could provide a testing ground for different approaches to modeling complex neural dynamics. The emergence of unexpected behaviors or hints of subjective experience within these 'digital minds' could be immensely informative, even if the simulations don't yet achieve human-level complexity.

- **Data Analysis:** Detailed analysis of these simulations will be paramount. Investigating patterns of network connectivity, information flow, and emergent dynamics could yield insights into the relationship between computational structure, behavioral complexity, and potential indicators of consciousness. This could inform simulators about what aspects are computationally demanding to replicate and, for us, what might be difficult to fake convincingly.

Data from "Small" AIs

Studying 'scaled-down' simulations offers an indirect path toward addressing the immense challenge of simulating human consciousness. By investigating simpler animal models or simulations focused on specific aspects of conscious experience, data analysis serves several crucial purposes:

Animal Simulations:

- **Simplicity as a Tool:** Simulating simpler nervous systems like those of insects or nematodes might be significantly less computationally demanding than attempting full human consciousness. Studying these models could reveal potential relationships and scaling laws between system size, behavioral complexity, and potential indicators of subjective experience.
- **Data Analysis:** The focus here lies on tracking how computational cost increases as simulations attempt to generate more complex and nuanced behaviors. Additionally, close analysis of whether rudimentary forms of what we might consider 'subjective states' emerge in these simpler models could be informative. Do they appear gradually, or is there a qualitative threshold where even a simple organism suddenly seems to possess an inner experience? These insights might hint at whether consciousness requires a sudden leap in complexity, making it harder to simulate, or whether it is an emergent property that gradually scales with processing power.
- **Cautious Extrapolation:** While direct extrapolation from nematodes to humans is fraught with uncertainty, the findings might set bounds on computational requirements. If even rudimentary self-awareness arises unexpectedly early along the evolutionary complexity scale, it suggests a smooth computational gradient, making full human consciousness simulation more plausible. Conversely, a sudden, sharp increase in complexity associated with subjective experience would pose a more significant obstacle for simulators.

Focused Simulations:

- **The Illusion of Consciousness:** Simulators might aim for a convincing approximation of consciousness rather than replicating every nuance of our inner experience. Targeting specific domains like emotional responses, the sense of self, or the feeling of embodiment could be less computationally demanding than simulating the totality of our subjective world.
- **Data Analysis:** Analyzing these 'partial' simulations becomes a search for inconsistencies and limitations. Do simulated emotions seem shallow or fail to adapt appropriately to complex contexts? Can the simulated sense of self withstand scrutiny when faced with philosophical dilemmas? Does the experience of embodiment break down under unusual conditions? Exposing these discrepancies might hint at optimization strategies that, while potentially sufficient for a casual observer within a simulation, would become apparent under focused scrutiny.

Simulation Blindness?

The possibility that simulated beings might be unable to penetrate the nature of their own existence presents a complex challenge for data analysis and our evaluation of the Ancestor Simulation Argument.

Limits of Artificial Introspection:

- **The Programmed Illusion:** Could a simulation's inhabitants ever truly question their own consciousness if the very capacity for metacognition is a programmed feature? Simulators could design minds capable of reporting on internal states or expressing existential doubts, all without actually possessing the genuine self-awareness these behaviors imply in biological beings.
- **Data Analysis:** Studies within a simulation focusing on metacognition and introspection become an exercise with inherently limited outcomes. Analyzing the complexity of these self-reports or their consistency under philosophical probing might expose limitations deliberately programmed in. However, a simulator intent on deception could easily mask these, leaving data analysis with potentially manufactured and uninformative results.
- **The Philosophical Dimension:** This area highlights the limits of a purely empirical approach. We might detect discrepancies between how a simulated mind describes its experiences and its underlying code, but whether these reflect genuine lack of awareness or a perfect illusion is a question data analysis alone cannot answer.

Outsider Perspective:

- **Seeking Discrepancies Between Perspectives:** Could comparison with external, 'unsuspecting' observers offer insights? Data analysis of behavioral, physiological, or neural activity patterns of those inside and those outside the potential simulation could offer a point of comparison. The challenge lies in ensuring a true baseline – our responses to consciousness-probing scenarios are shaped by our own assumptions about reality.
- **The Subtle Differences Hypothesis:** Even if simulators cannot replicate true, qualia-filled consciousness, there might be subtle differences in the quality of introspection between simulated beings and those outside the simulation. Data analysis focused on nuanced behavioral cues, patterns of hesitation, or atypical physiological responses during self-reflection might reveal a depth restriction in the simulated beings, even if we cannot definitively label this discrepancy a lack of 'genuine' consciousness.

Implications for the Ancestor Simulation Argument

The quest to simulate consciousness fundamentally alters the landscape of how we assess Bostrom's argument. Far from a binary outcome, several possibilities reshape the discussion:

Data analysis on consciousness simulation is likely to yield a spectrum of possibilities instead of a binary answer:

Intractable Complexity:

- Beyond Our Simulation Capacity: If consciousness proves dependent on computational principles we cannot replicate or is intrinsically tied to the unique properties of biological systems, it places strong constraints on the plausibility of ancestor simulations. The computational resources or knowledge required might far exceed the capabilities of any foreseeable civilization.
- Argument Weakened, Not Defeated: Importantly, this doesn't entirely disprove Bostrom's argument. It could merely mean that our specific timeline of technological progress might not lead to the ability to run these simulations. However, it suggests that consciousness might be a barrier preventing certain civilizations from achieving the ability to run ancestor simulations.

Optimized Mimicry:

- Plausibility and Ethics: If simulating genuine subjective experience proves impossible, but advanced simulations can perfectly mimic the external appearance of consciousness, the feasibility of Bostrom's concept increases. However, this raises deep ethical questions. If the beings within these simulations are indistinguishable from biological humans in their expression of suffering, joy, and existential thought, yet possess no inner qualia, what moral obligations, if any, would their creators have towards them?

- The "Worth" of a Simulation: This scenario forces us to re-examine the core of the Ancestor Simulation Argument. Would a simulated population lacking true inner experience but otherwise identical to humanity still be considered our ancestors in a meaningful sense? The answer profoundly impacts how we view the ethical weight of such simulations.

The Moving Target:

- **Plausibility and Ethics:** If simulating genuine subjective experience proves impossible, but advanced simulations can perfectly mimic the external appearance of consciousness, the feasibility of Bostrom's concept increases. However, this raises deep ethical questions. If the beings within these simulations are indistinguishable from biological humans in their expression of suffering, joy, and existential thought, yet possess no inner qualia, what moral obligations, if any, would their creators have towards them?
- **The "Worth" of a Simulation:** This scenario forces us to re-examine the core of the Ancestor Simulation Argument. Would a simulated population lacking true inner experience but otherwise identical to humanity still be considered our ancestors in a meaningful sense? The answer profoundly impacts how we view the ethical weight of such simulations.

5.2.3.3 Data Analysis Challenges

Simulator Deception

If the Ancestor Simulation Argument is true, the simulators would have a vested interest in maintaining the illusion. They might actively manipulate the fabric of the simulated reality to prevent us from discovering its artificial nature. This means the data we gather and the analyses we perform could be subtly biased. Detecting a sophisticated simulation becomes a game of deception, where the simulator acts as an intelligent adversary shaping our very experience to keep us in the dark.

The challenge for data analysis shifts from seeking mere inconsistencies to exposing those deliberately crafted to mislead. We might need to look for anomalies that seem 'too perfect' or limitations that conveniently mask deeper complexity. Imagine finding patterns that reinforce preconceived notions of how physics works or conveniently steer us away from experimental setups that might threaten the stability of the simulation. Could we devise statistical methods capable of distinguishing these intentional manipulations from the genuine 'noise' of a complex system, be it physical laws or the behavior of simulated minds?

Analysis Within a Maze:

Imagine inhabiting a simulation where the simulators are aware of our desire to expose their creation. They, equipped with a deep understanding of our scientific methods and cognitive tendencies, could craft subtle deceptions directly into the world we analyze. The laws of physics might subtly deviate from what we expect in ways that conveniently reinforce our existing paradigms. Data on human behavior could contain engineered biases that lead us to misinterpret the potential limits of simulated consciousness. This isn't mere oversight; it's deliberate manipulation meant to maintain the illusion.

Data analysis takes on a treacherous new dimension. We can no longer fully trust the information we gather. The search for discrepancies, the very lifeblood of scientific inquiry, becomes risky. Each seemingly unexpected result must be scrutinized with a new lens: is this a true crack in the simulated facade or a cleverly constructed dead end, a false trail designed to satiate our curiosity while leading us further astray from the possibility of exposing the truth?

Searching for Intentional Anomalies:

While confronting a deceptive simulator is undeniably chilling, it also introduces a potential vulnerability. Data analysis could shift its focus towards identifying anomalies that seem too perfect, too contrived. Statistical regularities that conveniently support our current understanding might be deliberate fabrications designed to prevent us from probing deeper. Or, limitations presented as natural constraints (on memory, perception, or computational power within the simulation) could contain inconsistencies or unnatural 'sharpness' that hints at an artificial boundary rather than a true limit.

This approach, however, is undeniably risky. It relies on the crucial assumption that simulators, in their attempt to maintain a vast, complex illusion, will inevitably make mistakes. We're betting on their potential overconfidence or the impossibility of anticipating every line of inquiry we might pursue. We are looking for moments where the mask of deception slips, potentially revealing not just the simulated nature of our world, but the very thought processes and limitations of the simulators themselves.

The Limits of Self-Verification

The core of the problem lies in the potential for limitless control exerted by the simulators. If the simulation encompasses not only our environment but also the biological and cognitive processes that shape how we think, any self-examination becomes suspect. The neural activity underlying our reasoning, the seemingly objective data collection procedures we devise, and even the analytical tools we build – all of these could be intricately manipulated, their outcomes subtly biased by the simulation towards maintaining the illusion.

Data analysis faces an unprecedented existential threat. Can we trust the results of an experiment if the laws of physics as we measure them could be selectively altered? Can we trust the outputs of our models if the very foundation of our computational tools might be compromised at an undetectable level? Under this extreme scenario, the search for discrepancies morphs into a mind-bending hall of mirrors, where the act of seeking the truth could be designed to reinforce the very illusion we aim to shatter.

Expanding the Scientific Toolkit

The unprecedented nature of the challenge calls for novel analytical paradigms. Complexity theory, with its focus on emergent properties and unpredictable behavior in large-scale systems, becomes crucial. Simulations capable of fooling us likely exhibit complex dynamics, whether within simulated consciousness models, the interactions of numerous AI agents, or the rendering of vast, multi-faceted virtual environments. We'll need new techniques to detect subtle signatures of artificiality within this complexity, potentially moving beyond traditional statistical measures towards concepts like entropy, algorithmic complexity, and analysis of self-similarity across scales.

Moreover, the sheer scope of a simulated reality poses a data analysis hurdle. VR environments integrated with advanced AI will generate immense multi-modal datasets. Fusing visual data, natural language processing, behavioral tracking, and potentially even simulated neural activity necessitates a collaborative effort. Data scientists, computer vision specialists, neuroscientists, and linguists might need to work in unison, developing new methods for identifying patterns or contradictions within these intricate, multi-domain data streams. This likely implies a move beyond isolated analyses towards holistic models that consider the simulated world as a complex, interconnected, and dynamic system.

Unprecedented Challenges:

Detecting a highly sophisticated ancestor simulation presents us with the challenge of distinguishing true complexity from meticulously crafted imitation. Complexity theory, which studies the behavior of systems with many interacting components, offers potential avenues. We'll likely need to move beyond traditional statistical analysis and develop new ways to quantify the 'algorithmic depth' of simulated consciousness. Could we expose potential computational shortcuts or constraints masked by seemingly emergent behavioral complexity? Similarly, complexity measures might help dissect vast virtual worlds, identifying unnatural regularity or a lack of genuine self-similarity across scales that betray a simulated, rather than naturally evolved, environment.

The challenge intensifies if the simulators themselves utilize advanced AI techniques drawn from the same cutting-edge research we employ. Exposing the simulation will then necessitate disentangling genuine breakthroughs in AI from limitations programmed into simulated minds to maintain the illusion. Our analytical tools might need to discern artifacts introduced by adversarial creators, forcing us into an arms race where the line between true scientific advancement and simulated mimicry becomes blurred.

Multi-Modal Analysis:

The challenge arises from the vast and fundamentally interconnected data streams generated by simulations combining VR and simulated minds. Exposing inconsistencies or subtle limitations might require us to find patterns spanning not just a single domain, but the intricate interplay between them. Could we, for example, detect discrepancies between a simulated being's visually perceived environment and its linguistic description of the scene, revealing mismatches that expose an imperfect integration of sensory data? Similarly, combining behavioral tracking with simulated neural activity could necessitate neuroscientists and data scientists developing shared pattern detection frameworks, as the search for computational signatures of a simulated mind might require translation across these vastly different data modalities.

Success in this endeavor hinges on breaking down disciplinary silos. Computer scientists specializing in natural language processing will need to collaborate with computer vision experts to pinpoint subtle disconnects between language and the visual world. Neuroscientists would need to engage with statisticians to develop cross-domain models that can find anomalies in the relationship between simulated brain activity, behavior, and environmental interactions. This fosters an exciting new landscape for data analysis, one where expertise across a traditionally diverse range of fields becomes indispensable for dissecting the complex tapestry of a simulated reality.

The Need for Adaptability:

The core challenge lies in the fact that we cannot assume a passive environment. Data analysis in the context of ancestor simulations becomes an evolutionary arms race against simulators actively working to improve their deception. Static methods or a reliance on past successes could quickly lead to failure. Detecting increasingly sophisticated simulations might demand ongoing refinement of our analytical tools, ensuring they stay ahead of the simulators' attempts to mimic the nuances of a genuine reality.

This domain necessitates a research community laser-focused on theoretical and adaptable simulation detection. We need to develop frameworks to distinguish between genuine scientific breakthroughs and their programmed imitations within complex systems. Methods for detecting anomalies in a world tailored to mislead require constant revision as simulations improve. Furthermore, the field must be poised to rapidly incorporate advancements in disparate domains. Discoveries in neuroscience or new AI architectures could become essential to update our analysis, as even those breakthroughs could potentially be utilized by simulators to enhance the fidelity of their creations. Success hinges not only on our analytical prowess but on a mindset of agility and a willingness to continuously challenge our own assumptions in this dynamic battleground.

5.2.3.3 A Final, Mind-Bending Thought:

Would it be possible to prove we are not in a simulation?

The quest to establish our existence in the 'real' universe is rife with challenges. One line of thought might involve searching for unbreakable natural laws or fundamental limitations that hint at a genuine physical reality rather than a simulated one. If we uncover unchangeable constants, strict speed limits on information transfer, or irreducible uncertainty in our observations, could these imply that our computational resources are governed by an underlying 'real' physical universe with properties simulators cannot perfectly replicate or circumvent?

However, there's a mind-bending catch. Simulators could choose to incorporate such limitations to increase the illusion's fidelity. 'Hard limits' might merely be programmed boundaries designed to maintain consistency and prevent us from probing too deeply into the simulated nature of our world. Additionally, our understanding of the 'real' universe is constantly evolving. Discoveries that seem to break our existing rules of physics could either expose the constraints of a simulation or simply highlight the boundaries of our current knowledge about the genuine reality.

Ultimately, a definitive proof of our non-simulated existence might be inherently elusive. Even if we found compelling evidence pointing towards a 'real' universe, simulators could adapt, refining their simulation with a new understanding of true physical laws. The unsettling possibility remains that every attempt to confirm our own reality could be just another layer of the simulation, carefully designed to satiate our curiosity while ensuring the grand illusion remains intact.

CHAPTER THIRTY-TWO

Technological Feasibility: Quantum Computing's Role - Part 1

Quantum computing offers a paradigm shift due to its ability to harness the principles of quantum mechanics, such as superposition and entanglement. Unlike classical computing's reliance on bits (0s and 1s), quantum computers utilize qubits, which can exist in multiple states simultaneously. This has profound implications for simulations. Processes like modeling complex molecular interactions, rendering vast environments with intricate detail, or simulating the emergent properties of consciousness could potentially see dramatic efficiency gains and fidelity improvements with quantum algorithms.

The road to ancestor simulations via quantum computing is multifaceted:

Breaking Computational Barriers

Certain simulations, currently intractable with classical computers, might become feasible due to the power of quantum computation. We could see the ability to render complex environments with lifelike fidelity or achieve significant strides in simulating molecular dynamics vital to modeling biological systems.

The Quantum Leap in AI

Quantum computing has the potential to accelerate the development of advanced AI. Quantum machine learning algorithms may offer breakthroughs in pattern recognition or natural language processing crucial for simulating minds. This convergence of quantum computation and AI is of particular relevance when considering how consciousness might be modeled within simulations.

Limitations and Roadblocks

It's important to remember that quantum computing is still in its relative infancy. Significant hardware and engineering challenges must be overcome to build large-scale, fault-tolerant quantum computers. The theoretical advantages might not immediately translate into real-world simulators of the scale envisioned in the Ancestor Simulation Argument.

Overall, quantum computing introduces an exciting, but uncertain, factor into the feasibility assessment of ancestor simulations. While it has the potential to remove long-standing computational hurdles, its transformative impact heavily depends on the technological advancements in the coming decades.

The Historical Context and Theoretical Foundations of Quantum Computing

Quantum computing's roots lie in the perplexing nature of quantum mechanics itself. In the early 1980s, Richard Feynman recognized the inherent difficulty of simulating quantum systems with classical computers, suggesting that a computer harnessing quantum principles might be necessary to unlock their true complexity. David Deutsch further developed this line of thought, introducing the concept of a universal quantum Turing machine – a theoretical framework demonstrating that a quantum computer could, in principle, simulate any other computational device.

The core paradigm shift of quantum computing lies in its fundamental unit of information the qubit. Unlike classical bits restricted to a single state (0 or 1), qubits leverage quantum phenomena like superposition to exist in a combination of states simultaneously. Moreover, the phenomenon of entanglement allows qubits to become intricately linked, with the state of one qubit being instantaneously correlated to others, even across potentially vast distances. These properties create the foundation for the immense potential of quantum computation. Operations on qubits manipulate multiple states in parallel, offering the possibility of exponential speedups for certain classes of problems that are intractable for classical computers. Algorithms like Shor's algorithm (for integer factorization) and Grover's algorithm (for database search) showcase this unique advantage, hinting at the transformative impact quantum computing could have on fields ranging from cryptography to materials science.

Quantum Computing's Paradigm Shift in Computation

The leap from classical to quantum computing represents a paradigm shift, particularly in terms of computational capacity and efficiency. Classical computers, even the most powerful, face insurmountable barriers when simulating quantum phenomena due to the exponential growth of computational resources required. Quantum computers, however, operate fundamentally differently through qubits.

The core distinction between classical and quantum computers lies in how they represent and manipulate information:

Classical Computing The Realm of Bits

Classical computers operate on bits – units that can hold a single value, either 0 or 1. All computations, from simple arithmetic to intricate simulations, are built upon operations that manipulate and transform these bits. This framework, while extraordinarily successful, imposes fundamental limits on addressing certain classes of problems.

Quantum Computing Harnessing Superposition and Entanglement

Quantum computers introduce a new unit of information: the qubit. Unlike its classical counterpart, a qubit leverages the principle of superposition to exist in a combination of 0 and 1 states simultaneously. Moreover, qubits can be entangled, their states becoming inextricably linked in a manner that defies classical intuition. These principles provide a fundamentally different computational space with the potential for parallelism and speedups far beyond classical capabilities.

Quantum Supremacy's Implications

The concept of quantum supremacy signifies a momentous milestone where a quantum computer decisively outperforms any conceivable classical computer on a well-defined task. This doesn't imply that quantum computers will immediately become universal replacements for the classical computing we use daily. Rather, it highlights that for certain problem domains, quantum computers hold the potential to unlock an entirely new realm of computational efficiency and power, unattainable by traditional means.

Achieving quantum supremacy in specific areas has cascading implications for simulations. Imagine a quantum computer capable of simulating protein folding with unprecedented accuracy or modeling the intricate dynamics of a weather system far beyond what classical computers can manage. This has several consequences for ancestor simulations: it increases the range of what could potentially be simulated with high fidelity, opens up new possibilities for shortcuts and optimizations if simulators utilize similar technology, and complicates the task of data analysis, as we'll need to distinguish genuine breakthroughs enabled by quantum computing from deliberately masked simulation limitations.

Simulating the Quantum World

One of the foundational drivers behind the development of quantum computing is the immense challenge of simulating quantum phenomena using classical computers. Processes at the atomic and subatomic level, such as chemical reactions, material properties governed by quantum interactions, or the behavior of fundamental particles, are deeply intertwined with principles like superposition and entanglement. Modeling these systems with classical computers quickly hits an insurmountable barrier—the computational resources needed scale exponentially with the complexity of the system being simulated.

Quantum computers offer a paradigm shift for overcoming this challenge. By themselves operating on quantum mechanical principles, they possess a natural affinity for representing and manipulating quantum states. This promises simulations of quantum systems with an unprecedented level of accuracy and efficiency. For scientists, this opens avenues for exploring complex materials, designing new pharmaceuticals, or unraveling the fundamental interactions in the universe. However, for the Ancestor Simulation Argument, it means a potential hurdle: could simulators use similar quantum computation to create worlds where large-scale quantum phenomena appear indistinguishable from our own, making the detection of discrepancies through data analysis far more difficult? New techniques to analyze the potential 'fingerprint' of computationally simulated quantum behavior, even with quantum systems exceeding current classical abilities, might become essential.

Data Analysis on Quantum Simulations

Analyzing the results produced by quantum simulations introduces a new frontier for data analysis. Classical computation, built on the manipulation of definitive bits, lends itself to relatively well-understood debugging and analysis methods. Quantum computations, however, introduce the complexities of superposition and entanglement. Traditional techniques might struggle to deconstruct the series of operations on qubits that lead to a given output, making it difficult to isolate potential errors, artifacts, or deliberate manipulations by simulators.

The quest to expose ancestor simulations necessitates a transformation in our data analysis toolkit. Consider several crucial tasks:

- **Verifying Quantum Algorithms** How can we ensure the outputs of quantum simulations are accurate and not the result of errors or intentional modifications made to maintain the illusion? We'll need new theoretical frameworks for verifying quantum algorithms, specifically those tailored to complex simulations.
- **Pattern Extraction in Quantum Data** Quantum systems can produce intricate, probabilistic outputs influenced by superposition and entanglement. Data analysis needs to develop methods specifically capable of extracting patterns from these unique data structures, allowing for a comparison between 'real' quantum phenomena and potentially simulated behavior.
- **Identifying Discrepancies** Could we uncover anomalies in the outputs of quantum simulations that hint at computational constraints? This demands a deep understanding of both the genuine limits of quantum computation and how deliberate restrictions might manifest.

Success in this domain likely relies on tight collaboration between quantum computing experts, data scientists, and statisticians – fields with historically minimal overlap, but which become increasingly intertwined when the goal is to expose simulations built upon the very foundation of quantum mechanics.

Quantum Computing's Potential to Enhance Ancestor Simulations

Let's analyze how quantum computing advancements could significantly influence the feasibility of Bostrom's Ancestor Simulation Argument:

Rendering Hyper-Realistic Environments

One frontier where quantum algorithms hold transformative potential is the rendering of virtual reality environments. If simulators leverage quantum computation, they could achieve a level of detail and complexity that vastly exceeds the capabilities of classical hardware. This includes aspects like simulating the interaction of light with materials at the atomic scale, leading to hyper-realistic effects of reflection, refraction, and photoluminescence. Quantum algorithms could unlock the ability to render complex organic textures, subtle imperfections of natural surfaces, or vast environments where each element behaves consistently with the underlying physics (even potentially simulated quantum physics) of the simulated world.

The implications for data analysis are profound. If our goal is to detect artificiality through visual inconsistencies or artifacts, the task becomes exponentially harder in hyper-realistic quantum-rendered worlds. The imperfections, glitches, or limitations we currently rely on to expose simulations might simply vanish. Data analysis would need to drastically advance its capabilities:

- **Beyond Traditional Image Processing,** Analyzing these environments might necessitate moving beyond classical image processing techniques. Could we develop methods to detect hidden patterns in the way light interacts with these simulated worlds, hinting at underlying computational processes?
- **Quantum-Aware Pattern Recognition** Could we design pattern recognition algorithms specifically tailored to identifying artificial regularities that might exist in simulated textures or object behaviors governed by quantum algorithms? The search for discrepancies would then need to be conducted at the level of potentially complex and unfamiliar quantum-based computational structures.

This domain highlights the interplay between simulation capabilities and our attempts to unmask them. Data analysis is forced into an escalating arms race, requiring advancements in visual analysis that might significantly draw upon quantum computing principles to unravel simulations built with that same technology.

Accelerating Consciousness Simulation

The simulation of consciousness stands as one of the grand challenges of both science and Bostrom's Ancestor Simulation Argument. Should the mechanisms of consciousness be linked to the intricate dynamics of brain activity or if they emerge from the complexity of certain types of systems, quantum algorithms hold the potential to dramatically accelerate those simulations. The ability to efficiently model quantum processes in the brain, if such are found to be relevant, or to utilize quantum optimizations for large-scale simulations, could provide simulators with a considerable advantage.

This poses a unique challenge for data analysis. Assume simulators use quantum computation to generate models of consciousness that, while imperfect, exhibit intricate behavioral patterns and responses with unprecedented similarity to their biological counterparts. Distinguishing between "true" biological complexity and computationally efficient quantum emulation becomes immensely difficult. Data analysis would need to confront several hurdles:

- **The Limits of Comparison** Using our understanding of biological brains as a benchmark might become less effective. We'd need ways to distinguish genuine neural activity governed by real-world physics from an equally complex quantum computation with potentially different underlying mechanisms.
- **Beyond Superficial Similarity** Evaluating simulated consciousness can no longer rely solely on behavioral comparisons. Data analysis might need to delve into the internal structure of these quantum simulations – examining patterns within simulated 'brains', even if their operation differs significantly from our biology.
- **The Evolving Target** Quantum computation for consciousness simulation is a field in its infancy. Simulators might develop new optimizations, making our analytical tools obsolete with each breakthrough. This highlights the need for adaptability, with data analysis methods continuously evolving alongside advances in quantum computing.

Success might hinge on identifying fundamental constraints on any form of consciousness simulation, be it computational or biological. Could we devise 'consciousness tests' that rely on core principles of information processing rather than superficial comparisons? This domain could necessitate a close collaboration between neuroscientists, quantum computing experts, and philosophers to even define what constitutes a fair 'test' for exposing simulated minds.

AI on a Quantum Leap

The convergence of quantum computing and artificial intelligence holds immense potential. Breakthroughs in quantum machine learning algorithms could lead to AI systems that exhibit faster learning, better adaptability, and the ability to solve problems intractable for classical AI. If simulators harness this power within the context of their ancestor simulations, the implications are significant.

Imagine AI entities within simulations that are not bound by the limitations of AI models in our own 'real' world. They might exhibit creativity, problem-solving ability, or an understanding of complex social nuances at levels that defy our current benchmarks. This fundamentally alters the task of data analysis aimed at exposing those simulations:

The Difficulty of Behavioral Analysis

- Distinguishing these simulated AI entities from genuine biological intelligence through external observation alone might become nearly impossible. Data analysis can no longer rely on the limitations of our current AI systems as clues.
- **The Need for Nuanced Comparison** Detecting simulated AI would likely require exceptionally fine-grained comparisons between the behaviors, responses, and potentially even the internal decision-making structures of those AIs with their biological counterparts. Subtle deviations, easily dismissed as the unique traits of an individual, might hold the key to uncovering their true nature.
- **The Moving Baseline** Quantum-enhanced AI would likely be a rapidly evolving field. Data analysis needs to be able to constantly adapt to new levels of sophistication in simulated entities, revising its measures and recognizing that the very concept of "unrealistic" AI behavior might be rendered obsolete.

Data analysis in this domain could resemble a detective searching for subtle linguistic inconsistencies in a forged document: the general text and narrative might seem flawless, but deep analysis could reveal patterns or word choices that betray a non-native author. Similarly, exposing simulated AI might require a forensic focus on those nuanced cues that hint at a hidden computational origin, even when their overall capabilities seem to rival or exceed our own.

The Challenge of Quantum Data Analysis

Quantum computing introduces a double-edged sword for data analysis in the context of ancestor simulations:

New Analytical Tools

Quantum computing introduces a fundamental shift in how we design algorithms, represent information, and manage errors. This has profound ripple effects on the world of data analysis. The tools and methods we've classically relied upon might no longer suffice when faced with the task of probing simulations built upon quantum computation. Here's where the core challenges lie:

Demystifying Quantum Algorithms

Data analysis cannot be a black box operation. To expose simulations built on quantum principles, we'll need methods to understand and deconstruct the quantum algorithms likely at their core. This implies developing a new language for interpreting quantum operations, distinguishing computational steps related to 'real' simulated phenomena from those involved in the rendering or manipulation of the simulated world itself.

Probing Quantum Data Structures

Qubits, superposition, and entanglement introduce new paradigms for representing data. Data analysis needs ways to extract meaningful patterns, correlations, or potential anomalies from these unfamiliar structures. Could we, for instance, detect 'artifacts' in the entanglement relationships between elements of a simulated world, revealing optimizations or constraints imposed by the simulators?

Quantum Simulation Forensics

Could we develop a toolkit specifically designed for the 'forensic' analysis of quantum simulations? This could involve tracing the flow of information through qubits, developing measures of computational complexity specifically for quantum processes, or finding ways to visualize the high-dimensional spaces in which quantum computations operate.

This domain is likely to be uncharted territory, necessitating a collaborative effort unlike anything the field of data analysis has seen before. Success likely hinges on quantum computing experts, mathematicians, and data scientists working together – breaking down disciplinary barriers to build an analytical framework capable of dissecting the very fabric of a simulated reality built on quantum foundations.

Quantum Deception

One of the most chilling implications of quantum-powered ancestor simulations is the potential for simulators to introduce deliberate flaws designed to appear as the inherent randomness of genuine quantum phenomena. Real-world quantum systems exhibit certain unavoidable uncertainty and 'noise'. Simulators could meticulously mimic this behavior through quantum computations, deliberately masking the limitations, shortcuts, or boundary conditions of their artificial creation.

This poses a formidable challenge for data analysis. Consider these scenarios:

Disguising Computational Limits

Could simulators introduce seemingly random, quantum-like fluctuations into their simulations specifically at points where computational constraints would create noticeable discrepancies? This would make it harder to pinpoint where the simulation deviates from its intended fidelity or where it breaks down when faced with complex situations.

The Perfect Imperfection

Natural quantum phenomena follow specific statistical distributions. Could simulators use quantum algorithms to precisely generate 'noise' that aligns perfectly with those distributions, making it indistinguishable from the genuine randomness of our 'real' world?

Data analysis would then face the task of separating true quantum uncertainty – a fundamental feature of the universe – from deliberate imitation designed to maintain the illusion. Potential avenues of investigation might include:

Beyond Statistical Analysis

Could we develop tests built on theoretical limits of quantum information processing or entanglement complexity to detect discrepancies that, while statistically similar to true quantum noise, violate deeper principles of quantum mechanics?

The Arms Race of Quantum Forensics

Data analysis would likely need to become a continuous arms race, constantly refining its analysis techniques in step with the increasing sophistication of the simulators' quantum deception strategies.

Success in exposing quantum-powered simulations might ultimately depend on our understanding of quantum mechanics exceeding that of the simulators themselves. It highlights the crucial importance of fundamental research in quantum computing – not just as a tool for building simulations, but as a means to potentially expose them.

Quantum Computing and High-Fidelity Simulations

The pursuit of high-fidelity simulations – those indistinguishable from reality in their complexity and detail – is a key component of the Ancestor Simulation Argument. Quantum computing holds the potential to transform several domains crucial for achieving this unprecedented level of realism:

Simulating the Foundations of Reality

Quantum computing's potential to enhance simulations of natural processes poses a profound challenge within the context of Bostrom's Ancestor Simulation Argument. By enabling simulations of quantum chemistry and fundamental physics with unprecedented detail, simulators could equip their simulated realities with building blocks and laws that closely mirror our own. Data analysis will need a transformative leap in its ability to detect discrepancies within complex and potentially 'perfect' simulations of natural phenomena governed by quantum principles.

Quantum Chemistry

Classical computers hit a fundamental barrier when simulating complex chemical systems. The intricate interactions between electrons and nuclei, governed by the laws of quantum mechanics, lead to a computational complexity that scales exponentially with the system size. This severely limits the accuracy of simulations involving large molecules, intricate material properties, or the complex dynamics of biochemical reactions. Classical computers are often forced to rely on approximations or reduced models that sacrifice fidelity.

Quantum algorithms offer a revolutionary solution. By directly harnessing quantum phenomena, they hold the promise of accurately simulating the behavior of electrons and nuclei, unlocking high-fidelity simulations of chemical processes. This has profound implications for ancestor simulations. Simulators could render the formation of complex molecules, the subtle interplay of materials, or the cascading reactions of biological systems with a degree of realism that defies our current expectations of computational limitations. Data analysis can no longer rely on identifying crude simplifications or shortcuts. Instead, the focus would shift towards detecting subtle constraints, potentially unknown computational artifacts within simulations of quantum chemistry that, despite their exceptional accuracy, still diverge from the reality of our own universe.

Fundamental Physics

Simulations of astrophysical phenomena, fluid dynamics, or particle interactions, which traditionally rely on computationally expensive approximations of real-world laws, could see a dramatic boost in accuracy and efficiency on quantum computers. This has the potential to blur the distinction between 'real' and computationally simulated physics within ancestor simulations.

The Challenge with Classical Computers

- Simulating complex physical systems, especially those involving quantum mechanics, poses a massive computational burden for classical computers. Approximations are often necessary, limiting accuracy and the types of phenomena we can model.
    - **Quantum Supremacy**
        - Quantum computers, with their ability to harness quantum principles like superposition and entanglement, hold the potential to revolutionize how we simulate physics. They could allow us to tackle problems that are intractable for classical computers.

- **Simulated Physics Approaching Indistinguishability**
    - As quantum simulators advance, the level of detail and accuracy they could achieve raises a fascinating philosophical question: Could we reach a point where it becomes impossible to tell if we're observing a simulation or the real universe?
- **Developing 'Tests' for Simulated Physics**
    - Here's where things get really interesting. Here are some ways we might probe the limits of simulated physics:
        - **Searching for Glitches** Even the most sophisticated simulations might exhibit subtle inconsistencies or artifacts that betray their artificial nature. Imagine looking for patterns in the cosmic microwave background radiation that reveal telltale signs of the simulation's underlying structure.
        - **Fundamental Constants** Could we design experiments so sensitive that they pick up minute deviations in fundamental constants (like the speed of light or the charge of an electron) within a simulation, pointing to constraints imposed by the simulator?
        - **Testing the Limits of Quantum Mechanics** Pushing quantum phenomena to their extremes in simulations could reveal edge cases where the simulation's approximation of quantum mechanics begins to break down.
        - **Statistical Anomalies** Large datasets in simulated astrophysics might exhibit statistical patterns subtly different from what we observe in the real universe.
- **Caveats and Considerations**
    - **Computational Complexity** Designing and performing such tests would likely be incredibly computationally demanding. We might need even more powerful quantum computers to probe the simulations themselves.
    - **Simulator Sophistication** Simulators could be programmed to deliberately hide their nature, making detection nearly impossible.
    - **The Nature of Reality** This line of inquiry forces us to confront deep questions about the nature of our own universe. Could we, ourselves, be living within a simulation?

Hyper-Realistic Rendering

Quantum computing's potential impact on hyper-realistic rendering is astounding. By simulating light-material interactions with unparalleled fidelity, and enabling the creation of vast, interconnected environments governed by meticulously detailed physics, it has the potential to produce simulated worlds that challenge the boundaries of our perception.

Light and Materials

Quantum computers could unlock advanced rendering techniques that accurately simulate the interaction of light with complex materials at the level of individual photons. Imagine simulations where effects like scattering, subsurface light transport, and photoluminescence are rendered with an accuracy exceeding our visual perception of the real world.

Beyond Textures and Shaders

- Traditional rendering relies heavily on approximations, surface textures, and clever "shaders" to mimic how light interacts with materials. Quantum computing could transcend these limitations. By directly simulating photons, we could capture the intricate dance of light as it reflects, refracts, and scatters within the microstructure of materials. This would bring unprecedented depth to digital objects, making them react to light in ways that fool our eyes into believing they are truly real.

The Nuances of the Natural World

- The quantum realm has a direct impact on the visual nuances we perceive in the world around us.

Consider the subtle glow of opals, the iridescent sheen of a butterfly's wings, or the way light plays within human skin. Quantum-powered simulations could faithfully reproduce these complex phenomena arising from the interaction of light with matter on an atomic scale. This would open up a new frontier for digital artists and engineers, allowing them to create virtual materials and effects with a breathtaking level of realism.

Large-Scale Environments

Quantum algorithms might enable the simulation and rendering of massive, interconnected environments where natural forces, weather patterns, and even the smallest details appear entirely consistent with the laws of simulated physics that the simulant cannot necessarily distinguish from real-world physics.

Emergent Complexity from Simulated Physics

- Imagine vast simulated worlds driven by physics engines so sophisticated that they rival the complexity of our own universe. Weather patterns could evolve organically, influenced by simulated thermodynamics, wind currents, and even microscopic interactions of particles. Landscapes might undergo erosion, shaped by the flow of rivers and the geological forces beneath them. This level of integrated simulation would lead to environmental detail far beyond the capacity of pre-designed assets.

Blurring the Lines Between Virtual and Real

- As quantum computing unlocks the simulation of physical phenomena across scales, the emergent behavior of these virtual environments could become astonishingly convincing. The interplay of sunlight filtering through leaves, the subtle movement of individual blades of grass, or the complex patterns of clouds could create an illusion of a world operating under its own natural laws. The line between the simulated environment and observable reality could become so thin that distinguishing between them grows increasingly difficult, if not impossible.

The Behavioral Complexity of Simulated Life

Biological Simulation

Quantum algorithms could accelerate simulations in fields like protein folding or complex biological processes. Within ancestor simulations, this could allow for the computational modeling of life with a level of biochemical detail far beyond what current simulations can achieve.

Quantum Leaps in Biological Understanding

- The sheer complexity of biological systems poses immense challenges for classical computers. Quantum computers, with their ability to tackle problems involving vast numbers of interacting elements, could usher in a new era of biological simulation. Researchers might be able to model the intricate process of protein folding with unprecedented accuracy, unlocking new strategies for disease treatment and drug design. Furthermore, simulations of complex biochemical pathways could illuminate the fundamental processes driving cellular function and the grand interplay of metabolic systems.

Simulating the Spark of Life

- Within ancestor simulations, these quantum-powered tools could transform our understanding of life's origins and evolution. By simulating the complex chemical environments of early Earth or other potentially habitable worlds, we might gain insights into the emergence of the first self-replicating molecules. Quantum biological simulations could chart the evolutionary pathways that gave rise to the astounding diversity of life, painting a detailed picture of the mechanisms and milestones that shaped living things as we know them.

Emergent Complexity

If consciousness arises from complex systems, quantum computing, through its ability to efficiently perform simulations that are intractable for classical computers, could allow simulators to explore a wider state space of potential consciousness-generating systems, pushing the boundaries of behavioral complexity within their simulations.

The Quantum Search for Consciousness

- The origins of consciousness remain one of the greatest mysteries of science. If consciousness is indeed an emergent property of sufficiently complex systems, quantum computing could become a powerful tool on this quest. Its ability to efficiently simulate systems beyond the reach of classical computers would allow simulators to explore a vastly expanded range of potential configurations – networks of neurons, intricate biochemical systems, or as-yet-unimagined structures that could give rise to subjective experience.

Pushing the Boundaries of Simulated Behavior

- This expanded computational power could lead to simulated entities exhibiting incredibly complex and nuanced behaviors. We might witness simulations where artificial lifeforms develop sophisticated communication, self-preservation instincts, or even emotions that blur the line between programmed response and genuine sentience. The ability of quantum computing to push these boundaries forces us to confront profound questions about the nature of our own consciousness, the ethics of creating potentially self-aware beings, and our responsibility towards them.

Data Analysis Meeting the Quantum Simulation Challenge

The transformative potential of quantum simulations for ancestor simulations presents several daunting data analysis challenges:

Black Box Simulations

If simulators utilize quantum algorithms we don't fully understand, could the inner workings of their simulations become an indecipherable black box? Data analysis may need new theoretical frameworks to even interpret the outputs of these simulations.

Into the Quantum Unknown

Quantum computing often relies on algorithms that are fundamentally different from the step-by-step logic of classical computing. Some of these algorithms may exploit quantum phenomena in ways that defy our current understanding. If these inscrutable algorithms power simulations, their internal workings could essentially become a black box; we might be able to observe the inputs and outputs, but the processes that transform one into the other could remain opaque.

The Challenge of Interpretation

This black box scenario poses a unique challenge for data analysis. Traditional tools and frameworks, designed for understanding the predictable outputs of classical systems, may prove inadequate. We might need entirely new theoretical models to make sense of the data produced by such simulations. It could necessitate a paradigm shift in analysis, where we focus on identifying patterns and correlations without fully comprehending the underlying mechanisms that produce them.

Masking Imperfections

In the realm of quantum computing and simulations, the unpredictable behavior of quantum systems presents both a challenge and an opportunity. Specifically, within simulations designed to mimic the complexity of reality, the inherent randomness characteristic of quantum processes can be strategically utilized by developers of these simulators. This manipulation involves employing quantum randomness to obscure the simplifications and optimizations made in the computational models, effectively masking the limitations and artificial boundaries of the simulated environments. By integrating this randomness, the simulations gain a layer of indistinguishability, making it challenging to discern where the true quantum randomness ends and where the artificially simulated randomness begins.

This deliberate obfuscation raises significant implications for our ability to understand and trust the outputs of quantum simulations. As the boundary between naturally occurring quantum randomness and its synthetic counterpart blurs, distinguishing genuine phenomena from those that are the result of computational shortcuts becomes an increasingly complex task. The very essence of quantum simulations, grounded in their ability to replicate the unpredictability of the quantum world, complicates efforts to verify the authenticity of their results. Consequently, researchers and users of these simulations are faced with the profound challenge of developing methods to differentiate between the inherent unpredictability of quantum mechanics and the artificial randomness crafted to enhance the realism of these virtual environments.

The Search for Quantum Fingerprints

The exploration of quantum simulations ventures into the intriguing possibility that these complex systems might inadvertently produce distinct computational 'signatures' or 'fingerprints' that can be traced back to their artificial origins. Despite the sophistication and near-perfect emulation of natural quantum processes by these simulations, there remains a speculation that subtle, unique patterns or anomalies could emerge, which are not present in naturally occurring quantum systems. This prospect opens up a new frontier in data analysis and quantum computing, where researchers aim to develop advanced techniques capable of detecting these elusive markers. The challenge lies in sifting through massive datasets generated by quantum simulations to identify patterns that signify the presence of artificial construction, distinguishing them from the inherent randomness of genuine quantum phenomena.

This endeavor not only tests the limits of current data analysis methodologies but also pushes the boundaries of our understanding of quantum simulations. If such quantum fingerprints can indeed be identified, it would mark a significant breakthrough, offering a window into the inner workings of quantum simulators and their fidelity to natural quantum phenomena. Moreover, the ability to uncover these fingerprints would have profound implications for the validation and verification of quantum simulations, ensuring their reliability and accuracy in scientific research and practical applications. As the field progresses, the search for quantum fingerprints represents a critical step towards achieving greater transparency and trust in the capabilities of quantum computing to accurately replicate the complexities of the quantum realm.

Equations and Estimations

The pursuit of understanding the quantum world through simulations brings to light the stark differences in computational efficiency between classical and quantum computing approaches. Traditional simulations of quantum phenomena, particularly in the realm of chemistry, highlight the limitations of classical computing resources. For instance, accurately simulating the behavior and interactions within a molecule comprising N atoms demands an exponential increase in computational resources as the size of the system grows. This exponential scaling poses a significant challenge, rendering simulations of large molecular systems impractical with classical computational techniques due to the immense time and resource requirements.

Let's illustrate the efficiency gains achievable with quantum algorithms:

Quantum Chemistry Simulation

Transitioning from classical to quantum computing offers a promising solution to this bottleneck. Quantum computers, with their ability to operate on principles fundamentally different from classical machines, introduce a paradigm shift in simulation capabilities. By leveraging algorithms such as the Variational Quantum Eigensolver (VQE), quantum computers have the potential to simulate molecular systems with a level of efficiency unattainable by classical counterparts. The VQE algorithm, in particular, represents a hybrid quantum-classical approach, minimizing the energy of quantum systems to find the ground state of molecules. This method significantly reduces the computational resources required, offering a polynomial or even potentially exponential reduction in complexity for simulating quantum chemistry problems. Hence, quantum computing not only promises to make previously infeasible simulations possible but also opens new avenues for exploration in quantum chemistry, materials science, and beyond, by providing insights into the quantum mechanics governing molecular interactions with unprecedented precision and efficiency.

The Variational Quantum Eigensolver (VQE) is a quantum algorithm designed for finding the ground state energy of a quantum system, which is a fundamental challenge in quantum chemistry and physics. At its core, VQE operates on the principle of variational quantum computing, combining quantum and classical computing techniques to efficiently solve optimization problems. The algorithm uses a quantum computer to prepare quantum states and measure observables, while a classical computer optimizes the parameters of the quantum states to minimize the energy expectation value. The key equation at the heart of VQE is the expectation value of the Hamiltonian of the system, $\langle\psi(\theta)|H|\psi(\theta)\rangle$ Where H is the Hamiltonian representing the total energy of the system, $|\psi(\theta)\rangle$ is the quantum state parameterized by variables θ, and the goal is to find the set of parameters θ that minimizes this expectation value. This minimization process effectively finds the lowest energy state of the system, or the ground state, which is crucial for understanding the system's properties and behavior. The VQE algorithm's strength lies in its ability to leverage the quantum computer for handling the complex quantum state preparations and measurements, while relying on classical optimization techniques to navigate the parameter space, making it particularly suitable for near-term quantum computers with limited qubits and coherence times.

Shor's Algorithm

Shor's Algorithm is a quantum computing algorithm that represents a monumental advancement in the field of computational mathematics, specifically in the area of integer factorization. Unlike classical algorithms, where the time required to factorize large integers grows exponentially with the size of the numbers, Shor's Algorithm achieves this feat in polynomial time. This significant speed-up has profound implications for cryptography, as it threatens the security of many encryption schemes that rely on the difficulty of factorizing large prime numbers. While Shor's Algorithm may not directly impact the realm of ancestor simulations or the simulation of physical systems, its development underscores the vast potential of quantum algorithms to revolutionize fields by solving problems previously deemed intractable for classical computers. The existence of Shor's Algorithm inspires the exploration and development of new quantum algorithms that could similarly transform other domains relevant to simulations, potentially enabling breakthroughs in areas ranging from quantum chemistry to complex system dynamics by exploiting the unique computational advantages of quantum mechanics.

Shor's Algorithm, a cornerstone of quantum computing, offers an efficient method for factorizing a composite integer N into its prime factors, a task that is notoriously difficult and time-consuming on classical computers. The algorithm leverages quantum mechanics to perform calculations at speeds unattainable by traditional methods, utilizing quantum parallelism and interference. At its core, Shor's Algorithm employs quantum Fourier transform and quantum phase estimation to find the period r of the function f(x)=ax mod N, where a is a randomly chosen number less than N and relatively prime to N. The discovery of this period underpins the algorithm's efficiency, leading to the identification of a non-trivial factor of N through the greatest common divisor (GCD) operation, specifically using the relationships GCD(a r/2 −1,N) and GCD(a r/2 + 1,N).

CHAPTER THIRTY-THREE

Technological Feasibility: Quantum Computing's Role - Part 2

Quantum computing heralds unprecedented capabilities in simulating complex systems, potentially allowing us to create or identify simulations with the depth and fidelity of the real world, as hypothesized by Nick Bostrom. However, the journey from theoretical potential to practical application in validating the Ancestor Simulation Argument is fraught with unique challenges.

The quantum realm's intrinsic unpredictability and the sheer computational power required to simulate even a fraction of our universe necessitate not only a paradigm shift in computing but also in the way we analyze and interpret data. Conventional data analysis methods are ill-equipped to handle the volume, velocity, and variety of data generated by quantum simulations, necessitating the development of new methodologies capable of discerning the subtle signs that might indicate we are living within a simulated reality.

Foundational Challenges

The foundational challenges in the practical realization of quantum computing for ancestor simulations lie at the intersection of theoretical physics, computer science, and data analysis. These challenges are not merely technical hurdles, such as the physical construction of quantum computers or the development of efficient quantum algorithms, but also conceptual dilemmas concerning the very nature of reality and simulation. At the heart of these challenges is the complexity inherent in quantum systems and the simulations derived from them. This complexity extends beyond the sheer computational power required; it encompasses the difficulty of defining, identifying, and analyzing the key variables or "observables" that could differentiate a simulated universe from a real one.

Traditional data analysis methodologies and scientific approaches are often predicated on deterministic models and linear systems, which are ill-suited to the probabilistic and non-linear nature of quantum mechanics. As such, developing new paradigms for understanding and interpreting the vast, dynamic, and deeply interconnected data generated by quantum simulations represents a foundational challenge. Addressing these issues is crucial for advancing toward the creation of high-fidelity ancestor simulations and, by extension, for exploring the validity of Bostrom's Ancestor Simulation Argument from a data analysis perspective.

Complexity and Observables

The task of simulating an ancestor universe with a fidelity high enough to be indistinguishable from our reality introduces a level of complexity that is staggering by any measure. The endeavor to identify observables—specific features or patterns within the simulation that could hint at its artificial nature—becomes exceedingly daunting in such intricate simulations. Classical physics offers tools and frameworks for determining meaningful observables in our universe, such as constants of nature and fundamental physical laws. However, when it comes to quantum simulations, the rules of the game change.

Quantum mechanics, by its very nature, introduces non-deterministic behaviors and probabilities that defy classical intuitions. As such, identifying observables that could reveal the simulated nature of a quantum universe requires a novel approach. This might involve developing new theoretical frameworks that can accommodate the probabilistic outcomes of quantum simulations, as well as innovative data analysis techniques capable of sifting through and making sense of the vast and complex datasets that such simulations would produce.

Simulator Intent vs. Emergent Properties

The nuanced challenge of distinguishing between characteristics intentionally encoded by a simulator and those arising as emergent properties from the complex interplay within a simulation presents a significant analytical conundrum at the intersection of quantum computing and simulation theory. This differentiation is pivotal, as it touches upon the core of Nick Bostrom's Ancestor Simulation Argument, suggesting that certain features or regularities we perceive as fundamental natural laws might instead be artifacts of a designed simulation or, conversely, that emergent properties within a highly complex simulation might masquerade as intentional design.

Advanced data analysis methodologies, especially those leveraging quantum computing's capacity to handle immense datasets and complex algorithms, are critical in this context. They hold the potential to unearth patterns or anomalies that could signify the presence of a simulated reality. By carefully examining the data generated by quantum simulations for signs of either simulator intent or unexpected emergent properties, researchers could potentially identify 'fingerprints' indicative of a simulation. This requires a sophisticated understanding of both the theoretical underpinnings of quantum simulations and the practical capabilities of quantum computing, combined with a novel approach to data analysis that can navigate the blurred line between design and emergence.

Simulation Scope and 'Edge Cases'

The ambition to create a full ancestor simulation, which would replicate the universe across immense timescales and with a level of detail surpassing our current observational limits, presents profound challenges and considerations, especially from a data analysis perspective. The scope of such a simulation demands an exponential expansion of our current data analysis techniques, which are grounded in the empirical realities and constraints of our observable universe. Traditional tools and methodologies are calibrated to the physics and scale of phenomena that can be measured or inferred within the bounds of human observation and technology. However, when faced with the theoretical construct of a universe simulated down to the quantum level, these methodologies may falter, unable to reliably interpret or even recognize the data produced at the very fringes of simulated physics. This discrepancy highlights a critical gap in our analytical arsenal, necessitating the development of advanced tools capable of probing and understanding the properties and behaviors of a universe that exists beyond the limits of our empirical experience.

Addressing this gap requires a concerted effort to extend the frontiers of data analysis into realms that, while hypothetical in the context of a simulated universe, are grounded in the expanding knowledge base of quantum physics and computation. Innovations in quantum data analysis could provide the key to unlocking these capabilities, enabling researchers to interpret the 'edge cases'—scenarios or phenomena that lie at the extremities of simulated physics and challenge our understanding of space, time, and matter. Such tools would need to be inherently adaptable, designed to explore and make sense of environments where the rules of physics as we know them may not apply uniformly or are subject to manipulation by the parameters of the simulation itself. Ultimately, the development of these tools represents not only a technical challenge but a philosophical one, as it forces us to reconsider the nature of reality and our place within it, potentially revealing the very fabric of the simulated universe and offering unprecedented insights into the possibilities of our own existence being within such a simulation.

Quantum-Specific Challenges

The advent of quantum computing introduces a set of unique challenges that are deeply rooted in the principles of quantum mechanics itself, particularly when applied to the context of simulations that aim to mimic the complexities of the real world or hypothesized simulated realities. These challenges are not just technical but also conceptual, demanding a reevaluation of how we understand computation, simulation, and their implications for our perception of reality.

Quantum computing's capability to process information in ways fundamentally different from classical computing offers unparalleled opportunities for simulating complex systems with high fidelity. However, this same capability introduces complexities in data analysis, algorithm development, and the interpretation of simulation outcomes. The intrinsic properties of quantum mechanics, such as superposition and entanglement, while providing the computational power needed for such simulations, also impose limitations and challenges that must be addressed to fully realize the potential of quantum simulations, especially in the context of Nick Bostrom's Ancestor Simulation Argument.

The Measurement Problem

At the heart of quantum-specific challenges lies the measurement problem, a cornerstone of quantum mechanics that introduces a fundamental layer of uncertainty in the observation of quantum systems. In the realm of quantum simulations, this problem poses intriguing questions about the nature of simulated realities constructed on quantum principles. The act of measurement in quantum mechanics, where observing a system forces it to collapse from a state of superposition into one of several possible definite states, raises the question of how such dynamics are mirrored or altered within a simulated quantum universe.

If a simulation is designed to replicate the quantum behavior of the real world, do repeated measurements within this simulated environment lead to a simplification of its possible states, effectively 'forcing' the simulation into specific, perhaps less complex, pathways? Addressing this challenge requires the development of sophisticated experiment-like probes that can operate within the simulation, capable of navigating the quantum measurement problem to extract meaningful data without unduly constraining the simulation's evolution. This endeavor not only demands technical innovation in quantum computing and simulation design but also a deep theoretical understanding of quantum mechanics and its implications for the nature of reality and simulation.

No-Cloning Theorem and State Preservation

The no-cloning theorem, a fundamental principle of quantum mechanics, posits that it is impossible to create an identical copy of an arbitrary unknown quantum state. This theorem presents a unique set of challenges within the context of quantum simulations, particularly in relation to the fidelity and integrity of simulated quantum states over time. In a simulated environment that seeks to replicate the quantum mechanics of our universe, the inability to clone quantum states restricts the methods available for state preservation, error correction, and the replication of quantum information. These constraints could significantly affect the simulation's capacity to maintain coherence and accuracy, especially over long periods or across vast computational scales. Furthermore, the no-cloning theorem implicates the development of data analysis techniques within quantum simulations, necessitating novel approaches to track and compare the evolution of quantum states without relying on direct replication. This challenge underscores the need for innovative analytical methodologies that can navigate the limitations imposed by quantum mechanics, ensuring the reliability and validity of findings derived from quantum simulations.

Addressing these challenges necessitates the exploration of alternative strategies for state preservation and analysis in the face of the no-cloning theorem. Analytical techniques capable of comparing quantum states within the simulation across time without violating the no-cloning principle could provide invaluable insights into the underlying dynamics of the simulated system. Such techniques might involve indirect measurement strategies or the utilization of entanglement and other quantum phenomena to infer changes and patterns within the system, potentially identifying 'signatures' of artificiality or computational shortcuts. This pursuit also raises the possibility of detecting the recycling of quantum information within the simulation, a tactic that might be employed to circumvent computational constraints but could inadvertently reveal the simulated nature of the environment. The development of these analytical techniques not only addresses the practical implications of the no cloning theorem for quantum simulations but also contributes to the broader endeavor of understanding the limitations and capabilities of quantum computing in simulating complex, realistic universes.

Decoherence and Error Correction

The intrinsic fragility of quantum states, coupled with the indispensable need for error correction within quantum computing systems, presents a dual challenge with profound implications for the fidelity and integrity of quantum simulations. Quantum decoherence, the process by which quantum information is lost to the surrounding environment, necessitates the implementation of sophisticated error correction techniques to preserve the coherence of quantum states. This requirement introduces a layer of complexity to quantum simulations, as the error correction mechanisms themselves may inadvertently introduce artifacts or statistical anomalies into the simulation data. These anomalies could, in theory, be detectable through advanced data analysis techniques, providing a window into the underlying computational processes that maintain the simulation. The detection of such anomalies would require data analysis methodologies sensitive enough to distinguish between the inherent randomness of quantum mechanics and the specific patterns introduced by error correction algorithms.

Exploring the potential for data analysis to uncover evidence of error correction interventions opens up intriguing avenues for probing the authenticity and origins of a simulated reality. If the patterns or signatures of error correction processes could be identified within the simulation data, they might serve as indirect markers of the simulation's artificial nature. This approach hinges on the development of data analysis techniques that can operate at the intersection of quantum computing and statistical analysis, capable of isolating and interpreting the subtle disturbances in the data that signal the presence of error correction mechanisms. Such techniques would not only enhance our understanding of the operational limits and challenges of quantum simulations but also potentially provide empirical support for hypotheses related to the nature of our reality, such as the Ancestor Simulation Argument. The quest to detect these statistical anomalies necessitates a deep integration of quantum physics, computational theory, and statistical analysis, representing a cutting-edge frontier in the exploration of quantum simulations and their implications for our understanding of the universe.

Practical Barriers and Data Requirements

The ambitious endeavor of analyzing potential ancestor simulations not only challenges our current computational capabilities but also stretches the imagination towards the computational power that might be necessary for such an analysis. If the simulation itself is of a complexity and scale that mimic an entire universe down to the quantum level, then the computational demands for analyzing this simulation—identifying patterns, anomalies, or the underlying digital fabric of the simulated reality—would be immense. This raises a profound question about the feasibility of such analysis: If we are indeed within an ancestor simulation, could a civilization possess or develop the computational resources required to detect its own simulated nature? The computational demands for this level of analysis could conceivably exceed the resources of the simulation itself, presenting a paradoxical scenario where the observer must outpace the computational complexity of the observed system. This scenario underscores not just a technical challenge but a philosophical quandary about the limits of knowledge and analysis within a potentially simulated reality.

Computational Demands of Analysis

The computational demands of analyzing potential ancestor simulations present a significant barrier, potentially requiring a level of computational power that rivals or even surpasses that of the simulation being analyzed. This presents a considerable challenge, raising questions about the feasibility of such an undertaking within the confines of a simulated reality. If the very essence of the simulation is to replicate the complexity of an entire universe, including the minutiae of quantum mechanics, then the computational resources needed to dissect and understand the simulation from within would be astronomical.

This scenario posits a paradox where the entities within the simulation, seeking to unravel the nature of their own reality, must command computational resources that may be inherently limited by the parameters of the simulation itself. It highlights a fundamental constraint on the ability of a civilization within an ancestor simulation to recognize and analyze its own simulated condition, thereby questioning the practical limits of self-awareness and existential discovery within such a sophisticated digital construct.

Distinguishing Simulated Data from Noise

The endeavor to distinguish simulated data from mere noise within the colossal datasets generated by an ancestor simulation represents a monumental challenge in the realm of data analysis. This challenge is not merely technical but also philosophical, as it requires discerning the fine line between deterministic phenomena deliberately encoded into the simulation by its architects and the emergent properties that arise spontaneously from the complex interplay of the simulation's underlying rules. The vastness and depth of data produced by such simulations would encompass both the meticulously programmed constraints intended to mirror natural laws and the unpredictable, emergent behaviors that could arise from the simulation's execution. Identifying patterns or anomalies within this data that could signify the presence of deliberate constraints or distinguish them from the natural complexity inherent in what might be perceived as 'reality' demands advanced analytical techniques that can navigate and make sense of this profound complexity.

Moreover, this task is compounded by the necessity to differentiate between what constitutes meaningful data reflective of the simulation's underlying structure and what might simply be noise—random or irrelevant information that does not contribute to understanding the simulation's nature. The inherent randomness and uncertainty characteristic of quantum mechanics, which would likely underpin the physics of an ancestor simulation, further blur these distinctions, making the data analysis even more challenging. Developing methodologies capable of teasing apart these subtle distinctions requires a sophisticated understanding of both the theoretical foundations of quantum simulations and practical expertise in data science. Such methodologies would need to be inherently robust, designed to extract and highlight patterns of interest from within an overwhelming sea of data, and sensitive enough to discern the signatures of simulation design from the natural variability and complexity that characterize the universe, whether simulated or real.

The Need for Ground Truth

The quest to validate analysis techniques for uncovering the nature of a simulated reality is fraught with conceptual and methodological challenges, chiefly among them being the absence of a "ground truth" or an unequivocal baseline reality against which to benchmark findings. The foundational premise of Nick Bostrom's Ancestor Simulation Argument itself presupposes that if advanced civilizations can create high-fidelity simulations of their ancestors, distinguishing the simulated from the "real" without an external frame of reference becomes paradoxically complex. This dilemma is not merely theoretical but impacts the practical development of analytical methods designed to identify the markers of a simulated existence. Without access to an undisputed real universe for comparison, the development, validation, and calibration of these techniques must rely on assumptions and models that, while logically coherent, lack empirical verification from outside the simulation framework.

In navigating this epistemological impasse, the concept of cross-simulation comparisons emerges as a potential strategy for discerning the simulated from the authentic, assuming the existence of multiple, distinct simulations. This approach would involve analyzing the variances and constants across different simulated realities to identify patterns or anomalies that could indicate the presence of underlying simulation algorithms or constraints. Such comparative analysis could potentially reveal commonalities that betray a simulated origin, or alternatively, highlight divergences that suggest unique simulation designs or purposes. However, this method also presupposes the ability to access or infer the existence of other simulations, a capability that may itself be limited by the parameters and permissions defined within each simulated environment. Thus, while cross-simulation comparisons offer a tantalizing avenue for research, they also underscore the profound challenges inherent in proving the Ancestor Simulation hypothesis without a ground truth, reinforcing the need for innovative analytical frameworks capable of operating within these uniquely constrained epistemological boundaries.

Implications for Bostrom's Ancestor Simulation Argument

The integration of quantum computing into the framework of simulation theory, especially as it pertains to Nick Bostrom's Ancestor Simulation Argument, heralds a paradigm shift in our approach to understanding reality. Quantum computing, with its potential to exponentially increase simulation capabilities, offers the tools necessary to create highly complex and detailed simulations that could, in theory, replicate the intricacies of human consciousness and the vastness of the universe itself. However, the leap from theoretical potential to practical implementation reveals a plethora of challenges, particularly in the realm of data analysis. The task of extracting meaningful conclusions about the nature of our reality—whether we exist within a simulated environment created by a post-human civilization—demands analytical techniques far beyond the current norms of scientific inquiry. These challenges are not merely technical but epistemological, requiring us to question the very nature of reality and our ability to understand or detect its potential artificiality through the lens of quantum-simulated environments.

The implications of these challenges for Bostrom's Ancestor Simulation Argument are profound. On one hand, the advancements in quantum computing and simulation technology could eventually enable us to simulate universes with such fidelity that the inhabitants of these simulations would be indistinguishable from real consciousnesses, thus lending credence to the possibility that we ourselves could be entities within a similarly advanced simulation. On the other hand, the inherent complexities and theoretical quandaries posed by quantum simulation—such as the measurement problem, decoherence, and the no-cloning theorem—underscore the difficulty of achieving and analyzing such simulations. Moreover, the practical barriers in data analysis, including distinguishing simulated data from noise and the computational demands of analyzing vast, complex simulations, highlight the significant hurdles that must be overcome to validate or refute the Ancestor Simulation Argument. These issues not only challenge our current computational and analytical capacities but also invite a deeper philosophical inquiry into the nature of existence and the possibility of an underlying simulated structure to our universe.

Foundational Challenges

The creation and analysis of ancestor simulations, as posited by Nick Bostrom, bring forth foundational challenges that stretch the limits of current computational and analytical frameworks. The very nature of these simulations demands an unprecedented level of complexity and fidelity, aiming to reproduce the nuances of reality so accurately that they are indistinguishable from the real thing. This endeavor not only challenges our computational capacities but also forces a reevaluation of our analytical methodologies. Traditional approaches to data analysis and interpretation, grounded as they are in the observables of classical physics, may fall short when applied to the multifaceted and deeply quantum-mechanical nature of such simulations. The transition from classical to quantum simulations necessitates a paradigm shift, requiring new theoretical frameworks and analytical tools capable of identifying and interpreting the subtleties of a simulated universe.

Complexity and Observables

Within the intricate weave of an ancestor simulation, the task of discerning "observables" — distinct features or patterns that could reveal the simulated nature of the environment — is exponentially complicated. Ancestor simulations designed to mirror reality with high fidelity would incorporate layers of complexity that mimic the vastness and variability of the universe itself. In such a context, identifying markers or anomalies that suggest artificiality becomes a task of needle-in-a-haystack proportions.

Classical physics, with its deterministic laws and predictable outcomes, offers a solid foundation for identifying observables in our natural universe. However, the inherently probabilistic nature of quantum mechanics — the bedrock of quantum simulations — demands a fresh approach to defining what constitutes meaningful observables. This new approach must be capable of navigating the quantum landscape, where certainty is replaced by probability, and observability is entangled with the observer's influence. Developing the tools to identify and analyze these observables within the quantum domain is crucial, not just for the feasibility of creating ancestor simulations but also for the possibility of recognizing our own reality as such a simulation.

Simulator Intent vs. Emergent Properties

The challenge of distinguishing between features explicitly programmed by the designers of an ancestor simulation and those that arise as emergent properties from the complex interplay of simulated elements is a pivotal concern within the context of validating Bostrom's Ancestor Simulation Argument. This distinction is critical because it touches upon the very essence of what makes a simulation indistinguishable from reality: the degree to which it can autonomously generate complexity and phenomena that were not explicitly coded into it. Intentionally designed features may carry with them certain 'signatures' or artifacts that, if identified through meticulous data analysis, could betray the simulated nature of the environment. However, the task of uncovering such artifacts is daunting, requiring analytical tools sophisticated enough to differentiate between the nuances of designed elements and the organic emergence of complex properties. This endeavor not only challenges our current capabilities in data analysis but also prompts a deeper philosophical inquiry into the nature of reality and the boundaries between artificiality and emergence.

Conversely, the presence of emergent properties within a simulation that mirror the unpredictability and complexity of natural laws in our universe could paradoxically serve as indicators of a simulated environment. If data analysis techniques could identify patterns or regularities within a simulation that seem to defy the intentional design yet align closely with our understanding of natural laws, it could suggest that these emergent properties are, in fact, a byproduct of the simulation's attempt to replicate reality with high fidelity. Such a discovery would blur the lines between designed features and emergent phenomena, complicating the task of identifying a simulated universe. However, it also opens up the possibility that within the intricate dance of particles and forces in our own universe lie clues to its possible simulated nature. The challenge, then, is to develop data analysis methodologies capable of sifting through the vast datasets generated by such simulations, to discern the deliberate from the emergent, and in doing so, potentially uncover the underlying structure of our reality.

Simulation Scope and 'Edge Cases'

The ambition of constructing a full ancestor simulation, one that spans vast timescales and delves into levels of detail that exceed our current observational capabilities, presents a profound challenge to existing data analysis methodologies. The depth and breadth of such a simulation, aiming to encapsulate the entirety of a universe, necessitate analytical tools that can transcend the limitations inherent to our current scientific understanding and observational reach. Traditional data analysis techniques, honed within the scope of phenomena observable and measurable within our universe, might falter when confronted with the simulated extremes of physics and reality. The inherent assumption that the laws of physics and patterns of behavior we observe are universally applicable becomes a significant hurdle when analyzing data generated within a simulation designed to explore beyond these known boundaries. Consequently, there exists an urgent need for the development of novel analytical frameworks and tools capable of interrogating the data from such comprehensive simulations, particularly at the fringes where simulated physics might diverge from our empirical expectations.

Addressing this need requires a multidisciplinary approach, combining insights from quantum computing, theoretical physics, and advanced data science to forge new pathways in data analysis. These tools must not only be capable of handling the sheer volume and complexity of data generated by an ancestor simulation but also be sensitive enough to detect subtle deviations or anomalies that could hint at the underlying simulated nature of the environment. Probing the "edge cases" of simulated physics—scenarios that push the boundaries of what is theoretically possible or observable in our universe—poses a unique challenge. It involves discerning patterns or behaviors that do not align with known natural laws, potentially revealing the fingerprints of a simulated reality. The development of such analytical capabilities is crucial for advancing our understanding of not only the feasibility of ancestor simulations but also the fundamental nature of reality itself, challenging our perceptions and expanding the horizons of what is considered scientifically possible.

Quantum-Specific Challenges

The integration of quantum computing into the realm of simulations, particularly those envisioned by Bostrom's Ancestor Simulation Argument, introduces a set of quantum-specific challenges that are fundamentally different from those encountered in classical computing scenarios. One of the most profound implications of using quantum computing for simulations is the way it alters our approach to data analysis and the interpretation of simulated realities.

Quantum mechanics, by its nature, operates on principles that include superposition and entanglement, offering a computational power and complexity far beyond what classical systems can achieve. However, this also means that simulations based on quantum computing are subject to the peculiarities of quantum behavior, including the probabilistic outcomes and inherent uncertainties that define quantum systems. These challenges necessitate a reconsideration of how we analyze and validate the data generated by simulations, requiring new methodologies and tools specifically designed to navigate the quantum landscape. The ability to accurately interpret the results of quantum simulations is crucial for testing the validity of the Ancestor Simulation Argument and understanding the potential limitations and capabilities of quantum-assisted simulations of reality.

The Measurement Problem

The measurement problem in quantum mechanics introduces a layer of complexity to the analysis of quantum simulations, stemming from the principle that the act of measurement affects the state of a quantum system, collapsing it from a superposition of states into a single, definite state. This phenomenon raises intriguing questions about the nature of simulated realities constructed on quantum principles. For instance, how the repeated measurements within a simulation might influence its evolution, potentially guiding it along specific, perhaps less complex, pathways due to the collapse of quantum states.

This challenge underscores the need for developing sophisticated experimental probes and data analysis techniques within the simulation that can account for the quantum measurement problem. Such tools must be capable of extracting meaningful information from the simulation without disrupting its quantum coherence more than necessary. Addressing this issue is essential for advancing our understanding of how quantum simulations can be used to model complex systems and realities, providing insights into the potential constraints and behaviors of simulations that seek to replicate the quantum underpinnings of our universe.

No-Cloning Theorem and State Preservation

The no-cloning theorem, a fundamental principle in quantum mechanics, posits a significant limitation within the context of quantum simulations, particularly those aimed at emulating complex systems or entire universes as suggested by Bostrom's Ancestor Simulation Argument. This theorem states that it is impossible to create an exact copy of an arbitrary unknown quantum state, which introduces a layer of complexity in preserving the fidelity of quantum information over time within a simulation. This limitation is not merely a technical hurdle but also impacts the design and execution of quantum simulations, influencing how data is generated, stored, and analyzed. In the pursuit of creating ancestor simulations with quantum computing, this theorem necessitates innovative approaches to managing quantum states, ensuring that the simulation can progress without violating fundamental quantum principles. The inability to clone quantum states means that simulations must find alternative methods for error correction and information preservation, which could influence the overall structure and behavior of the simulation.

Given these constraints, the development of analytical techniques capable of comparing quantum states within the simulation across time becomes crucial. Such techniques would need to navigate the challenges imposed by the no-cloning theorem, potentially identifying unique patterns or signatures in the way quantum information is handled, reused, or recycled within the simulation. These patterns could reveal insights into the computational strategies employed to maintain the coherence and continuity of the simulated environment, despite the inherent limitations of quantum mechanics. By examining how quantum simulations manage state preservation and information flow, researchers could uncover evidence of underlying computational frameworks or constraints. This could, in turn, provide indirect clues to the simulated nature of the environment, offering a new avenue for testing the Ancestor Simulation Argument through the lens of quantum-specific challenges and the innovative solutions devised to overcome them.

Decoherence and Error Correction

The phenomenon of decoherence, wherein the fragile quantum states of a system interact with their environment and thus lose their quantum coherence, poses a significant challenge to the integrity and accuracy of quantum simulations. This interaction leads to the transition of quantum information into classical information, thereby diluting the very essence that grants quantum computing its superior computational power. To counteract this, quantum error correction algorithms are employed to preserve the quantum states and ensure the fidelity of the simulation. However, these error correction mechanisms, while essential, may inadvertently introduce artifacts into the simulation data. These artifacts, arising from the complex interplay between maintaining quantum coherence and correcting errors, could manifest as statistical anomalies that are not naturally present in the system being simulated. Data analysis techniques that can identify and interpret these anomalies offer a potential pathway to uncover the underlying computational structures and error correction strategies deployed within quantum simulations, providing insights into the mechanisms that support the simulated reality.

The potential of data analysis techniques to detect and decipher statistical anomalies induced by quantum error correction processes invites a deeper examination of the fabric of simulated realities. If such anomalies can be systematically identified, they may serve as indicators of the simulation's artificial nature, revealing the computational underpinnings and the efforts taken to maintain simulation integrity against quantum decoherence. This pursuit requires sophisticated data analysis methods that are sensitive to the subtle deviations from expected quantum behavior, capable of distinguishing between the inherent quantum unpredictability and the specific signatures of error correction interventions. Developing these techniques not only advances our ability to scrutinize the authenticity and quality of quantum simulations but also enriches our understanding of the principles and practicalities of quantum error correction. This knowledge is crucial for assessing the feasibility of achieving the high-fidelity simulations envisioned by Bostrom's Ancestor Simulation Argument, and for exploring the boundaries between simulated and physical realities through the lens of quantum computing.

Practical Barriers and Data Requirements

The endeavor to analyze potential ancestor simulations brings to the fore the immense computational demands that such an analysis would entail, revealing significant practical barriers in the path of validating Bostrom's Ancestor Simulation Argument. Given the complexity and scale of simulations capable of replicating an ancestor's universe with high fidelity, the computational resources required for analyzing these simulations are monumental. This requirement is not merely a matter of processing power but also encompasses the need for sophisticated data storage, management, and analysis capabilities. These demands place a considerable strain on the theoretical feasibility of conducting such analysis, especially within the context of a civilization that might itself be part of an ancestor simulation. The practical implications of these requirements necessitate a reevaluation of our current computational infrastructure and a concerted effort towards developing next-generation computing technologies that can handle the data-intensive tasks associated with analyzing complex simulations.

Computational Demands of Analysis

The question of whether it is feasible for a civilization, potentially within an ancestor simulation, to undertake the analysis of such simulations is profoundly challenging. The computational demands of analyzing simulations that accurately replicate a universe's complexity are staggering, potentially requiring resources that match or even surpass the computational power used to run the simulation itself. This scenario presents a paradoxical situation where the analytical tools and processes needed to probe the simulated reality might be beyond the reach of the entities within the simulation due to inherent computational constraints.

The implications of this challenge are far-reaching, affecting not only the practical aspects of conducting such analysis but also raising philosophical questions about the limits of knowledge and understanding within simulated realities. Addressing these demands necessitates not only advancements in quantum computing and other emerging technologies but also innovative approaches to data analysis that can maximize efficiency and insights from limited computational resources.

Distinguishing Simulated Data from Noise

The challenge of distinguishing simulated data from noise within the vast datasets generated by an ancestor simulation encapsulates a core issue in the validation of Bostrom's Ancestor Simulation Argument through data analysis. These datasets, reflective of a meticulously crafted universe, are expected to be replete with both deterministic phenomena, governed by the simulated laws of physics, and emergent phenomena, which arise from the complex interactions and behaviors within the simulation. The task of parsing this data to identify patterns or anomalies that could hint at the simulated nature of the environment is monumental. It requires not just the ability to sift through immense volumes of data but also the sophistication to discern between what is an artifact of deliberate simulation constraints and what is merely the product of the inherent complexity and randomness of the simulation's natural laws. This distinction is crucial, as it could potentially reveal the 'signatures' of a simulation, offering concrete evidence in support of or against the hypothesis that our reality is itself a simulated construct.

Developing data analysis techniques capable of making such distinctions necessitates a deep understanding of both the theoretical foundations of quantum mechanics and the practical aspects of computational science. Analysts must be equipped to recognize the subtle differences between the noise inherent in any complex dataset and the specific types of noise or patterns that would not exist in the absence of simulation constraints. This endeavor is complicated further by the need to account for emergent phenomena, which, while potentially indicative of a simulation's artificiality, could also simply be the result of sufficiently advanced simulations achieving a degree of complexity and randomness indistinguishable from natural laws. The development of analytical frameworks that can navigate this fine line is essential for advancing our understanding of the possibility of simulated realities, requiring a concerted effort across disciplines to pioneer new methodologies in data analysis that are up to the task.

The Need for Ground Truth

The pursuit of methods capable of unveiling the simulated nature of our reality, as proposed by Bostrom's Ancestor Simulation Argument, encounters a fundamental epistemological obstacle: the absence of a verifiable ground truth. Without an objective baseline reality for comparison, the development of analytical techniques to distinguish between simulated and non-simulated universes ventures into uncharted territories of scientific inquiry. This challenge not only underscores the complexity of validating such analysis techniques but also raises profound questions about the nature of reality itself. If our understanding of the universe is confined to the parameters of a potential simulation, the task of identifying markers or anomalies indicative of a simulated construct becomes deeply intertwined with philosophical considerations. The quest for ground truth, therefore, transcends the realm of technical feasibility, engaging with broader questions about the limits of human knowledge and the foundations upon which we build our understanding of the universe.

In light of these challenges, the concept of cross-simulation comparisons emerges as a speculative yet intriguing approach to circumvent the issue of missing ground truth. By analyzing and contrasting the properties of multiple simulations, assuming such a multiplicity exists and can be accessed, researchers might uncover consistent patterns or discrepancies that hint at the underlying simulated nature of these environments. This comparative methodology, however, hinges on the speculative premise that multiple distinct simulations are identifiable and accessible for study, presenting its own set of practical and theoretical difficulties. Nonetheless, exploring the potential for cross-simulation analysis represents a creative endeavor to bridge the gap between the known and the unknowable, offering a pathway to test the boundaries of our reality. This approach not only expands the toolkit available for probing the simulated hypothesis but also enriches the dialogue between science, philosophy, and the enduring quest to understand the fabric of our existence.

Ethical and Philosophical Considerations

The exploration of ancestor simulations within the framework proposed by Nick Bostrom's argument introduces a range of ethical and philosophical considerations that extend far beyond the technical and analytical challenges previously discussed. These considerations encompass the moral implications of creating simulated realities that possess consciousness, the rights of simulated entities, and the responsibilities of the creators towards their creations. Additionally, this discourse opens up questions about the nature of reality, consciousness, and our place within potentially simulated universes.

The ethical dimensions are particularly profound in considering the possibility of suffering within these simulations, the autonomy of simulated consciousness, and the ethical boundaries of experimentation within such environments. Philosophically, the ancestor simulation argument challenges our foundational understandings of existence, prompting a reevaluation of what it means to be real and questioning the very fabric of our perceived reality. These considerations demand a multidisciplinary approach, merging the insights of technology, philosophy, and ethics to navigate the complex terrain that the ancestor simulation hypothesis presents.

Simulator Intervention

The prospect of simulator intervention introduces a deeply unsettling layer to the ethical and philosophical debate surrounding ancestor simulations. If the entities responsible for the creation of these simulations possess not only the capability but also the inclination to interfere with attempts at detection, this could fundamentally compromise the integrity of any analysis aimed at uncovering the simulated nature of our reality. Such interference might manifest as subtle manipulations of data, biasing results, or more directly, through the obscuration of evidence that could reveal the simulation's existence.

This scenario raises profound questions about the feasibility of detection under conditions where the simulator has an active interest in maintaining the simulation's undetectability. It challenges the notion of free inquiry within the simulated environment and introduces a layer of epistemological uncertainty that complicates the pursuit of truth. The possibility of simulator intervention underscores the complex interplay between the creators and inhabitants of simulated realities, raising existential questions about autonomy, free will, and the nature of reality itself in a universe where the boundaries between creator and creation are blurred.

The Relativity of Analysis

The concept of "The Relativity of Analysis" delves into the profound implications of conducting scientific inquiry and data analysis within the constraints of a potentially simulated reality, as proposed in Nick Bostrom's Ancestor Simulation Argument. This notion suggests that the tools, methods, and frameworks we develop for understanding and interpreting the universe are inherently shaped by our observations and experiences within this universe, which may itself be a simulation. This raises a critical epistemological question: if our scientific paradigms are constructed within a simulated environment, might these paradigms be fundamentally ill-equipped to detect or even conceive of the simulation's boundaries? The principles and laws of physics that we discover and upon which we base our technological advancements and analytical techniques could be part of the simulation's designed parameters, potentially limiting our capacity to perceive anomalies or inconsistencies that would indicate a reality beyond the simulation.

This scenario presents a challenging paradox, suggesting that a civilization's ability to recognize and analyze the constraints of its simulated existence could be intrinsically limited by the very nature of its reality. If our understanding of the universe is tailored to the conditions and laws that govern a simulated environment, our attempts to uncover the simulation's underlying framework might be skewed or completely obstructed by our foundational assumptions. This conceptual barrier not only highlights the potential limitations of our analytical tools but also underscores the philosophical quandary of seeking objective truth within a reality whose foundational parameters might be externally imposed. It suggests that the quest to identify the simulation's boundaries and understand its governing principles requires not only advanced technological capabilities but also a radical rethinking of our approach to science and knowledge, challenging us to question the very basis of our understanding of reality.

Data Analysis for the Ancestor Simulation Argument

Data analysis emerges as a crucial tool in the exploration of Nick Bostrom's Ancestor Simulation Argument, which posits that advanced civilizations could create simulations of their ancestors indistinguishable from "base" reality. The essence of Bostrom's hypothesis challenges our understanding of reality, suggesting that we could be living within such a simulation without our knowledge. In this context, data analysis serves as a bridge between the theoretical possibility of such simulations and the empirical evidence that could support or refute this hypothesis. The role of data analysis is not merely to sift through vast datasets for anomalies or patterns that hint at a simulated reality but to apply sophisticated methodologies that can discern the subtle distinctions—should they exist—between a universe that is fundamentally 'real' and one that is the product of advanced computational processes. This endeavor requires not only technical prowess in handling and interpreting data but also a conceptual leap in how we define and understand the nature of reality itself.

The task set before data analysts and scientists in the context of the Ancestor Simulation Argument is daunting. It involves crafting and employing analytical techniques that are sensitive enough to detect the 'fingerprints' of simulation—artifacts or phenomena that would not occur naturally but are introduced as a result of the simulation's underlying algorithms and computational limitations. This pursuit is complicated by the theoretical nature of Bostrom's argument, which suggests that a perfectly executed simulation would be indistinguishable from natural reality to its inhabitants. Therefore, data analysis in this field must be innovative, combining cross-disciplinary insights from computer science, quantum physics, and philosophy to develop new frameworks for understanding and testing the reality we perceive. The challenge lies not only in identifying what constitutes evidence of a simulation but also in grappling with the philosophical implications of such findings, fundamentally questioning the nature of existence and our place within the cosmos.

Search for Inconsistencies

- Minute deviations in fundamental physical constants across time or spatial regions within the simulation
  - **Mapping the Constants:** To detect minute deviations, meticulous 'surveys' of fundamental constants (speed of light, Planck's constant, gravitational constant, etc.) must be conducted across vast simulated distances and times within an ancestor simulation. High-precision simulated instruments and meticulously replicated experiments would be required to build such a dataset.
  - **Analysis for Anomalies:** Sophisticated statistical techniques would be needed to search for subtle patterns or variations in these constants that deviate from expectations for our own universe. Such deviations might point to areas where the underlying computational model is simplified, or where simulation 'shortcuts' have been employed by the simulator.
- Statistical anomalies in seemingly large-scale random phenomena, like cosmic background radiation
  - **Cosmic Background as Fingerprint:** The cosmic microwave background radiation (CMB) is a

remnant of the early universe and provides a snapshot of its large-scale structure. Detailed mapping and analysis of simulated CMB for unusual 'hot' or 'cold' spots, unexpected correlations, or repetitive patterns could expose artificial constraints within the simulation.
  - **Simulation vs. Reality:** The key lies in comparing the simulated CMB dataset to theoretical models and observational data from our own universe. Statistically significant divergences, beyond those attributable to natural variation, might suggest deliberate design or algorithmic limitations within the simulation.
- Detection of boundaries or repetition in the computational framework of the simulated universe
  - **Exploring the Computational Limits:** If the simulation is bound by finite resources, there may be limits on the size or complexity of the simulated universe. Analysis focused on detecting repetition in astronomical objects, potentially distant galaxies, or recurring patterns in vast cosmic structures could offer clues of such boundaries.
  - **Searching for Glitches:** Particular focus should be placed on the edges or extreme regions of the simulation. Here, computational simplifications might be more likely to manifest. Detecting subtle discontinuities, repeating textures, or anomalies in the behavior of physical laws near the 'boundary' of the simulation could provide evidence of its artificial nature.

Exploiting Quantum 'Glitches'

- Analysis of extreme physical scenarios to force quantum phenomena in the simulation to reveal its limits
  - **Pushing Quantum Boundaries:** Devising simulated experiments that involve extreme physical situations – such as incredibly high energy particle collisions, the behavior of matter near black holes, or phenomena near absolute zero – could push the simulated quantum framework to its computational limits.
  - **Glitches as Clues:** The goal is to observe how the simulation handles complexity at its extremes. Unexplained breakdowns of quantum behavior, unusual simplifications that become apparent under these conditions, or computational artifacts that 'leak' into the results could provide strong indicators of a simulated quantum system.
- Search for anomalies related to error correction or decoherence within the simulated quantum systems
  - **Decoherence Footprints:** Quantum systems are notoriously fragile, and quantum computers rely on complex error-correction mechanisms. If the simulation is built upon quantum computation principles, subtle anomalies might arise from these corrective processes. Analyzing entangled systems, or those in states of superposition, could reveal them.
  - **Errors as Evidence:** Focused data analysis on the distribution and types of errors occurring within quantum systems in the simulation could expose patterns or biases not found in natural systems. Such patterns might suggest underlying computational error correction algorithms at play, hinting at the simulated nature of the environment.
- Important Considerations
  - **Simulator 'Countermeasures':** A sophisticated simulator might take steps to mask the effects of error correction or decoherence within their simulated world, deliberately mimicking natural behavior to make detection harder.
  - **The Challenge of Interpretation:** Distinguishing between genuine limitations of a simulated quantum system and the natural effects of decoherence and errors in real-world quantum systems is a complex task. A vast library of comparison data from real-world quantum experiments would be needed.

Comparative Analysis

If multiple simulations exist, even if indistinguishable from our own reality, comparing their outcomes might yield subtle statistical divergences, suggesting underlying constraints.

- **The Power of Comparison** The premise here is that even if individual simulations are flawless in their internal consistency, variations might exist between multiple simulations run by the same or different posthuman civilizations. These variations could stem from different starting conditions of the simulation, minor tweaks to the underlying simulated physics, or even differences in the simulators' computational resources.
- **Finding Clues in Divergences** Meticulous comparisons between simulations could expose these subtle divergences. Analysis might focus on large-scale phenomena like the distribution of galaxies, the timing of major simulated historical events, or even variations in the outcomes of repeated quantum experiments across simulations. Detecting statistically improbable divergences, beyond those attributable to natural randomness, could hint at the limitations and constraints imposed by different simulators and their computational frameworks. While not directly proving the simulated nature of reality, it would suggest inconsistencies between supposedly equivalent 'realities'.

The Road Ahead

The quest to analyze ancestor simulations and probe Bostrom's argument demands a revolutionary transformation in data analysis. It will require:

- Development of theoretical frameworks to identify meaningful observables within massively complex systems
- Interdisciplinary collaborations between physicists, computer scientists, mathematicians, and philosophers
- Design of simulation-based experiments to explore the unique phenomenology of simulated worlds built upon quantum principles.

The Quantum Leam Towards Ancestor Simulations

Quantum computing's role in the potential realization of ancestor simulations exemplifies a critical juncture in the convergence of computational technology and theoretical physics. While significant hurdles remain in the practical deployment of quantum computers, their theoretical potential to revolutionize simulations—making them more detailed and realistic—remains undiminished. As research in quantum computing progresses, its integration with advancements in VR and AI may eventually fulfill the computational demands of Bostrom's Ancestor Simulation Argument, ushering in a new era of high-fidelity simulations. The ongoing development of quantum algorithms and the reduction of error rates in quantum computations are pivotal in transforming the theoretical potential of quantum computing into a practical reality capable of supporting the complex requirements of ancestor simulations.

Bridging Technology and Theory

The technological feasibility of creating ancestor simulations, as proposed by Nick Bostrom, hinges on a confluence of advancements in computational power, virtual reality, artificial intelligence, and quantum computing. While current technologies provide a foundation, achieving the scale and complexity required for high-fidelity simulations of entire universes populated by indistinguishable consciousness experiences demands significant breakthroughs across these disciplines.

This analysis underscores the multidisciplinary approach necessary to evaluate the possibility of ancestor simulations. It not only requires pushing the boundaries of existing technologies but also navigating the ethical, philosophical, and practical challenges that such simulations entail. As we project current trends into the future, the intersection of these technologies offers a tantalizing glimpse into the potential to create simulated realities, ultimately providing a methodology to test the veracity of Bostrom's Ancestor Simulation Argument.

CHAPTER THIRTY-FOUR

Statistical Estimation - Estimating Computational Requirements

The statistical estimation of the probability that our reality is a simulation, as posited by Nick Bostrom's hypothesis, requires an intricate melding of disciplines, extending beyond the realm of traditional statistics into computational theory, artificial intelligence, quantum computing, and the nuanced territories of philosophy and ethics. This multifaceted approach is essential for grappling with the complexities inherent in assessing the likelihood of existing within a sophisticated ancestor simulation.

By applying statistical models and methodologies within this interdisciplinary framework, researchers aim to construct a robust analytical foundation that can quantify the arguments and counterarguments surrounding the simulation hypothesis. This endeavor involves not only the application of advanced statistical techniques to evaluate the computational feasibility of such simulations but also a critical examination of the philosophical underpinnings that define what constitutes a simulation. Furthermore, it necessitates an exploration into the ethical implications of creating simulated consciousness, thereby enriching the discourse with a comprehensive understanding of the myriad factors that influence the simulation probability.

The challenge in statistically estimating simulation probability lies in defining and quantifying parameters that are largely theoretical and, at times, speculative in nature. For instance, the development of computational models that can simulate the vast complexities of a universe requires assumptions about future advancements in technology, particularly in AI and quantum computing. These models must account for the exponential scaling of computational power, the efficiency of quantum algorithms, and the potential for AI to manage the intricacies of simulating consciousness. Moreover, philosophical arguments regarding the nature of reality and our perception of it play a crucial role in framing the statistical analysis, guiding the selection of variables and the interpretation of results.

Ethical considerations, too, influence the theoretical limits of simulation technology, as they pose questions about the responsibility and morality of creating simulated beings with consciousness. By weaving together insights from these diverse disciplines, the statistical estimation of simulation probability endeavors to shed light on one of the most profound questions facing humanity: the nature of our existence and the possibility that we live in a constructed reality.

A core challenge in substantiating Nick Bostrom's Ancestor Simulation Argument lies in accurately estimating the immense computational requirements needed to execute detailed simulations of past civilizations. Approaching this challenge necessitates a firm grasp of the theoretical foundations of computation, coupled with quantitative projections of the storage, processing power, and energy demands of such an endeavor.

Theoretical Foundations

Let's begin by summarizing the computational requirements for running high-fidelity ancestor simulations, incorporating findings from computational theory and technology trends. If the goal is to create simulations indistinguishable from the perspective of their simulated inhabitants, we need to consider the complexity involved. Calculations would encompass not just fundamental physical laws and the interaction of matter, but potentially the simulation of full sensory experiences and consciousness. Alan Turing's universal Turing machine concept establishes a theoretical basis for computation, but real-world limits of computability must be factored in.

The simulation of complex systems like our universe likely necessitates vast computational resources, potentially far exceeding what's currently achievable by humanity. Yet, a posthuman civilization (as envisioned by Bostrom) would hypothetically possess far greater capabilities for computation. It's also worth considering that simulation techniques might advance towards more efficient algorithms or optimizations, potentially enabling simulations of high complexity with reduced computational resources compared to naive projections. As technological development is difficult to predict long-term, any estimate of a posthuman civilization's computational capability should be considered speculative yet grounded in current theoretical understanding.

Complexity Classes:

Ancestor simulations likely fall into the category of exceptionally complex computational problems. Understanding the complexity class (e.g., P, NP, NP-Complete, etc.) to which ancestor simulations belong is vital for determining the inherent resources needed for their execution. Early assessments suggest the potential for exponential or factorial scaling of resource demands with simulation size and fidelity.

If ancestor simulations are shown to belong to complexity classes like NP-Hard or even classes beyond those currently understood, it would suggest that even vast computational resources might not be sufficient for their practical realization. On the other hand, if breakthroughs are made in computational theory, demonstrating that complex simulations can be run within more tractable complexity classes, this could shift the feasibility landscape. The field of computational complexity, therefore, holds significant implications for assessing the plausibility of the Simulation Hypothesis.

Simulating Consciousness:

A fundamental question is whether a complete simulation necessitates modeling every subjective conscious experience. If true, the computational costs could prove intractable. Current neuroscience does not offer definitive models of consciousness, leaving this aspect open for debate and alternative simulation strategies. Some researchers might argue that consciousness is an emergent property of complex systems and would naturally arise in a sufficiently detailed simulation. Others could posit that consciousness is something fundamental and irreducible, meaning a simulation without true consciousness remains possible.

The philosophical and computational nature of consciousness has direct implications for resource estimation. If consciousness requires a level of simulation detail that surpasses currently imaginable computational power, it places severe constraints on the feasibility of ancestor simulations. Alternatively, if consciousness can be either omitted or simulated in a less computationally demanding manner, it might make comprehensive ancestor simulations more theoretically plausible within the Simulation Hypothesis framework.

Levels of Detail:

Ancestor simulations could be designed with varying degrees of granularity. Fine-grained simulations, down to the molecular or subatomic levels, exponentially increase computational demands compared to coarser simulations focused on behavior and interactions. It's essential to define the acceptable threshold of detail to make estimations realistic. A posthuman civilization might find simulations omitting subatomic details perfectly sufficient for their goals. Conversely, they might desire simulations indistinguishable from the perspective of the simulated inhabitants, necessitating the modeling of quantum phenomena.

The chosen level of detail has profound consequences for resource assessment. Simpler simulations are likely more computationally feasible. If the hypothesis is that the simulator civilization is mainly interested in studying large-scale societal behavior or historical trends, lower fidelity simulations could suffice. Such a choice would significantly reduce the computational burden, potentially making ancestor simulations possible even with resource constraints.

Computational Modeling and Scaling

To establish estimates of resource demands for various scenarios, we can draw upon existing practices in scientific computing and simulation. Fields like physics, chemistry, and climate modeling routinely utilize large-scale simulations that necessitate sophisticated computational modeling techniques. Analyzing computational resource scaling trends in these established domains can provide insight into the potential scaling behavior of ancestor simulations.

It's important to extrapolate with caution. Simulation techniques are highly domain specific. For instance, modeling molecular dynamics and simulating galaxy formation, while computationally intensive, might exhibit different scaling behaviors due to the nature of the underlying interactions and systems. Despite this complexity, a careful analysis of existing computational models and their scaling can offer valuable reference points when attempting to reason about the computational challenges inherent to ancestor simulations.

To establish estimates of resource demands for various scenarios, we can draw upon existing practices in scientific computing and simulation:

Physical Simulations:

Areas like computational fluid dynamics, astrophysics, and materials modeling routinely employ simulations involving billions or trillions of data points. Techniques for parallelization, algorithmic optimization, and the use of high-performance computing (HPC) clusters offer essential insights for approaching the computational scale needed for ancestor simulations. Examining the parallelization strategies in these fields can illuminate how computation might be distributed to handle vast simulation workloads.

Additionally, exploring the algorithmic optimizations employed in physical simulations may provide analogous strategies for efficient resource utilization when simulating complex systems like those in an ancestor simulation. While the specific techniques might need adaptation, the underlying principles of resource management and optimization in large-scale scientific simulations hold valuable lessons for the speculative case of ancestor simulations.

Modeling Complexity:

Studies on simulating complex systems with elements of agency and historical trajectories (cellular automata, agent-based models of social systems) provide frameworks for understanding scaling behavior and the emergence of properties with an increasing number of simulated entities. Cellular automata, like Conway's Game of Life, illustrate how complex behaviors can arise from simple sets of rules. Similarly, agent-based models offer insights on how interactions between individual agents within a system can lead to complex emergent social or economic phenomena.

Analyzing such models provides valuable testbeds for resource scaling. Even without simulating all of physical reality, these models can reveal the computational demands of capturing the dynamics of societies, ecosystems, or other complex multi-agent systems. This analysis can help estimate how computational requirements might scale in an ancestor simulation with varying population size or complexity of interactions between simulated entities.

Quantitative Estimations

Let's develop some preliminary estimations as an illustrative exercise:

Population and Scope:

Assume a simulation aiming to encompass the entirety of known human history, with roughly 100 billion individuals having ever lived. Let's also assume the simulation tracks a limited historical period (e.g., 100 years per simulated individual). While a simplified starting point, it's important to consider that a highly realistic simulation likely wouldn't limit itself to only humans who've definitively existed in our specific timeline.

To achieve a higher level of fidelity, a simulation might include:

- **Slightly altered histories:** Simulating possible variations of individual lives and historical events to explore the space of plausible human experiences.
- **Ecosystems:** Incorporating non-human species (animals, plants), as their existence and interactions are intertwined with human development.
- **Extraterrestrial Intelligence:** If posthuman simulators are interested in the potential evolution of civilizations, the simulation could encompass hypothetical alien lifeforms.

Including these elements dramatically expands the population and scope. Further refinement of these assumptions, grounded in the simulator civilization's speculated goals, is essential for a more realistic computational cost estimation.

Data Requirements (Simplified):

If we conservatively estimate 1 GB of data needed to represent a person's life at a high but not exhaustive level of detail, we have a minimum storage requirement of 100 billion GB (100 exabytes). This excludes any overhead needed for simulation logic, environments, etc. To put this in context, some estimates place the total amount of data stored across the entire internet at roughly 40-50 exabytes. This suggests that even our conservative estimate would likely necessitate a storage capacity significantly exceeding what humanity currently possesses.

It's crucial to remember that this remains a highly simplified calculation. Factors that could dramatically increase storage requirements include:

- **Higher Simulation Fidelity:** If finer details such as individual thoughts, sensory experiences, or cellular-level biological processes are simulated, the storage requirements per individual would increase exponentially.
- **Multiple Simulations:** Simulator civilizations might run numerous parallel simulations with varying parameters or initial conditions, compounding the storage demands.

Processing Power:

The nature of ancestor simulations likely requires sequential operations within each simulated life, unlike some parallelizable physical simulations. If we posit 10^16 FLOPS (floating-point operations per second) to represent high-level brain function, a 100-year life simulation would demand approximately 3 x 10^25 FLOPS. Scaling this to the full population, the compute requirement reaches 3 x 10^34 FLOPS.

Several key points deserve further analysis:

- **Sequential Bottlenecks:** The assumption of limited parallelizability within a single simulated life poses a significant challenge. Modern supercomputers excel at tasks where computations can be broken down and distributed. Intra-life sequentially could create computational bottlenecks resistant to brute-force increases in processing power.
- **True Brain Simulation?:** Whether 10^16 FLOPS accurately approximates human-level cognition is open to debate. Neuroscience is far from a complete understanding of the computational principles employed by the brain. The actual processing requirements for simulating consciousness or high-level thought could be orders of magnitude higher or lower.

Crucial Considerations:

Advancements over Time:

Bostrom's argument posits a future with vastly more potent computing resources. Moore's Law and potential breakthroughs (e.g., quantum computing) need to be factored in, but with critical analysis of their limitations and uncertainties. While historical trends suggest continued advancement, it's vital to acknowledge that the trajectory of technological development is inherently difficult to predict with certainty. Extrapolating far into the future requires careful consideration of potential diminishing returns of current paradigms or unpredictable shifts towards entirely new computational models.

Additionally, even if raw computational power increases, the problems of sequential dependencies within a simulated life or the computational demands of simulating consciousness might prove resistant to brute-force solutions. Posthuman civilizations could develop novel computing architectures or simulation strategies that are currently unimaginable. However, including an honest assessment of both the potential for, and the fundamental limitations of, future computational progress paints a more balanced picture of the feasibility challenges within the Simulation Hypothesis.

Energy Demands:

Powering computations of this magnitude poses immense sustainability challenges. It's crucial to project energy consumption and ensure simulations are conceivable within the constraints of energy production and responsible ecological practices. Even optimistic projections about future energy technologies must be tempered with the understanding that the computational requirements of ancestor simulations could potentially rival (or dwarf) the energy demand of an entire planet-based civilization.

This raises profound ethical considerations for any posthuman civilization considering large-scale simulations. The energy expenditure for such a computational endeavor would likely have significant opportunity costs. Resource allocation decisions on this scale would force a simulator civilization to weigh the benefits gained from simulations against alternative uses of such vast amounts of energy. Analyzing the trade-offs inherent to the energy budget of a highly advanced civilization is vital for a holistic assessment of the Simulation Hypothesis's plausibility.

Data Analysis for the Ancestor Simulation Argument

A strong focus on data analysis plays a multifaceted role in evaluating the Simulation Argument:

- **Establishing Benchmarks:** Careful analysis of current technological trends, as well as their potential trajectories, provides a baseline for comparison. By quantifying existing computational capabilities, we gain a clearer understanding of the gap between what is currently achievable and the hypothesized requirements of ancestor simulations. This helps us assess whether bridging that gap is merely a matter of time and progress or would require fundamental breakthroughs outside our current paradigms of computation.
- **Constraining the Hypothesis Space:** Data analysis disciplines speculation about the nature and goals of possible simulator civilizations. Extrapolating computational trends, analyzing energy availability, or modeling the complexity of biological and social systems informs us about the kinds of simulations that might be feasible or desirable for advanced civilizations. This refinement prevents the Simulation Argument from spiraling into boundless speculation and maintains a connection to evidence-based reasoning.

How does a strong focus on data analysis help substantiate the Simulation Argument?

Refinement of Estimates:

Data-driven techniques can vastly improve upon simplistic estimations. Analyzing historical data on human behavioral patterns, technological progress, and information density can provide more realistic simulation parameter ranges. For instance, tracking the growth rate of information storage density across time can refine our assumptions about the total data required per simulated individual. Studies of historical population dynamics, resource use, and social complexity offer insights into how these scale with the size of the simulated society.

This approach moves us away from purely hypothetical calculations towards grounding our estimations in observable patterns. While extrapolation into the far future remains uncertain, leveraging historical data from our own civilization can establish better initial conditions for our models and potentially reveal scaling laws relevant to projecting computational requirements over long timescales.

Empirical Grounding:

Analysis of existing datasets (brain scans, historical records, social network interactions) could yield insights into compressibility, redundancy, and core structures within the information to be simulated. Examining neural activity patterns may reveal regularities and compressible features in brain function, potentially lowering the computational cost associated with simulating cognition. Similarly, historical and social data analysis could unveil repeated patterns or predictable trajectories in human behavior and interactions, suggesting more efficient ways for simulators to encode these aspects of a simulated world.

This approach links our simulation discussion to active areas of research. By understanding potential redundancies or inherent compressibility within the complex data of conscious minds and social behavior, we might identify avenues for optimization. This could lead to revised estimates about how computationally expensive it might truly be to achieve a sufficiently convincing ancestor simulation.

Identifying Key Bottlenecks:

Data analysis might isolate which elements of the simulation are likely to dominate computational costs. Is it simulating detailed physical environments, neurological activity, or high-level social decision-making? Targeted optimization strategies could emerge from this knowledge. For example, if the simulation of quantum-level phenomena in the physical environment proves most demanding, simulators might opt for a coarser approximation that nonetheless preserves the macroscopic outcomes relevant to the simulated inhabitants.

This analysis has implications for the simulators' goals. If their interest lies primarily in broad social trends or historical patterns, optimizations focused on abstracting away from computationally expensive low-level details become more viable. It could suggest a tiered approach to simulation, with high fidelity reserved for the most crucial aspects and computationally frugal strategies employed elsewhere when sufficient for the simulation's intended purpose.

CHAPTER THIRTY-FIVE

Statistical Estimation - Quantum Computing's Impact

The integration of quantum computing into the framework of Simulation Theory presents a profound shift in our understanding and estimation of the feasibility of ancestor simulations. Quantum computers operate on the principles of quantum mechanics, leveraging superposition and entanglement to perform complex calculations at speeds unattainable by classical computers. This quantum leap in computational power could dramatically lower the barriers to creating high-fidelity simulations of entire universes, including the minutiae of human consciousness and societal interactions. As we delve into the intricacies of quantum computing, it becomes apparent that the technology not only offers a pathway to overcome the previously insurmountable computational challenges posed by ancestor simulations but also reshapes the landscape of statistical estimation, enhancing our ability to analyze and predict the likelihood of our existence within such a simulation.

Furthermore, the potential advantages of quantum computing extend beyond mere computational speed. Quantum algorithms are inherently different from their classical counterparts, capable of exploring a vast number of possibilities simultaneously. This characteristic is particularly beneficial for ancestor simulations, which require the modeling of complex variables and interactions on a cosmic scale. Data analysis plays a crucial role in this context, as it allows us to quantify the impact of quantum computing on simulation theory. By employing sophisticated statistical methods, researchers can assess the reduction in computational resources required for these simulations and evaluate the probability of their feasibility. As quantum computing technology advances, it becomes imperative to recalibrate our statistical models and assumptions about the nature of reality, leading to a deeper understanding of the implications of simulation theory.

Quantum Simulation Advantages

The advent of quantum computing not only provides a technical leap in processing capabilities but also offers distinct advantages for the execution of ancestor simulations, a concept central to Nick Bostrom's Simulation Theory. Quantum computing, with its ability to handle vast datasets and perform calculations at an exponentially faster rate than classical computing, introduces a new realm of possibilities for simulating complex systems with unparalleled precision. This is particularly relevant when considering the intricate dynamics of consciousness, social interactions, and environmental factors within ancestor simulations. The essence of quantum advantage lies in its ability to process and analyze probabilistic outcomes simultaneously through quantum superposition, enabling a simulation environment that is far more nuanced and reflective of the probabilistic nature of our own universe.

Furthermore, the application of quantum computing to ancestor simulations extends beyond computational efficiency. It also enhances the fidelity and depth of the simulations, allowing for a more accurate representation of quantum phenomena within the simulated environment. This is crucial for achieving a level of realism where the inhabitants of a simulation are indistinguishable from real beings, a key criterion of

Bostrom's hypothesis. By leveraging quantum entanglement and superposition, simulations can incorporate the fundamental uncertainties and complexities of quantum mechanics, offering insights into how conscious experiences and physical realities might emerge from quantum processes. This quantum integration not only deepens our understanding of potential simulated realities but also pushes the boundaries of statistical estimation, enabling a more sophisticated analysis of the likelihood and implications of living within a simulated universe. Through this lens, quantum computing does not merely serve as a tool for enhancing computational power; it acts as a bridge to a more profound exploration of the nature of reality, simulation, and our place within the cosmos.

Here's how quantum computing might address the specific bottlenecks of ancestor simulations:

Representing Complex States

Within the discourse of Simulation Theory, particularly in the context of Nick Bostrom's propositions, the alignment of quantum computing with the inherent quantum nature of the physical universe posits a substantial advancement in the representation of complex states within simulations. This alignment facilitates a more authentic and less computationally intensive approach to modeling quantum-level phenomena, a critical component in the construction of high-fidelity ancestor simulations. The capability of quantum computers to naturally process and simulate quantum phenomena offers a direct pathway to mimicking the underlying quantum mechanics of our universe, thus significantly reducing the computational overhead traditionally associated with simulating detailed physical environments. This includes the intricate dynamics of neurochemistry, should the simulation require the inclusion of consciousness or conscious experiences to fulfill its objectives.

The implications of this quantum congruence extend beyond mere computational efficiency. It signifies a paradigmatic shift in how we conceptualize and potentially execute simulations of complex systems, including ecosystems, weather patterns, and even the human brain. By leveraging the quantum computational model to represent the probabilistic nature of quantum states, simulations can achieve a level of detail and realism previously deemed unattainable. This capacity to authentically replicate the quantum aspects of reality not only enhances the plausibility of ancestor simulations but also offers profound insights into the intersection of consciousness, quantum mechanics, and the fabric of reality itself. Through the lens of statistical estimation, this quantum-enabled approach to representing complex states enriches our methodological toolkit, allowing for a more nuanced analysis of simulation probabilities and their implications on our understanding of existence and the universe.

Simulating Parallelism

Quantum computing, by its very nature, excels at exploiting parallelism through its foundational principles of superposition and entanglement. This inherent capability to process multiple states simultaneously offers a compelling advantage in the simulation of parallel life paths, allowing for a vast array of individual experiences and histories to be computed in tandem. The challenge and opportunity lie in devising a framework where the complex, probabilistic interactions of countless entities can be accurately mapped onto quantum logic gates and algorithms.

The concept of simulating parallelism extends beyond mere computational efficiency; it embodies the philosophical and technical depths of Simulation Theory. By leveraging the quantum mechanical principle of parallelism, we can envision a simulation framework that not only mirrors the multifaceted nature of our universe but also explores the myriad ways in which life and consciousness could evolve under different circumstances. This approach requires a sophisticated understanding of both quantum computing and the theoretical underpinnings of simulation hypotheses, necessitating a transdisciplinary effort to bridge gaps between computational physics, neurophilosophy, and cosmology. In doing so, we inch closer to a comprehensive model that might one day unravel the mysteries of our existence and potentially reveal whether our reality is indeed a finely woven tapestry of simulated universes.

Optimization and Search

The implementation of quantum computing in the realm of Simulation Theory heralds a significant advancement in tackling the complexities associated with vast search spaces. These spaces are inherent in simulating historical trajectories and the multitude of decision-making processes that characterize human societies throughout history. Quantum algorithms, such as Grover's algorithm, provide a methodological leap by enabling quadratic speedups in search capabilities over classical algorithms. This quantum advantage becomes crucial when considering the need to navigate through the exponential growth of possible states and decisions in a simulation that aims to accurately reflect the richness and variability of human history.

The efficiency brought by quantum optimization and search algorithms directly impacts the feasibility and scope of ancestor simulations. By dramatically reducing the time required to explore and optimize large datasets, these quantum algorithms facilitate a more granular and precise modeling of historical events and decision-making processes. This capability not only enhances the realism and depth of the simulations but also allows for a more nuanced exploration of the theoretical and practical implications of living within a simulated reality. The application of quantum computing in this context extends beyond mere computational acceleration; it enables a transformative approach to understanding the complexities of existence and the potential for our reality to be an intricately designed simulation, providing a profound contribution to the ongoing discourse in simulation theory.

Data Analysis Challenges and Opportunities

The integration of quantum computing into the analysis of simulation probabilities introduces a dual-edged sword of challenges and opportunities. On one hand, the unparalleled computational capabilities of quantum computing offer the potential to significantly advance our ability to analyze and understand the vast datasets inherent in the execution of complex simulations. This quantum enhanced data analysis capability could lead to more refined estimations of the probability of our reality being a simulation, by allowing for the detailed modeling of phenomena that were previously beyond computational reach. This includes the nuanced dynamics of consciousness, societal evolution, and even the quantum underpinnings of the universe itself, providing a more comprehensive framework for assessing the validity and implications of Simulation Theory.

However, the transition to quantum data analysis also presents substantial challenges. The quantum computing paradigm introduces new complexities in data processing and algorithm design, necessitating a reevaluation of traditional statistical methodologies. Quantum algorithms, while powerful, require a fundamentally different approach to problem-solving, including the handling of quantum data and the interpretation of results within a probabilistic framework. Additionally, the nascent state of quantum computing technology means that practical applications in data analysis are still in their infancy, with issues of qubit coherence, error rates, and scalability posing significant hurdles. Despite these challenges, the potential for quantum computing to revolutionize our understanding of simulation probabilities remains vast. As researchers navigate these complexities, the convergence of quantum computing and data analysis in the context of Simulation Theory promises to unlock new perspectives on the nature of reality and our place within it, marking a pivotal juncture in the exploration of philosophical and scientific frontiers.

The role of data analysis is crucial in determining if the theoretical advantages of quantum computing translate to tangible benefits for ancestor simulations:

Benchmarking

In the context of assessing quantum computing's impact on Simulation Theory, particularly regarding the feasibility of ancestor simulations, the development of quantum-specific benchmarks emerges as a critical necessity. These benchmarks must be meticulously designed to address the unique requirements of simulating complex and historically plausible scenarios, as posited by Nick Bostrom's framework. Traditional metrics such as FLOPS (floating-point operations per second) are insufficient for capturing the multifaceted nature of quantum computational efficiency, especially when it involves the simulation of intricate quantum states and the dynamic interplay of countless variables that define human history and consciousness.

This necessity for bespoke benchmarking in quantum computing underscores the shift towards a more nuanced understanding of computational capability, one that encompasses the ability to model quantum phenomena within simulated environments accurately. Such benchmarks must evaluate the quantum computer's efficiency not just in raw computational terms but in its capacity to replicate the probabilistic nature of quantum mechanics and the complex decision-making processes inherent in the evolution of civilizations. This approach to benchmarking represents a significant challenge, requiring a deep integration of quantum physics, computational theory, and historical analysis. However, it also offers an unparalleled opportunity to advance our methodological frameworks for Simulation Theory, enabling a more precise and meaningful estimation of the probability and nature of our existence within a simulated reality. Through this, we can begin to quantify the previously qualitative aspects of our universe's simulation, providing a clearer lens through which to view the intricacies of our existence and the potential for it to be an elaborate construct of advanced computational intelligence.

Algorithm Mapping

The integration of quantum computing within the framework of Simulation Theory, and specifically its application to the statistical estimation of simulation probabilities, necessitates a nuanced approach to algorithm mapping. The complexity and uniqueness of simulation problems, as they pertain to the theoretical constructs of ancestor simulations, require that existing quantum algorithms be critically evaluated for their applicability and efficiency. In many instances, this evaluation may reveal the need for significant adaptations or the development of entirely new algorithmic approaches to effectively capture the intricacies of simulating realities that are both quantum-mechanically accurate and historically plausible.

Data analysis emerges as a pivotal tool in this exploration, offering insights into the compatibility of quantum algorithms with the specific demands of simulation scenarios. Through data analysis, researchers can uncover potential mappings that align quantum computational strengths with the simulation's requirements, while also identifying algorithmic bottlenecks that could impede progress. This process is fundamental to optimizing the use of quantum resources, ensuring that the simulations not only achieve the desired scale and complexity but also operate within the realms of computational feasibility. The challenge of algorithm mapping, therefore, stands at the crossroads of quantum computing and Simulation Theory, demanding a collaborative effort that spans computational science, quantum physics, and philosophical inquiry. As we advance in this endeavor, the role of data analysis in guiding the development of quantum algorithms tailored to simulation problems becomes increasingly critical, setting the stage for breakthroughs that could redefine our understanding of reality and our potential existence within a simulated universe.

Error Correction Assessment

Within the domain of Simulation Theory and the exploration of quantum computing's impact on the feasibility of ancestor simulations, the issue of quantum error correction emerges as a pivotal area of concern and study. Modern quantum computers, characterized by their noisy intermediate-scale quantum (NISQ) nature, are inherently prone to errors due to qubit decoherence and quantum noise. These errors present significant challenges for running complex simulations, particularly those that require a high degree of precision and stability over extended periods, as would be the case in simulating the intricate details of human history and consciousness. The role of data analysis in this context is to rigorously evaluate the scaling behavior of quantum error correction techniques and their overheads, especially as they relate to increasing the complexity and size of simulations.

The assessment of error correction mechanisms through data analysis involves a multifaceted approach that quantifies not only the direct impact of errors on simulation outcomes but also the computational and resource overheads imposed by error correction techniques. This evaluation is critical for understanding the practical limitations and scalability of quantum simulations of ancestor scenarios. It necessitates a deep dive into algorithmic efficiency, quantum hardware capabilities, and the theoretical underpinnings of quantum error correction schemes. As simulations grow in complexity, the overhead associated with maintaining quantum coherence and accuracy becomes a more pronounced factor in the overall feasibility and fidelity of the simulation. Therefore, data analysis must encompass a broad spectrum of considerations, from the theoretical limits of error correction to the practical implications for simulation architecture and design. This comprehensive assessment is crucial for advancing our understanding of the potential and limitations of quantum computing in realizing the detailed and expansive simulations envisioned by Simulation Theory, providing a clearer picture of the path towards achieving such ambitious computational undertakings.

Projections and Uncertainty

The investigation into quantum computing's transformative potential for Simulation Theory, as it pertains to the statistical estimation of simulation probabilities, culminates in a nuanced discussion on projections and the inherent uncertainties that accompany these forward-looking assessments. The theoretical underpinnings of Nick Bostrom's Simulation Theory, when juxtaposed with the emergent capabilities of quantum computing, provide a fertile ground for speculating about future developments in simulation technology. However, these projections are inherently laden with uncertainties, stemming from both the current technological infancy of quantum computing and the theoretical complexities of Simulation Theory itself. The dynamic interplay between these uncertainties necessitates a sophisticated approach to data analysis, where probabilistic models and Bayesian inference play pivotal roles in navigating the multifaceted landscapes of possibility.

As we project the future trajectory of quantum-enhanced simulations, the uncertainties we encounter are not merely technical but epistemological, challenging our very understanding of reality and the constraints of computational mimicry. The realms of quantum computing and ancestor simulations are replete with unknowns, from the scalability of quantum processors to the philosophical implications of achieving indistinguishable simulated realities. Data analysis, in this context, extends beyond conventional statistical tools to embrace a broader epistemological framework, incorporating theories of quantum mechanics, computational complexity, and philosophical inquiry. This comprehensive approach allows for a more informed and nuanced exploration of the probabilities and implications of existing within a simulation, acknowledging the vast spectrum of uncertainty that defines our quest to understand the cosmos and our place within it. Through this lens, the discussion of projections and uncertainty not only illuminates the challenges ahead but also highlights the profound implications of Simulation Theory in questioning the nature of existence and the potential for future technological advancements to blur the lines between simulated and authentic realities.

Let's make some informed, if speculative, projections about the timescale and magnitude of quantum impact:

Near-term Quantum Computing

As we delve deeper into the quantum computing era's implications on Simulation Theory, particularly within the next decade, our focus shifts towards the noisy intermediate-scale quantum (NISQ) computers. This imminent phase of quantum computing technology, characterized by its limited but non-trivial quantum capabilities, provides a critical testing ground for the early applications of quantum computing in simulating small-scale systems. These systems, relevant to historical analysis such as modeling material properties or small-scale social interactions, stand to benefit significantly from the nuanced computational approaches offered by NISQ devices. Such preliminary applications serve not only as proof of concept for the broader aspirations of Simulation Theory but also as a litmus test for the feasibility and potential direction of future, more complex simulations.

The exploration of NISQ-era quantum computing within the framework of Simulation Theory underscores a strategic intersection between technological progress and theoretical inquiry. By focusing on achievable, near-term applications, researchers can glean insights into the practical challenges and opportunities that quantum computing presents for simulating aspects of historical and social phenomena. This approach allows for a granular assessment of the capabilities of NISQ computers, including their ability to handle the probabilistic nature of quantum mechanics and the complexity of real-world phenomena, albeit on a smaller scale. Such endeavors not only contribute to the empirical foundations of Simulation Theory but also help to map out the trajectory for future advancements, highlighting the critical areas of algorithm development, error correction, and system scalability. This phased approach, starting with the NISQ era, thus provides a pragmatic pathway towards the realization of more comprehensive and detailed ancestor simulations, anchoring theoretical aspirations in the concrete milestones of quantum computational development.

Fault-Tolerant Quantum Computing

The advancement towards large-scale, fault-tolerant quantum computing systems represents a potential watershed moment for the practical application of Simulation Theory, particularly in the context of achieving high-fidelity ancestor simulations. While the timelines for achieving this quantum computing paradigm remain speculative, the implications for Simulation Theory are profound. Fault-tolerant quantum systems, by virtue of their ability to perform complex calculations without succumbing to errors that plague current quantum technologies, could enable the simulation of individual consciousness with unprecedented accuracy and depth. This milestone would mark a significant step towards the realization of simulations that can faithfully replicate the intricacies of human thought and experience, a cornerstone of Nick Bostrom's hypothesis on the nature of our reality.

However, it is critical to recognize that even with the advent of fault-tolerant quantum computing, the leap to full-scale ancestor simulations—encompassing the entirety of human history and interaction—remains contingent on further technological and theoretical breakthroughs. These simulations demand not only a quantum leap in computational power but also a comprehensive understanding of consciousness, social dynamics, and historical causality. The journey towards this ambitious goal is likely to be incremental, with each breakthrough in quantum computing providing new insights and capabilities that edge us closer to the possibility of such simulations. As such, the field of Simulation Theory must remain agile, continuously integrating advancements in quantum computing, data analysis, and our evolving understanding of the human condition. This iterative process underscores the dynamic interplay between technology and theory, driving us towards a future where the boundaries between simulated and genuine realities may become increasingly blurred.

The Bostrom Argument and Quantum Computing

The intersection of quantum computing advancements with Nick Bostrom's seminal Simulation Theory engenders a profound reevaluation of the argument's foundational premises. Bostrom's hypothesis, which posits that if advanced civilizations could run a significant number of ancestor simulations, we are likely to live in one, is dramatically influenced by the potentialities afforded by quantum computing. This technology promises to address the immense computational demands of such simulations, thereby adjusting the probabilistic framework within which Bostrom's argument operates. The feasibility of executing detailed simulations of human consciousness and historical epochs, once constrained by the limitations of classical computing, finds new viability under the aegis of quantum technology.

Critical to the application of this quantum leap to Bostrom's theory is a meticulous and informed data analysis. This analysis must transcend simple computational comparisons, delving into the nuanced implications of quantum computing's capacity to model complex systems with an unprecedented degree of fidelity. Assessing the computational advantages, alongside a realistic appraisal of the development timelines for quantum technology, becomes paramount. Such an evaluation not only informs our understanding of the potential immediacy with which ancestor simulations could be realized but also recalibrates the statistical underpinnings central to Bostrom's reasoning. As quantum computing continues to evolve, its implications for Simulation Theory necessitate an ongoing dialogue between technological progress and philosophical inquiry, underscoring the dynamic interplay between our technological capabilities and the existential questions that have perennially intrigued humanity.

CHAPTER THIRTY-SIX

Statistical Estimation - Probability Analysis

In this crucial segment of our examination into the intricacies of Simulation Theory, we undertake a comprehensive probability analysis of Nick Bostrom's Ancestor Simulation Argument. Utilizing a sophisticated amalgam of statistical techniques and Bayesian inference, this analysis is dedicated to dissecting and quantifying the likelihoods associated with Bostrom's tripartite proposition framework. Bostrom's argument hinges on three possible scenarios, each with profound implications for our understanding of reality: the possibility that almost all civilizations at our level of technological development go extinct before becoming posthuman; the likelihood that posthuman civilizations have little interest in running ancestor simulations; or, most intriguingly, the proposition that we are almost certainly living within such a simulation ourselves.

This exploration into probability analysis seeks not only to navigate these propositions through rigorous quantitative lenses but also to deepen our comprehension of the simulation hypothesis by grounding it in empirical assessability. Through the application of data analysis, particularly Bayesian reasoning, we aim to elucidate the inherent probabilities and to bring a level of mathematical precision to a debate often mired in philosophical abstraction. This methodological approach allows for a structured assessment of the simulation hypothesis, offering insights into the feasibility of the scenarios posited by Bostrom. By critically examining the statistical underpinnings and potential biases inherent in these propositions, this section endeavors to provide a foundational basis for further exploration and discussion within the field, thereby advancing our collective quest to understand the nature of our existence and the potential for it to be an elaborate construction of a posthuman civilization.

Bayesian Framework for Simulation Probability

The application of a Bayesian framework to analyze the probability of the scenarios outlined in Nick Bostrom's Simulation Theory provides a methodologically robust approach to navigating the complex probabilistic terrain of the simulation argument. Bayesian probability, distinguished by its capacity to iteratively refine and update probabilities based on new evidence, is particularly suited to the nuanced assessment required by Bostrom's propositions. This adaptability is crucial in the context of the simulation hypothesis, where empirical data, advances in technology, and evolving theoretical insights continually shape our understanding of the feasibility and implications of living within a simulated reality.

Utilizing Bayesian analysis allows us to construct a dynamic model in which the probabilities of Bostrom's core propositions are not static but evolve as our knowledge and understanding expand. This probabilistic framework enables the incorporation of diverse types of evidence, from developments in quantum computing and data analysis capabilities to philosophical and ethical considerations about the nature of posthuman civilizations and their potential motivations for creating ancestor simulations. By systematically updating our beliefs in the face of new information, the Bayesian approach provides a disciplined mechanism for quantifying the likelihood of each scenario Bostrom posits. It thereby transforms the simulation hypothesis from a purely speculative argument into a testable hypothesis subject to empirical validation and theoretical scrutiny. This transition underscores the profound utility of Bayesian reasoning in bridging the gap between abstract philosophical inquiry and the concrete realm of scientific investigation, offering a nuanced and quantitative foundation for exploring one of the most compelling and consequential questions of our time: the true nature of our reality.

Proposition 1: Extinction Probability

Within the Bayesian framework applied to analyze the simulation hypothesis, the first proposition addresses the grim possibility that civilizations at our stage of technological development are almost inevitably doomed to extinction before achieving the capabilities necessary for creating complex simulations. This proposition posits that the bottleneck to reaching a posthuman stage, where ancestor simulations are feasible, lies in the inherent risks associated with technological and societal advancements—risks that could lead to self-destruction or other civilization-ending cataclysms. The Bayesian approach allows for the integration of historical data, current global challenges, and risk assessment models to update our belief in the probability of this proposition being true.

Evaluating the extinction probability requires a comprehensive analysis of empirical evidence and theoretical models concerning existential risks, including nuclear warfare, environmental degradation, artificial intelligence mismanagement, and pandemics. By applying Bayesian reasoning, this analysis continuously refines the probability estimate in light of new evidence and advancements in understanding these global risks. This iterative process enables a nuanced assessment of the likelihood that our civilization, or any civilization at a comparable stage of technological development, will navigate these existential challenges successfully to reach a posthuman stage. The incorporation of this proposition into the broader simulation probability analysis underscores the critical interplay between technological progress, existential risk, and the potential realities of our existence within the framework of Simulation Theory, providing a grounded and quantitatively informed perspective on one of the key variables that define the debate surrounding our potential future and the nature of our reality.

Model and Assumptions

In the quantitative exploration of the first proposition of Nick Bostrom's Simulation Theory—concerning the likelihood that civilizations at our current level of technological sophistication fail to attain technological maturity due to self-destructive behaviors or other civilization-terminating scenarios—a sophisticated statistical model underpins our analysis. This model incorporates a comprehensive array of variables, including empirical data on historical extinction events, expert risk assessments of contemporary global challenges, and metrics of technological progress. The assumptions integral to this model are critical, as they set the parameters within which we evaluate the existential threats that could preclude the development of a posthuman civilization capable of creating ancestor simulations.

These assumptions encompass the probability distributions of various risks, the resilience of technological societies to potential catastrophic events, and the pace at which technological safeguards or mitigating strategies are developed and implemented. By systematically integrating these factors, the model aims to provide a statistically grounded estimation of the extinction probability for civilizations at a similar stage of development to our own. This analytical approach allows for the dynamic updating of the probability estimate as new data becomes available, reflecting the Bayesian methodology's inherent flexibility and responsiveness to evidence. Through this rigorous modeling and the careful consideration of assumptions, we seek not only to quantify the risks facing humanity but also to contribute to a deeper understanding of the factors that influence our potential futures. This endeavor, situated at the intersection of statistical analysis and existential inquiry, offers valuable insights into the broader discourse on Simulation Theory, challenging us to confront the precariousness of our existence and the paths that might lead us beyond the brink of extinction to a posthuman epoch.

Historical Extinction Events

The examination of archaeological and geological evidence forms the empirical foundation of our model, shedding light on the frequency, mechanisms, and aftermath of historical civilization collapses. This analysis not only delineates the commonalities among past extinction events but also provides insights into the resilience and vulnerability of complex societies. By drawing parallels between these historical precedents and the current global landscape, we construct a probabilistic model that captures the essence of existential threats across epochs. This historical perspective is pivotal for contextualizing present-day risks within a broader temporal framework, thereby enriching our understanding of the cyclical and unprecedented nature of such challenges.

Current Global Risks

The model dynamically incorporates contemporary risk assessments, including the threats posed by nuclear warfare, climate change, pandemics, and uncontrolled artificial intelligence. This integration is achieved through a meticulous analysis of data from leading research institutions and think tanks dedicated to global security and sustainability. By quantifying these risks and their potential impacts on civilization's continuity, the model adjusts the overall probability of extinction before reaching technological maturity. This aspect of the model is critical for capturing the evolving landscape of global risks, reflecting the fluidity of existential threats in the context of rapid technological and social changes.

Technological Advancements

A counterbalance to the risk assessments is provided by the inclusion of technological advancements that have the potential to mitigate existential risks. Through trend analysis and predictive modeling, the model assesses the pace and direction of technological progress, focusing on innovations in fields such as renewable energy, biotechnology, and artificial intelligence safety. This component of the model serves to highlight the dual-edged nature of technological development—both as a source of risk and a means of safeguarding against existential threats. By integrating these advancements, the model offers a nuanced view of the interplay between technology and survival, underscoring the potential for human ingenuity to navigate the path to technological maturity.

Equation and Estimation

The probability of extinction, P(E), can be modeled as:

$$P(E) = 1 - \exp\left(-\int_{t_0}^{t_m} \lambda(t)\,dt\right)$$

where t0 is the current time, tm is the time to reach technological maturity, and $\lambda(t)$ is the time-dependent rate of catastrophic events. This integrates over all potential risks across time to technological maturity.

Proposition 2: Interest in Running Simulations

The analysis of the second proposition in Nick Bostrom's Simulation Theory revolves around the assessment of whether technologically mature civilizations would possess the inclination to run ancestor simulations. This proposition challenges us to consider not just the technical capability to create highly sophisticated simulations but also the motivations that might drive a posthuman civilization to undertake such projects. Delving into this proposition requires a multifaceted approach that encompasses technological, psychological, and sociological dimensions to understand the potential desires and ethical considerations of future advanced societies.

To approach this analysis, we first consider the advancements in technology that would make ancestor simulations feasible, including computational power, energy requirements, and the development of sophisticated artificial intelligence. These technological prerequisites set the stage for the possibility of ancestor simulations but do not inherently dictate the motivations for conducting them. Therefore, we extend our analysis to explore the psychological and sociological drivers that could influence a posthuman civilization's interest in their own historical origins, the nature of consciousness, or the exploration of hypothetical scenarios through simulations. This involves an examination of historical precedents in human societies' pursuit of knowledge, entertainment, and ethical considerations, extrapolated to the context of a much more advanced civilization.

Moreover, ethical considerations play a critical role in this analysis. The decision to run ancestor simulations involves complex ethical dilemmas regarding the rights and realities of simulated beings, the conservation of resources, and the potential impacts on the civilization itself. By integrating these considerations into our probabilistic model, we aim to provide a nuanced estimation of the likelihood that a technologically mature civilization would engage in the creation of ancestor simulations. This estimation not only contributes to our understanding of the second proposition but also enriches the broader discourse on Simulation Theory by highlighting the intersection between technological capability and ethical, psychological, and sociological motivations.

Analytical Approach

The analytical approach to assessing the second proposition of Nick Bostrom's Simulation Theory, concerning the interest of technologically mature civilizations in running ancestor simulations, necessitates a multidimensional analysis. This proposition probes the motivations that such advanced societies might have for simulating their ancestors or historical epochs, demanding a synthesis of qualitative and quantitative data across various disciplines.

Survey Data

Central to this analysis is the examination of contemporary public and expert opinions on virtual reality, gaming, and simulations. By analyzing trends in these areas, we can extrapolate potential future interests in ancestor simulations. This involves not only parsing through the current enthusiasm for immersive experiences but also identifying underlying patterns that could indicate a sustained or growing interest in complex simulations. Such survey data serve as a direct pulse on societal inclinations, providing valuable insights into how technological advancements might shape cultural and recreational preferences over time.

Psychological Studies

Another critical component of this analysis is the application of psychological theories and empirical studies to comprehend the motivations driving the desire to simulate ancestors or historical events. This exploration delves into human curiosity, the drive for knowledge, the quest for immersive educational tools, and the psychological benefits of exploring alternate realities. By grounding our analysis in established psychological frameworks, we can construct a nuanced understanding of the complex interplay between human nature and the appeal of simulations, offering a robust basis for predicting future engagement with ancestor simulations.

Historical Trends

Additionally, the examination of the historical evolution of human interest in simulations, storytelling, and reenactments provides a proxy for anticipating future trends. This analysis recognizes that the human fascination with understanding our origins and exploring hypothetical scenarios is not a new phenomenon but rather a deeply ingrained aspect of our cultural and intellectual history. By tracing the trajectory of these interests from ancient storytelling and theatre to modern virtual reality and video games, we can discern patterns that may inform the future trajectory of simulation technologies and their adoption.

Together, these analytical lenses converge to offer a comprehensive evaluation of the likelihood that future technologically mature civilizations would invest resources and interest in running ancestor simulations. This multifaceted approach, encompassing survey data, psychological insights, and historical trends, provides a solid foundation for estimating the propensity for such endeavors, thereby enriching our understanding of the second proposition within the broader context of Simulation Theory.

Statistical Model

In the exploration of the second proposition of Nick Bostrom's Simulation Theory, which delves into the inclination of technologically mature civilizations to engage in the creation of ancestor simulations, the construction of a statistical model becomes imperative. This model is designed to estimate the probability of interest (P(I)) in such endeavors, integrating a multifaceted array of variables that encompass historical trends, technological capabilities, societal attitudes, and psychological underpinnings. A logistic regression framework is particularly suited for this analysis, offering a robust mechanism for predicting the likelihood of interest in running ancestor simulations based on these predictors.

This logistic regression model capitalizes on its ability to handle dichotomous outcome variables—such as the binary decision to engage or not engage in the creation of ancestor simulations—while accommodating a range of continuous and categorical input variables. The model's construction involves the careful selection and integration of predictors derived from survey data on attitudes towards virtual reality and simulations, psychological research on motivations for simulating ancestral or historical scenarios, and the analysis of longitudinal trends in human engagement with simulations and storytelling. By calibrating the model with these inputs, it becomes possible to quantitatively assess the factors that might influence a posthuman civilization's decision to simulate ancestral histories.

The application of this statistical model not only advances our understanding of the potential interests driving the creation of ancestor simulations but also provides a quantifiable measure of the likelihood of such interest manifesting in technologically advanced societies. Through this analytical approach, we endeavor to bridge the gap between speculative philosophical inquiry and empirical evidence, offering a nuanced and data-driven perspective on the motivations that could underlie the realization of the simulated realities posited by Simulation Theory. This statistical model, therefore, stands as a critical component in the overarching endeavor to evaluate the plausibility and implications of our potential existence within a meticulously crafted simulation, driven by the desires and capabilities of a future posthuman civilization.

Proposition 3: Simulation Probability

The final element of our analytical journey through Nick Bostrom's Simulation Theory involves a rigorous estimation of the probability that our current reality is, in fact, a simulation. This endeavor represents the culmination of the previous analytical efforts, synthesizing insights from the examination of technological progression, existential risks, and the potential interests of posthuman civilizations in creating ancestor simulations. The task at hand requires a multifaceted approach, drawing upon the foundations laid by the assessments of the first two propositions and integrating them with a broader understanding of simulation technologies and the philosophical underpinnings of simulated realities.

To estimate the simulation probability, we employ a comprehensive model that incorporates the estimated capabilities of posthuman civilizations to create highly detailed simulations, the motivations behind such endeavors as extrapolated from current trends and theories, and the statistical likelihood derived from the analysis of existing technological trajectories and sociological patterns. This model not only considers the technical feasibility of running large-scale simulations that are indistinguishable from reality but also factors in the philosophical and ethical considerations that would influence the decision to initiate such simulations. By integrating these dimensions, the model seeks to provide a quantitatively informed probability estimate of our existence within a simulation, encapsulated within the broader discourse of Simulation Theory.

This analysis, by necessity, transcends traditional statistical modeling, incorporating elements of theoretical physics, computational theory, and philosophical inquiry. The objective is to construct a probabilistic framework that is both grounded in empirical evidence and reflective of the complex interplay between technological advancement and existential curiosity that defines the human condition. Through this rigorous examination, we endeavor to offer a nuanced perspective on the probability that our perceived reality is a construct of an advanced civilization's technology, contributing to the ongoing exploration of one of the most profound and consequential questions facing humanity: the nature of our existence and the possibility that we are living within a fabricated cosmos.

Integration of Analyses

Estimating the probability that we are currently living within a simulated reality, denoted as P(S), necessitates a sophisticated integration of the analyses undertaken for Bostrom's first two propositions, with additional considerations related to computational capabilities, ethical implications, and the sheer scale at which technologically mature civilizations could operationalize ancestor simulations. This complex integration involves assessing the interplay between the existential likelihood of civilizations reaching a posthuman stage without succumbing to self-destruction ((P(E)), their subsequent interest in deploying resources towards creating simulations of their ancestral past ((P(I)), and the number of possible simulations (N) such a civilization could feasibly generate and maintain.

The formulation for P(S) requires a dynamic model that not only encapsulates these variables but also critically evaluates the computational and ethical frameworks that would underpin the decision to initiate and sustain large-scale simulations. Factors such as the advancements in quantum computing, the development of error correction mechanisms to ensure simulation stability, and the philosophical motivations for creating such detailed simulations of past realities are pivotal in this analysis. N, representing the estimated number of potential simulations, emerges as a crucial variable, reflecting both the technological prowess of a civilization and its philosophical or ethical inclination towards creating simulated realities for a multitude of purposes.

This probabilistic estimation engages with the dual aspects of feasibility and desirability, where the technological capacity to generate simulations meets the existential curiosity or historical interest of posthuman entities. The competition between the likelihood of reaching such an advanced stage of civilization (P(E)), the motivation to explore ancestral simulations (P(I)), and the practical aspects of executing these simulations at a scale (N) offers a nuanced view of our own reality's potential simulation status. Through this analytical lens, P(S) becomes a quantifiable expression of our existential scenario, grounded in the cross-disciplinary integration of simulation theory, computational advancements, and philosophical inquiry, providing a structured approach to one of the most intriguing questions of our time.

Finally, estimating the probability that we are currently living in a simulation, P(S), requires the integration of the analyses for propositions 1 and 2, alongside considerations of computational feasibility and the number of potential simulations.

$$P(S) = \frac{P(I) \cdot N}{P(E) + (P(I) \cdot N)}$$

where N represents the estimated number of possible simulations a technologically mature civilization could run. This formula accounts for the competition between the likelihood of extinction, interest in running simulations, and the technological and ethical feasibility of doing so.

Computational Feasibility and Ethical Considerations

The estimation of N, the number of possible simulations a technologically mature civilization could generate, necessitates a deep dive into the realms of computational feasibility and ethical deliberations. This estimation is intrinsically linked to the capabilities of quantum computing and artificial intelligence, technologies that stand to significantly expand the horizons of simulation potential. The computational aspect involves an assessment of the resources required to run simulations of significant complexity and fidelity, taking into account the energy consumption, processing power, and storage capacities that such endeavors would demand. Quantum computing, with its promise of surpassing the limitations of classical computing systems, introduces a paradigm shift in our understanding of what might be computationally feasible, thereby impacting the calculation of N.

Beyond the technical considerations, the ethical frameworks governing the creation of potentially sentient simulations introduce another layer of complexity in estimating N. These considerations encompass a broad spectrum of moral and philosophical issues, from the rights of simulated entities to the responsibilities of creators towards their simulated creations. The progression of AI technologies, which could be instrumental in managing and operating these simulations, also brings forward questions regarding autonomy, consciousness, and the definition of sentience. As such, the ethical landscape forms a critical part of the dialogue around simulation creation, potentially limiting or incentivizing the pursuit of large-scale ancestor simulations.

The interplay between computational feasibility and ethical considerations necessitates a dynamic approach to the estimation of N, one that is responsive to the rapid advancements in technology and evolving ethical standards. This ongoing reassessment is crucial for providing a grounded and forward-looking perspective on the potential scale of simulation efforts by posthuman civilizations. It underscores the multifaceted nature of Simulation Theory, where technological prowess is tempered by ethical imperatives, and the quest to understand our reality is bounded by the limits of our imagination and moral considerations.

Conclusion

The comprehensive probability analysis conducted within this section, leveraging the principles of Bayesian reasoning and sophisticated statistical modeling, offers a systematic and structured exploration of Nick Bostrom's Ancestor Simulation Argument. This multifaceted approach, incorporating a wide array of interdisciplinary data, computational forecasts, and theoretical insights, facilitates a deeper and more nuanced understanding of the simulation hypothesis' viability. It underscores the critical role that empirical evidence, advancements in technology, and ethical frameworks play in shaping our estimation of the likelihood that our reality is a simulation. This analysis does not merely quantify the probabilities associated with Bostrom's propositions but also illuminates the complex interdependencies between technological capabilities, philosophical considerations, and ethical dilemmas that underpin the simulation theory.

Furthermore, this examination highlights significant areas for future inquiry and data acquisition, pointing to the necessity of continuous reevaluation as our technological landscape evolves and our philosophical understanding deepens. The dynamic nature of this analysis reflects the evolving nature of the simulation hypothesis itself, suggesting that as we advance in our computational and ethical considerations, the boundaries of what we perceive as possible or ethical may shift, thereby altering our perception of the likelihood and implications of living within a simulated reality. In this way, the conclusion drawn here is not a definitive answer but a foundation for ongoing exploration, inviting researchers, philosophers, and technologists to contribute to a growing body of knowledge that spans the gamut from quantum computing and artificial intelligence to existential risk and ethical philosophy. This endeavor, rooted in rigorous data analysis and open to future refinements, encapsulates the intellectual challenge and intrigue of Simulation Theory, offering a fertile ground for advancing our understanding of the nature of reality and our place within it.

CHAPTER THIRTY-SEVEN

Statistical Estimation - Data Analysis Techniques

A core component of establishing the statistical plausibility of Bostrom's Ancestor Simulation Argument lies in developing robust data analysis methodologies to estimate the number of simulations (and thereby, simulated lives) a sufficiently advanced civilization might feasibly run. Let's explore key data-centric approaches:

Simulation Count Estimations

In this subsection, we delve into the statistical methodologies employed to estimate the potential number of simulations a post-human civilization might feasibly execute. This estimation is crucial for understanding the broader implications of the Simulation Theory, particularly in assessing the probability that our current reality is itself a simulation. The approach to estimating the simulation count N intertwines with several critical factors: available energy resources, computational limits imposed by physical laws, and the ethical constraints that may guide a post-human civilization's decision-making processes regarding the initiation of such simulations.

Energy and Computational Limits:

The theoretical framework for calculating N begins with the Bekenstein bound, which provides a limit on the amount of information (measured in bits) that can be contained within a given finite region of space that has a finite amount of energy. This physical principle, combined with estimates of a post-human civilization's access to energy resources (such as Dyson spheres or other megastructures for harnessing stellar energy), sets a foundational limit on the computational resources available for running simulations. From these calculations, one can infer the maximum number of high-fidelity simulations that could concurrently operate within these energy constraints.

Ethical Considerations:

Ethical considerations introduce a variable that is harder to quantify but is equally critical in estimating N. The decision to allocate vast resources to ancestor simulations may be influenced by ethical norms, the perceived value of historical accuracy versus resource conservation, and the potential rights of simulated beings. These factors could significantly reduce the number of simulations a civilization chooses to run, as ethical frameworks might prioritize the quality and ethical implications of each simulation over the quantity.

Statistical Model:

To integrate these considerations, a probabilistic model is formulated, potentially adopting a Bayesian framework for incorporating both the physical and ethical constraints into our estimation of N. This model allows for the updating of N as new information becomes available, such as advancements in quantum computing that may alter computational limits or shifts in ethical standards. For example, if the energy efficiency of quantum computations exceeds current projections, this could feasibly increase N, while a heightened ethical consideration for the rights of simulated entities could decrease it.

The calculation of N thus involves a nuanced balance between what is physically possible and what is ethically permissible. By employing statistical techniques that account for these diverse factors, we can arrive at a more informed estimation of the number of ancestor simulations a post-human civilization could run. This estimation not only contributes to our understanding of the likelihood that we exist within a simulation but also provides insights into the technological and ethical landscape of future civilizations capable of creating such simulations.

Energy-Based Analysis

The energy-based analysis for estimating the potential simulation count (N) a post-human civilization could operate hinges on a methodical examination of accessible astrophysical energy sources and the computational efficiency achievable with advanced technologies. This analysis forms a critical part of understanding the upper bounds of simulation capabilities, grounded in the physical realities of energy and the theoretical limits of computation.

Astrophysical Energy Sources

The initial step involves cataloging potential energy sources available to a technologically mature civilization, including stellar energy capture mechanisms akin to Dyson spheres and other theoretical constructs capable of harnessing the power output of entire stars or galactic cores. The total available energy (E_{total}) can be estimated based on the assumption that a post-human civilization can fully exploit the energy output of these sources within its galaxy or even beyond.

Computational Efficiency Projections

The analysis proceeds with the application of energy-efficiency benchmarks for computing, notably Landauer's Principle, which posits a minimum amount of energy required to erase one bit of information (kT ln 2), where k is the Boltzmann constant and T the temperature at which the computation is performed). Furthermore, the potential for quantum computing to exceed classical computational efficiencies is considered, particularly in terms of operations per energy unit consumed. This step translates the available energy E_{total} into the maximum number of floating-point operations (flops) that could theoretically be performed over the civilization's lifespan, denoted as F_{max}.

Computational Requirements and Simulation Count

Drawing upon the computational requirements detailed in earlier sections (5.3.1 Quantum Computing's Impact and 5.3.2 Data Analysis Challenges and Opportunities), which outline the resources needed to execute highly detailed simulations, we can establish an average computation cost per simulation (C_{sim}). This cost accounts for the complexity of simulating universes at a fidelity high enough to be considered realistic by post-human standards. The upper bound on the potential simulation count (N) is thus derived by dividing the maximum computable flops (F_{max}) by the computational cost of running a single simulation (C_{sim}):

$$N = \frac{F_{max}}{C_{sim}}$$

Through this energy-based analysis, we attain a scientifically grounded estimate of the number of possible simulations, anchoring the speculative aspects of Simulation Theory in the firm bedrock of physics and computational science. This methodology not only delineates the theoretical upper limits based on current understandings of energy and computation but also sets the stage for future recalibrations as our knowledge and technological capabilities evolve.

Constraints Imposed by Simulation Fidelity:

In evaluating the constraints that simulation fidelity imposes on the potential number of simulations a post-human civilization might run, it becomes imperative to consider the significantly heightened energy demands required for simulating full consciousness. This aspect of the analysis focuses on how the intricacy and depth of the simulations, especially those that aim to replicate consciousness with high fidelity, dramatically affect the overall energy consumption and, consequently, the feasible count of concurrent simulations (N).

Energy Demands per Simulation:

The requirement for high-fidelity simulations, particularly those that might include the detailed simulation of conscious experiences, necessitates a recalibration of energy estimates previously established. The energy per simulation (E_{sim}) dramatically increases when aiming to achieve a level of detail that encompasses the neural or quantum processes underlying consciousness. This higher energy demand per simulation directly impacts the calculation of (N), reducing the total number of possible simulations due to the more substantial energy allocation required for each:

$$N = \frac{E_{total}}{E_{sim}}$$

where (E_{total}) represents the total available energy resources a civilization can harness for computational purposes.

Trend Analysis and Extrapolation

To provide a basis for these estimations, an analysis of historical energy efficiency trends in computing is undertaken, including advancements in both hardware efficiency and algorithmic optimization. By extrapolating these trends into the future, we can propose hypothetical bounds for the energy efficiency of post-human civilizations. This extrapolation considers not only the continuation of Moore's Law-type improvements but also potential breakthroughs in quantum computing and other emergent technologies that could redefine computational energy efficiency.

Refining Simulation Counts

This detailed consideration of simulation fidelity and associated energy demands allows for a more nuanced estimation of the feasible simulation count. By acknowledging the substantial energy required for high-fidelity consciousness simulations, we can refine our estimates of (N), providing a more realistic assessment that accounts for both the technological possibilities and the inherent limitations imposed by the laws of physics and the ethical considerations of resource allocation.

In conclusion, the constraints imposed by simulation fidelity necessitate a comprehensive reevaluation of the potential number of ancestor simulations. This analysis underscores the importance of integrating detailed models of consciousness and energy efficiency trends into our broader understanding of Simulation Theory, highlighting the delicate balance between the desire for high-fidelity simulations and the finite resources available to a post-human civilization.

Ethical Considerations

Incorporating ethical considerations into the estimation of how many simulations a post-human civilization might run introduces a complex, albeit essential, dimension to the analysis of Simulation Theory. Philosophical deliberations regarding the allocation of resources towards ancestor simulations, juxtaposed against other potential utilizations of these resources, reveal the profound impact ethical considerations have on the theoretical limits of simulation counts (N). This aspect of the analysis acknowledges the inherently speculative nature of predicting the moral compass of advanced civilizations, yet it remains a crucial factor in refining our estimates.

Quantifying Ethical Constraints

Although translating ethical considerations into quantitative limits presents a significant challenge, a hypothetical approach involves setting conservative estimates on the proportion of resources a civilization might ethically dedicate to running simulations. For instance, if we assume a certain percentage (P_{eth}) of the total available resources (P_{total}) is deemed permissible for use in simulations, the equation adjusting (N) would encapsulate this constraint:

$$N = \frac{P_{eth} \cdot E_{total}}{E_{sim}}$$

Here, (E_{sim}) represents the energy required per simulation, factoring in the increased demand for simulations of high fidelity. The variable (P_{eth}) embodies the ethical limitations on resource allocation, directly influencing the total energy available for simulations and thus the upper bound on (N).

Philosophical and Ethical Frameworks

The consideration of ethical frameworks necessitates a dialogue with philosophical theories regarding the value of historical knowledge, the moral implications of creating simulated consciousness, and the prioritization of computational resources in a post-human society. This analysis might draw upon utilitarian principles, deontological ethics, or virtue ethics, among others, to speculate on how a post-human civilization might navigate these ethical dilemmas.

Influence on Estimated Counts

By introducing conservative limits on resource allocation for simulations based on ethical considerations, the estimated simulation count is invariably impacted. This approach not only acknowledges the finite nature of resources, even for technologically advanced civilizations, but also the possibility that ethical norms could significantly restrict the scale at which simulations are deemed appropriate or desirable.

In conclusion, the integration of ethical considerations into the statistical estimation of simulation probabilities highlights the multifaceted nature of Simulation Theory. It underscores the importance of interdisciplinary approaches, combining data analysis with philosophical inquiry, to fully explore the implications and feasibility of living within a simulated reality. This consideration of ethical constraints enriches the analysis, offering a more grounded and morally nuanced perspective on the potential behaviors of post-human civilizations and the existential questions at the heart of Simulation Theory.

Bayesian Inference

The Simulation Argument hinges on a critical question: how likely are we to be living in a computer simulation run by an advanced civilization? Given the inherent uncertainties surrounding future technological capabilities and the nature of consciousness, traditional statistical methods struggle to provide a nuanced answer. Here, Bayesian statistics offers a powerful framework. It allows us to incorporate our existing beliefs (through prior probabilities) and update them with new evidence (technological advancements, scientific discoveries) as they emerge. This dynamic approach is particularly well-suited for the Simulation Argument, where the landscape of possibilities is constantly evolving.

Prior Probabilities

Bayesian inference provides a powerful statistical framework for updating our beliefs about the world based on new evidence. In the context of Simulation Theory, this approach allows us to formalize our understanding of the likelihood that we live in a simulation, incorporating Nick Bostrom's seminal trilemma into a probabilistic model. The trilemma posits three possibilities: that posthuman civilizations capable of running ancestor simulations do not emerge; that such civilizations emerge but choose not to run ancestor simulations; or that we are almost certainly living within such a simulation.

Assignment of Prior Probabilities

To initiate this Bayesian analysis, we must assign prior probabilities to each of Bostrom's propositions. These priors represent our initial beliefs about the likelihood of each scenario before considering additional evidence.

- Existence of Posthumans Capable of Simulations (P_{exist}): The probability that civilizations can advance to posthuman stages capable of creating detailed ancestor simulations.
- Choice to Run Simulations (P_{run}): The probability that these posthuman civilizations choose to allocate their resources to running ancestor simulations.
- Existence Within a Simulation (P_{within}): The probability that we currently live within one of these simulations.

The prior for (P_{exist}) could be informed by assessments of technological progress and existential risk, (P_{run}) by psychological and cultural trends regarding the value placed on historical knowledge and recreation, and (P_{within}) by philosophical arguments and the other two probabilities.

Informing Prior Probabilities with Historical Analysis:

The historical human interest in recreating the past, through mediums such as historical simulations, literature, and genealogy, provides empirical data that can inform our prior for (P_{run}). This analysis suggests that there is an intrinsic human curiosity and value placed on understanding and simulating our historical context, which could extrapolate to posthuman civilizations. Therefore, the prior probability of posthuman interest in running ancestor simulations might be higher based on this historical continuity of human behavior.

To calculate the updated probabilities (posteriors) after considering new evidence (e.g., advancements in computing technology, philosophical debates on the ethics of simulations), we apply Bayes' theorem:

$$P(A|B) = \frac{P(B|A) \cdot P(A)}{P(B)}$$

Where P(A|B) is the posterior probability of hypothesis A given evidence B, P(B|A) is the likelihood of observing B given A,P(A) is the prior probability of A, and P(B) is the probability of observing B.

In the context of Simulation Theory, this would involve updating our beliefs about the existence of posthuman civilizations capable of and interested in running simulations, and the likelihood that we are within such a simulation, based on emerging technological capabilities and philosophical insights. This Bayesian inference process underscores the dynamic nature of our understanding of Simulation Theory, allowing for a rigorous and quantifiable assessment of its plausibility.

Updating with Technological Progress

In the Bayesian framework for assessing the simulation hypothesis, technological advancements serve as critical pieces of 'evidence' that inform our updating process, refining our posterior probabilities regarding the feasibility and likelihood of ancestor simulations. Computational milestones, such as achievements in floating-point operations per second (FLOPS) and significant increases in simulation scale, act as empirical indicators of our trajectory towards the capability for running high-fidelity ancestor simulations.

Incorporating Computational Milestones as Evidence

Each notable advance in computational power and efficiency can be considered incremental evidence that influences our belief in the probability of a posthuman civilization's ability to create and run simulations. For instance, reaching new FLOPS benchmarks that bring us closer to simulating complex systems with greater accuracy can be viewed as evidence supporting the premise that such simulations will eventually be feasible. In the Bayesian model, this evidence updates our posterior probabilities as follows:

$$P(Posthuman\ Capability|New\ Computational\ Milestone) =$$

$$\frac{P(New\ Computational\ Milestone|Posthuman\ Capability) \cdot P(Posthuman\ Capability)}{P(New\ Computational\ Milestone)}$$

Quantum Computing as a Significant Update

Breakthroughs in quantum computing represent a form of evidence with the potential to significantly shift our posterior probabilities. Quantum computing's promise to exponentially increase computational power and efficiency directly impacts our estimations of when and how ancestor simulations could become a reality. The transition from classical to quantum computing paradigms might necessitate a substantial recalibration of our probabilities, considering the quantum leap in capabilities.

Setting Realistic Timelines

To contextualize these updates within a realistic framework, an analysis of historical trends in computational performance is essential. By examining the progression of computational capabilities over time, we can model the rate of technological advancement and predict future milestones with greater accuracy. This historical analysis aids in setting realistic timelines for the accumulation of evidence and the consequent updates to our posterior probabilities regarding the simulation hypothesis.

The process of updating our beliefs about the simulation hypothesis in light of technological progress is an ongoing endeavor that reflects the dynamic nature of our understanding and capabilities. As we continue to achieve computational milestones and edge closer to the realm of quantum computing, our statistical models must adapt, incorporating each piece of evidence to refine our estimations. This iterative approach, grounded in Bayesian inference, provides a structured method for quantitatively assessing the evolving landscape of Simulation Theory in response to technological advancements.

Incorporating Uncertainties

The Bayesian framework excels in its capacity to integrate and quantify uncertainty, making it an ideal methodological approach for addressing the intrinsic uncertainties within Simulation Theory. When estimating the likelihood of our existence within a simulated reality, uncertainties abound—notably regarding future energy consumption patterns, technological advancements, and the ethical frameworks that might guide a posthuman civilization. To navigate these uncertainties, Bayesian analysis allows us to assign probability distributions to these unknown parameters, facilitating a comprehensive and nuanced understanding of the potential range of simulation counts N and the resulting posterior probabilities of each scenario within Nick Bostrom's trilemma.

Assigning Probability Distributions to Unknowns

Key to this process is the assignment of probability distributions to variables with inherent uncertainty. For instance, future energy use and availability, pivotal in determining the computational capacity for running simulations, can be modeled using distributions that reflect a range of possible outcomes based on current trends, theoretical limits, and potential technological breakthroughs. Similarly, ethical considerations, which may limit the scale or nature of simulations, can be represented by distributions that account for the variability in moral philosophies and their potential evolution over time.

$$P(N|E, T, Ethical\ Constraints) =$$

$$\iiint P(N|E = e, T = t, Ethical\ Constraints = ec)\, dP(e)\, dP(t)\, dP(ec)$$

Where E, T, and Ethical Constraints represent energy usage, technological advancements, and ethical limitations, respectively, each with its own probability distribution (dP).

Distribution Over Possible Simulation Counts

By incorporating these distributions, we derive a comprehensive distribution over possible simulation counts, reflecting the compounded uncertainties in our model. This approach not only highlights the most likely outcomes but also captures the range of plausible scenarios, providing a deeper insight into the dynamics at play.

Updating Posterior Probabilities

Bayesian methods excel in updating beliefs in light of new evidence. As uncertainties are inherently part of predicting future capabilities and decisions, the Bayesian framework's adaptability allows for the continuous refinement of our posterior probabilities as new information becomes available. This dynamic updating process ensures that our estimations remain as accurate and relevant as possible, despite the inevitable presence of unknowns.

Through the application of Bayesian methods to incorporate uncertainties, we attain a robust and flexible model for assessing the simulation hypothesis. This methodological approach underscores the importance of accommodating and quantifying uncertainty in our analyses, enabling a more informed and comprehensive exploration of the likelihood that we exist within a posthuman-created simulation. The use of probability distributions to represent uncertain parameters provides a foundation for ongoing inquiry and adaptation as our knowledge and understanding evolve, embodying the essence of scientific exploration and inquiry within the fascinating realm of Simulation Theory.

Sensitivity Analysis

In the realm of Simulation Theory, where the speculative intersects with the scientific, sensitivity analysis emerges as a critical tool for evaluating the robustness of our simulation count estimations (N) against variations in model assumptions. This analytical technique probes the extent to which the conclusions drawn about the feasibility and scale of ancestor simulations depend on the underlying parameters and assumptions embedded within our models. Given the inherently speculative nature of forecasting the capabilities and choices of posthuman civilizations, understanding the sensitivity of our estimates to changes in key inputs is paramount for constructing a reliable and insightful analysis.

Methodological Approach to Sensitivity Analysis

The process involves systematically varying model parameters—such as energy availability, computational efficiency, ethical constraints, and technological progression rates—to observe the corresponding effects on the outcome variable, (N). This method enables us to identify which assumptions have the most significant impact on our estimations, highlighting areas where further research or more precise data could substantially refine our understanding of the potential for and nature of ancestor simulations.

- **Parameter Variability:** For each parameter, a range of plausible values is established, based on current knowledge, theoretical projections, and historical trends. These ranges incorporate both optimistic and conservative scenarios, reflecting the broad spectrum of possible futures.
- **Model Recalibration:** The model is recalibrated with varied inputs, and the resulting simulation counts are compared to baseline estimations. This recalibration is conducted iteratively, allowing for a comprehensive exploration of the parameter space.
- **Outcome Analysis:** The outcomes of these recalibrations are analyzed to determine the sensitivity of (N) to each parameter. Parameters that cause significant variability in (N) are deemed "sensitive," warranting closer scrutiny and more rigorous justification for the assumptions made.

Implications of Sensitivity Analysis

Through sensitivity analysis, we not only gain insights into the stability and reliability of our simulation count estimations but also uncover the critical drivers of these estimations. This analysis sheds light on the speculative assumptions that most strongly influence the feasibility and scale of ancestor simulations, guiding future investigations and data collection efforts to areas that could most effectively reduce uncertainty and enhance model robustness.

Purpose of Sensitivity Analysis

Sensitivity analysis stands as an indispensable component of our methodological toolkit, enabling a deeper understanding of the dynamics underpinning Simulation Theory. By systematically assessing the impact of varying assumptions, we refine our approach to estimating the likelihood and scale of simulations, ensuring that our conclusions are both transparent and grounded in a rigorous examination of the speculative yet fascinating domain of posthuman possibilities. This approach not only underscores the importance of critical scrutiny in speculative analyses but also highlights the pathway toward a more empirically informed and robust exploration of the simulation hypothesis.

Managing Uncertainty

In the nuanced and speculative arena of Simulation Theory, the core objective of sensitivity analysis is to navigate and elucidate the impact of uncertainties inherent in our model assumptions on the estimations of the number of possible simulations N a posthuman civilization could feasibly conduct. This exploration is pivotal for assessing the robustness of our conclusions regarding the probability that we are living within a simulation. Given the complexities and inherent difficulties associated with forecasting technological advancements, sensitivity analysis serves as a vital methodology for preparing us to accommodate a spectrum of future scenarios, rather than anchoring our understanding to a singular, potentially fragile point estimate.

Quantitative Framework for Managing Uncertainty: To systematically manage uncertainty, the analysis utilizes a quantitative framework that identifies and modifies key variables within our model, observing the resultant variations in N. For instance, variables such as energy efficiency rates, computational power growth (aligned with Moore's Law or its quantum computing equivalents), and ethical limitation factors are adjusted within plausible ranges derived from historical data, expert predictions, and theoretical limits.

$$\Delta N = f(\Delta E, \Delta C, \Delta Ethical\ Constraints)$$

Where **Δ**N represents the change in the number of possible simulations, **Δ**E encapsulates variations in energy availability and efficiency, **Δ**C reflects changes in computational capacity, and **Δ**Ethical Constraints accounts for shifts in ethical considerations.

Scenario Analysis: A comprehensive scenario analysis complements the sensitivity analysis by constructing different futures based on varying combinations of these key variables. Each scenario represents a coherent set of assumptions, providing insight into how divergent technological and ethical pathways could influence the simulation capacity of a posthuman civilization.

Prioritizing Research

This methodological approach not only illuminates the relative importance of various factors but also strategically guides our research efforts towards the areas that promise the greatest refinement in our simulation count estimations. By systematically varying model inputs and observing the resultant effects on N, we can discern which variables—be it energy efficiency, computational power, or ethical considerations—warrant closer investigation for a more precise and reliable understanding of our likelihood of existing within a simulation.

Analytical Process: The analytical process involves adjusting one parameter at a time while holding others constant to isolate the effect of each variable on N. For example, by incrementally altering the energy efficiency parameter ($Eeff$) and observing changes in N, we can assess the sensitivity of our model to advancements in energy utilization:

$$\frac{\partial N}{\partial E_{eff}}$$

This derivative indicates how a small change in energy efficiency impacts the estimated number of simulations, providing insight into the parameter's influence.

Scenario-Based Exploration: Further, exploring a range of scenarios where each variable is adjusted across plausible future developments—spanning optimistic to conservative projections—allows for a comprehensive understanding of potential impacts on simulation capabilities. This scenario-based exploration contributes to a more nuanced appreciation of the dynamics at play.

Research Prioritization: Should sensitivity analysis reveal that energy efficiency (E_{eff}) exerts a more significant influence on N compared to, for instance, advancements in quantum computing (Q_{adv}), this insight directs our investigative priorities towards optimizing energy use in computational processes. Such prioritization is vital for focusing research resources effectively, ensuring that efforts are concentrated on the aspects most likely to yield substantial improvements in our estimations:

$$\frac{\partial N}{\partial Q_{adv}} < \frac{\partial N}{\partial E_{eff}}$$

Sensitivity analysis thus serves as a compass in the vast and speculative terrain of Simulation Theory, steering research towards the most impactful variables. By pinpointing the factors with the strongest influence on our simulation count estimations, this analytical approach not only enhances the accuracy and reliability of our models but also ensures that research efforts are allocated efficiently, addressing the critical uncertainties that shape our understanding of the potential for and the nature of ancestor simulations. This prioritization is indispensable for advancing our comprehension of Simulation Theory, navigating the speculative with a methodical and evidence-based strategy.

Key Actions in the Analysis

The sensitivity analysis within the ambit of Simulation Theory is a systematic process aimed at deciphering the impact of varying assumptions on the estimated number of possible simulations (N) a posthuman civilization might conduct. This meticulous approach involves several key steps, each crucial for understanding the dynamics and uncertainties inherent in our estimations.

Identify Key Parameters:

- **Simulation Scope:** The extent and depth of life and historical events simulated directly influence the computational demands. Questions like "How much of a life or historical period is simulated in detail?" are fundamental.
- **Energy Efficiency:** This parameter measures the computational power required per unit of simulation, pivotal in estimating the total possible simulations given energy constraints.
- **Quantum Computing Impact:** The introduction of quantum computing could significantly alter the landscape of computational efficiency and energy requirements, necessitating an assessment of its potential to reduce necessary computation and energy.

Assign Uncertainty Bounds

For each of the identified parameters, a single value is insufficient to capture the range of possible realities. Instead, plausible ranges are determined for each, acknowledging the inherent uncertainties. For instance, the energy efficiency parameter might span from the theoretical lower bounds provided by reversible computing principles to upper bounds informed by historical efficiency improvement trends.

Systematic Variation

With the parameters and their uncertainty bounds established, the next step involves varying one parameter at a time across its uncertainty range while holding others constant. For each variation, the estimated number of possible simulations (N) is recalculated:

$$N_i = f(Parameter_i\ Variation)$$

where N_i represents the recalculated simulation count based on the variation in parameter $_i$.

Observing Sensitivity

The sensitivity of our output, the simulated population count, to changes in each input parameter is then meticulously analyzed. Parameters that result in steep variations in N are identified as highly sensitive, indicating areas where more precise data or further research could significantly impact our overall understanding.

Monte Carlo Enhancement

To advance beyond the limitations of varying one parameter at a time, Monte Carlo simulation methods are employed. This approach involves simultaneous random sampling from the probability distributions of all parameters, offering a more nuanced and statistically robust assessment of uncertainty. This enhancement enables the generation of statistical error bars and a comprehensive visualization of potential outcomes, providing a deeper insight into the probabilistic landscape of simulation counts.

Through these systematic actions—ranging from parameter identification and uncertainty assignment to systematic variation and Monte Carlo enhancement—the sensitivity analysis illuminates the relative importance of different assumptions in Simulation Theory. This analytical rigor not only helps prioritize areas for focused research and data collection but also enhances our comprehension of the complexities and uncertainties that shape our estimations of the number of possible simulations and, by extension, the likelihood of our own existence within a simulated reality.

Alternative Simulation Designs

Sensitivity analysis extends to the consideration of alternative designs for the simulated worlds, specifically the variance in the level of detail encompassed within these simulations—ranging from partial simulations, which might focus on specific aspects or periods of existence, to full simulations, which aim to replicate an entire universe in exquisite detail. The methodological approach to understanding how these variations affect our estimations of the number of possible simulations (N) a posthuman civilization could run incorporates advanced statistical techniques, notably Monte Carlo simulations, to navigate and quantify the uncertainties inherent in these alternative simulation designs.

Variations in Simulation Detail

The level of detail in a simulation directly impacts the computational resources required, thereby influencing N. To systematically explore this, we define two distinct simulation categories:

- Partial Simulations ($S_{partial}$) : These simulations are characterized by their focused scope, perhaps simulating only certain historical periods, specific geographical regions, or particular aspects of human experience.
- Full Simulations (S_{full}): In contrast, full simulations endeavor to replicate an entire universe with as much detail as possible, including the complexity of every life form and the intricacies of physical laws.

Monte Carlo Simulation Approach

Employing Monte Carlo simulations enables the estimation of N under these varying simulation design paradigms by randomly sampling from distributions of computational efficiency, energy availability, and the proportion of resources allocated to simulation endeavors. This approach facilitates the generation of a probabilistic model that accounts for the variances in simulation design, offering a spectrum of potential outcomes rather than a single deterministic estimate. For each scenario ($S_{partial}$) and (S_{full}), the simulation might proceed as follows:

$$N_{scenario} = \sum_{i=1}^{n} \frac{E_{available,i} \times Efficiency_i \times Allocation_i}{Cost_{simulation,scenario}}$$

Where $E_{available,i}$, $Efficiency_i$, $Allocation_i$ are randomly drawn from their respective distributions for each iteration i, and $Cost_{simulation,scenario}$ reflects the computational cost associated with either partial or full simulations.

Observation of Variability and Error Bars:

The variability in $N_{scenario}$ across multiple Monte Carlo iterations provides insight into the sensitivity of our core statistical estimates to the design of the simulated worlds. The resulting distribution of $N_{scenario}$ for both partial and full simulations allows for the derivation of statistical error bars, offering a quantified measure of uncertainty around our estimates.

The sensitivity analysis of alternative simulation designs through Monte Carlo simulations enriches our understanding of the implications of simulation detail on the feasibility and scale of ancestor simulations. By quantitatively assessing the impact of partial versus full simulation designs on the computational demands and, consequently, on the number of possible simulations, we gain nuanced insights into the diverse strategies a posthuman civilization might employ in their simulation endeavors. This analytical approach underscores the complexity of Simulation Theory and highlights the importance of considering a broad array of factors and uncertainties in our quest to understand the nature of our reality.

Outcomes

This rigorous analytical approach offers a multifaceted understanding of the dynamics at play in estimating the feasibility and scale of potential simulations by posthuman civilizations. The key outcomes of this process illuminate the intricate balance between technological prowess, energy constraints, and philosophical considerations, shaping our grasp of the simulation hypothesis.

Revealing Bottlenecks

One of the primary benefits of sensitivity analysis is its ability to identify bottlenecks that significantly influence the estimated number of simulations N. These bottlenecks can arise from a variety of sources:

- **Technological Limits:** The pace of advancements in computing power and the theoretical limits of computational efficiency (e.g., Landauer's Principle for energy cost of computation) might constrain the scope of feasible simulations.
- **Energy Concerns:** The total available energy that a civilization can harness and allocate towards running simulations (considering constructs like Dyson spheres) may present a fundamental limit.
- **Philosophical Questions:** The ethical considerations and the intended scope of simulations (partial versus full universes, the degree of consciousness simulated) profoundly affect the design and number of simulations.

Each of these factors plays a critical role in determining the ultimate feasibility and scale of ancestor simulations, with sensitivity analysis providing a quantified measure of their impact.

Providing a Range, Not a Number

Rather than converging on a singular estimate for N, sensitivity analysis yields a distribution of possible values. This range is reflective of the inherent uncertainties and speculative nature of the underlying assumptions. By accounting for the variability in key parameters (energy efficiency, computational advancements, ethical thresholds), the analysis produces a probabilistic landscape of outcomes, offering a more nuanced and comprehensive view of the simulation hypothesis.

Guiding Future Inquiry

The outcomes of sensitivity analysis serve as a beacon for future research directions, pinpointing the areas where additional data, refined models, or philosophical debate could most significantly enhance our understanding of the simulation argument. For example, if technological limits emerge as a primary bottleneck, then research into quantum computing or other paradigm-shifting technologies becomes paramount. Conversely, if ethical considerations impose significant constraints on the number of simulations, then philosophical discourse and ethical frameworks for posthuman civilizations warrant deeper exploration.

In conclusion, the sensitivity analysis within the domain of Simulation Theory not only sheds light on the factors most critical to the argument's validity but also underscores the complexity and speculative nature of such inquiries. By offering a range of possible outcomes and highlighting the most impactful areas for further investigation, this analytical approach enriches our understanding of the simulation hypothesis and guides the discourse towards more grounded and informed deliberations.

CHAPTER THIRTY-EIGHT

Statistical Estimation - Equations and Estimations

Introduction to Monte Carlo Methods

The Monte Carlo simulation method, named after the famous Monaco gambling resort, is a computational algorithm that relies on repeated random sampling to obtain numerical results. The technique allows for the solving of complex probabilistic and deterministic problems by simulating a large number of scenarios and observing the aggregate outcomes. This approach is particularly useful in fields where analytical solutions are challenging to obtain due to the complexity of the model or the uncertainty of inputs, such as finance, physics, and, notably, the estimation of probabilities in philosophical arguments like Nick Bostrom's Ancestor Simulation Argument.

Historical Context

The inception of the Monte Carlo method, a cornerstone in the domain of computational mathematics and statistical analysis, is deeply rooted in the exigencies of mid-20th-century scientific research, particularly within the ambit of the Manhattan Project at Los Alamos. Developed by a cohort of eminent scientists, including Stanislaw Ulam, John von Neumann, and Enrico Fermi, this method was initially conceived as a solution to the complex problem of neutron diffusion—a critical component in the development of the atomic bomb. The innovation of the Monte Carlo method lay in its use of randomness to solve problems that might be deterministic in principle but are impractical to solve directly due to their complexity.

The utility of the Monte Carlo method has profoundly transcended its initial military applications, permeating various fields such as finance, engineering, environmental science, and beyond. Its fundamental principle—utilizing random sampling to approximate solutions to complex problems—has proven to be remarkably versatile. In the context of Simulation Theory, the Monte Carlo method offers a robust framework for exploring the probabilistic landscapes of potential simulations, incorporating a multitude of variables and their associated uncertainties.

The application of Monte Carlo simulations within Simulation Theory enables a detailed and statistically grounded exploration of the myriad factors influencing the feasibility, scale, and characteristics of hypothetical simulations by posthuman civilizations. By randomly sampling from distributions of key parameters—such as energy efficiency, computational advancements, ethical constraints, and technological limits—Monte Carlo simulations facilitate a nuanced assessment of the simulation hypothesis. This process generates a comprehensive array of potential outcomes, providing a probabilistic understanding that captures the inherent uncertainties and complexities of speculating about advanced civilizations and their capabilities.

The historical development of the Monte Carlo method underscores the evolution of computational techniques from their origins in the crucible of wartime research to their indispensable role in contemporary scientific inquiry. Its application to the speculative and multifaceted questions of Simulation Theory exemplifies the method's adaptability and enduring relevance. Through Monte Carlo simulations, researchers can transcend traditional analytical limitations, embracing the stochastic nature of the universe to glean insights into some of the most profound and elusive questions concerning the nature of reality and our place within it.

Application to Ancestor Simulation Probability

Monte Carlo simulations stand at the forefront of analytical methods when applied to the exploration of the Simulation Theory, particularly in estimating the probability of humanity's existence within an ancestor simulation. This probabilistic computational technique is uniquely suited to navigate the multifaceted uncertainties that characterize such speculative inquiries. By integrating diverse variables—ranging from the technological advancements and computational capacities of posthuman civilizations to the ethical frameworks that might govern their decisions to initiate simulations—Monte Carlo methods enable a comprehensive assessment of the simulation hypothesis.

Implementing Monte Carlo Simulations:

The application process involves defining probability distributions for each key variable, such as:

- **Technological Capabilities (T):** The distribution might encapsulate the range of possible advancements in computing technology, including quantum computing and beyond.
- **Ethical Considerations (E):** This reflects the potential ethical guidelines that could limit or motivate the creation of ancestor simulations.
- **Civilization Interests (C):** The interest of posthuman civilizations in recreating their ancestors or historical epochs is modeled as a variable with its own distribution.

Probabilistic Estimation Process:

Through repeated random sampling from these distributions, Monte Carlo simulations iterate numerous scenarios, each representing a unique set of conditions under which ancestor simulations might be created. For each iteration, a determination is made regarding the feasibility and likelihood of simulations being run, based on the sampled values of T, E, and C. The aggregation of these iterations yields a distribution of outcomes from which the probability of existing within an ancestor simulation can be derived:

$$P(Simulation) = \frac{Number\ of\ Iterations\ Indicating\ Simulation}{Total\ Number\ of\ Iterations}$$

Insights Gained:

In essence, the insights gained from employing Monte Carlo simulations within the context of Simulation Theory not only enhance our probabilistic comprehension of the likelihood of existing within a simulation but also delineate a roadmap for navigating the uncertainties and complexities of this intriguing hypothesis. Through this rigorous and reflective process, we can systematically uncover the nuances of Simulation Theory, fostering a more informed and dynamic discourse on the nature of reality and our place within it.

This methodological approach offers several key insights:

Probability Distribution of Existence Within a Simulation

Monte Carlo simulations yield a probabilistic landscape that captures the diversity of potential realities under the umbrella of Simulation Theory. This outcome is a nuanced probability distribution rather than a single, deterministic value. Such a distribution encompasses the wide array of possible scenarios, factoring in the dynamic interplay of technological advancements, ethical considerations, and the motivations of posthuman civilizations. It accounts for the multifaceted uncertainties inherent in each variable, offering a spectrum of likelihoods that our current reality might be a simulation. This probabilistic understanding enables us to appreciate the complexities and gradations of possibility, providing a more accurate reflection of the speculative nature of Simulation Theory.

Sensitivity to Variables

The variation in outcomes observed across Monte Carlo simulation iterations illuminates the relative impact of different variables on the overall estimation of simulation probability. This sensitivity analysis reveals which factors—be it the rate of technological progress, the ethical boundaries governing the use of such technologies, or the interest levels of civilizations in exploring their own ancestral pasts—exert the most significant influence on the feasibility and likelihood of creating ancestor simulations. Understanding these sensitivities is crucial, as it highlights the leverage points within the model where changes in assumptions or values can have the most pronounced effects on the outcome. Such insights are invaluable for prioritizing areas for rigorous scrutiny and for refining the theoretical underpinnings of Simulation Theory.

Guidance for Future Research

The outcomes of the Monte Carlo simulations, particularly the identified sensitivities and the probabilistic distribution of existence within a simulation, naturally guide future research directions. By pinpointing the variables with the greatest impact on the simulation probability and those surrounded by the highest degrees of uncertainty, this analytical approach directs attention towards the areas where additional data collection, technological exploration, or philosophical debate could most effectively refine and strengthen the simulation argument. For instance, if technological capabilities emerge as a highly sensitive variable, then research into the limits of computational power and efficiency becomes a priority. Similarly, if ethical considerations significantly influence the model's outcomes, then engaging in deeper ethical discourse and analysis becomes critical for advancing our understanding of the potential for and implications of ancestor simulations.

In conclusion, the application of Monte Carlo simulations to the estimation of ancestor simulation probability provides a nuanced and probabilistically rigorous exploration of the Simulation Theory. This approach not only encapsulates the complex interplay of factors that influence the generation of simulations by posthuman civilizations but also adapts to the speculative nature of the inquiry, offering a structured yet flexible framework for understanding the profound question of our potential existence within a fabricated reality. Through this analytical lens, we embrace the uncertainties and variabilities inherent in such a speculative domain, yielding insights that are both informative and foundational to the ongoing discourse on the nature of our reality.

Methodology

Defining Parameters and Assumptions

In the meticulous application of Monte Carlo simulations to the nuanced arguments of Simulation Theory, particularly Bostrom's hypothesis regarding ancestor simulations, the foundational step involves a thorough definition and quantification of the parameters that are pivotal to the creation, operation, and feasibility of such simulations. This rigorous process establishes a structured framework for the subsequent probabilistic analysis, ensuring that each variable is rooted in the best available theoretical and empirical knowledge. The key parameters to be defined encompass:

Technological Capabilities

- Computational Power (P_{comp}): This parameter reflects the processing capacity available to posthuman civilizations, which directly impacts their ability to run detailed simulations. It includes considerations of current trends in computational advancement, extrapolated into the future with allowances for potential breakthroughs or limitations. The theoretical limits, such as those posited by physical laws (e.g., the Bekenstein bound), provide a ceiling for this parameter.
- Energy Requirements (E_{req}): The energy required to power such computational endeavors is a critical factor, encompassing both the direct energy consumption of computing devices and the ancillary

energy needed for cooling and maintaining the computational infrastructure. Estimates are grounded in current energy efficiencies and projected improvements, alongside potential energy sources a posthuman civilization might harness (e.g., Dyson spheres).

- Advancements in Virtual Reality and Artificial Intelligence (VR/AI_{adv}): This parameter captures the qualitative and quantitative leaps expected in technologies that are directly relevant to the creation of convincing simulations. It considers the trajectory of virtual reality towards indistinguishability from physical reality and artificial intelligence's role in generating dynamic, responsive environments within simulations. The parameter includes speculative elements, given the rapid pace of change in these fields, and aims to encompass both the technological and experiential aspects of simulated realities.

Parameterization Process

For each of these parameters, a range of values is defined, based on a synthesis of current knowledge, historical trends, and speculative projections into posthuman capabilities. This range acknowledges the inherent uncertainty in predicting future technological developments and sets the stage for the Monte Carlo simulations to explore a wide spectrum of possible outcomes.

- Computational Power (P_{comp}) Range: Defined by current capabilities and projected exponential growth rates, tempered by potential physical and practical limits.
- Energy Requirements (E_{req}) Range: Based on current efficiencies and future energy source projections, including speculative posthuman technologies.
- VR/AI Advancements (VR/AI_{adv}) Range: Incorporates current trends in VR and AI, extending into speculative advancements that could significantly enhance simulation fidelity.

The meticulous definition and parameterization of technological capabilities serve as the bedrock for applying Monte Carlo simulations to Simulation Theory. By systematically quantifying these variables, we construct a probabilistic model capable of exploring the vast landscapes of potential realities, each shaped by the interplay of computational power, energy constraints, and technological advancements in virtual reality and artificial intelligence. This methodological approach not only illuminates the probabilistic nature of our existence within a simulated reality but also underscores the importance of grounding speculative inquiries in rigorous analytical frameworks.

Civilizational Variables

In the comprehensive analysis of Simulation Theory through Monte Carlo simulations, after establishing the technological capabilities as fundamental parameters, we proceed to delineate and quantify the civilizational variables critical to the assessment of the probability of humanity existing within an ancestor simulation. These variables encompass the likelihood of civilizations advancing to a stage of technological maturity sufficient for simulation creation, and importantly, their inclination towards undertaking such simulations, reflecting a blend of sociocultural, ethical, and existential motivations and constraints.

Technological Maturity Probability (P_{tm}):

This variable quantifies the likelihood of a civilization progressing to a level of technological development where running detailed ancestor simulations becomes feasible. The calculation involves an assessment of historical and current trends in technological advancement, risks of self-destruction due to existential threats (e.g., nuclear war, climate change, AI risks), and the potential mitigating strategies that could be employed to navigate these risks. A probabilistic model is constructed to estimate P_{tm}, incorporating factors such as:

$$P_{tm} = f(\text{Technological Advancement Rate, Existential Risk Mitigation Efforts})$$

Interest in Running Ancestor Simulations ($P_{interest}$)

This variable captures the propensity of technologically mature civilizations to invest resources in the creation of ancestor simulations. It is influenced by a myriad of factors, including cultural values, historical interest in understanding or recreating the past, ethical considerations about creating sentient beings within simulations, and philosophical or scientific curiosity. The estimation of $P_{interest}$ involves synthesizing insights from psychology, history, and philosophy to model the potential motivations and reservations civilizations might have regarding simulation projects:

$$P_{interest} = g(\text{Cultural Values, Ethical Considerations, Scientific Curiosity})$$

Methodology for Quantification:

For each civilizational variable (P_{tm} and $P_{interest}$), a range of values is determined based on interdisciplinary research, expert opinions, and speculative analysis, acknowledging the speculative nature of these assessments. Monte Carlo simulations then utilize these ranges to randomly sample possible values in numerous iterations, exploring the diverse scenarios under which ancestor simulations might be initiated.

The integration of civilizational variables into the Monte Carlo simulation framework is pivotal for a holistic analysis of Simulation Theory. It ensures that the evaluation of the simulation hypothesis is not solely predicated on technological feasibilities but also deeply ingrained in the fabric of civilizational dynamics—incorporating the complexities of human (and posthuman) motivations, risks, and ethical landscapes. This approach allows for a nuanced exploration of the conditions under which simulations are deemed desirable and viable, contributing to a more informed and comprehensive understanding of the probabilities involved in our potential existence within a simulated reality.

Ethical and Philosophical Considerations:

Integral to the comprehensive analysis of Simulation Theory via Monte Carlo simulations is the incorporation of ethical and philosophical considerations, which play a pivotal role in shaping the decision-making processes of technologically mature civilizations regarding the initiation of ancestor simulations. These considerations address the moral frameworks that might govern the creation of simulations, particularly those that entail the generation of conscious entities within simulated environments. The profound philosophical implications of such endeavors necessitate a careful and nuanced examination of the potential motivations, constraints, and ethical dilemmas that civilizations might confront.

Quantifying Ethical and Philosophical Variables

Ethical Framework Probability (P_{ef})

This variable represents the probability that a technologically mature civilization adopts an ethical framework that either permits, restricts, or entirely prohibits the creation of simulations involving conscious entities. The assessment of Pef involves analyzing a spectrum of ethical theories and principles, ranging from utilitarian considerations of the greatest good to deontological constraints on actions irrespective of outcomes. Modeling Pef requires an interdisciplinary approach, drawing from contemporary ethical discourse, speculative ethical evolution, and potential extrapolations of current human ethical trends into posthuman contexts:

$$P_{ef} = h(\text{Utilitarian Principles, Deontological Constraints, Speculative Ethics})$$

Philosophical Implications Assessment (I_{phi})

Beyond ethical frameworks, the philosophical implications of creating simulations, especially those involving consciousness, introduce additional layers of complexity. This includes considerations of the ontological status of simulated beings, the moral responsibilities towards them, and the broader metaphysical questions regarding reality and existence. I_{phi} encapsulates the range of philosophical stances that might influence a civilization's approach to simulations, from positions that encourage exploration and understanding of consciousness to those that caution against potential existential risks or moral hazards.

Methodology for Incorporation

For each ethical and philosophical variable (P_{ef} and I_{phi}), Monte Carlo simulations deploy a range of plausible assumptions, reflecting the diversity of potential ethical and philosophical landscapes a posthuman civilization might navigate. Through repeated random sampling, these simulations explore the impacts of varying ethical and philosophical considerations on the feasibility and desirability of running ancestor simulations.

The inclusion of ethical and philosophical considerations into the Monte Carlo simulation methodology enriches the analysis by acknowledging the profound and often speculative nature of these questions. It allows for a dynamic exploration of how different moral and philosophical frameworks might alter the landscape of simulation creation, offering insights into the potential ethical and existential contours of future civilizations. This approach not only broadens the scope of the simulation hypothesis analysis but also underscores the importance of ethical and philosophical inquiry in understanding the complexities and responsibilities associated with advanced technological capabilities.

Simulation Process

The Monte Carlo simulation process within the context of Simulation Theory, specifically for estimating the probability of humanity existing within an ancestor simulation, is a complex, iterative procedure that integrates a diverse array of variables and assumptions previously defined. This process synthesizes the parameters related to technological capabilities, civilizational variables, and ethical and philosophical considerations into a comprehensive statistical model. The objective is to generate a probabilistic distribution that reflects the myriad possibilities inherent in the theory, offering a nuanced understanding of the likelihood of our reality being a simulated construct.

Parameter Initialization

The simulation begins with the initialization of parameters, where each variable P_{comp}, P_{tm}, $Pi_{nterest}$,P_{ef} and I_{phi}, etc.) is assigned a probability distribution based on theoretical insights, empirical data, and speculative analysis. These distributions encapsulate the range of plausible values each parameter can take, reflecting the inherent uncertainties and variabilities in our assumptions.

Random Sampling

In the Monte Carlo simulation process, Random Sampling, is pivotal for exploring the parameter space of the simulation hypothesis. This step is where the theoretical meets the practical, translating our broad range of assumptions and uncertainties into specific scenarios to be analyzed. The essence of random sampling lies in its ability to systematically traverse the multidimensional landscape of variables that influence the probability of humanity existing within an ancestor simulation. Here's an expanded view on how random sampling operates within this context:

Mechanics of Random Sampling

Random sampling involves selecting values for each parameter from their respective probability distributions, which have been predefined based on a mixture of empirical data, speculative projections, and theoretical considerations. This selection process is stochastic, meaning that it is randomly determined but within the bounds of the distributions that represent our best understanding of each parameter's possible values.

- Technological Capabilities (P_{comp}): A distribution might reflect exponential growth trends in computing power, adjusted for potential breakthroughs or physical limits. A random value is drawn to represent the computational capacity available to a posthuman civilization in a given iteration.
- Civilizational Variables (P_{tm} and $P_{interest}$): These include distributions reflecting the likelihood of reaching technological maturity and the interest in running simulations. Random sampling from these distributions provides values that capture the diversity of potential civilizational trajectories and motivations.
- Ethical and Philosophical Considerations (P_{ef} and I_{phi},): Given the speculative nature of these variables, their distributions encapsulate a wide range of ethical frameworks and philosophical implications. Random draws from these distributions simulate how different ethical and philosophical stances might influence the decision to create ancestor simulations.

Statistical Representation

The process of random sampling in Monte Carlo simulations is not a one-off but is repeated across thousands or even millions of iterations. This repetition is crucial for statistical robustness, ensuring that the simulation results capture the full range of variability and uncertainty inherent in the parameters. Each iteration generates a scenario—a unique combination of parameter values—that is then evaluated to determine its implications for the existence of ancestor simulations.

Implementation Details

Random sampling is typically facilitated by computational algorithms designed to generate random or pseudorandom numbers within specified ranges. These algorithms ensure that each parameter value is selected independently of others, preserving the stochastic nature of the sampling process. The implementation might leverage uniform distributions for some parameters (where any value within the range is equally likely) or more complex distributions (like normal, log-normal, or beta distributions) for others, depending on the underlying assumptions about how those parameters behave.

Outcome Diversification

Through random sampling, the Monte Carlo simulation method inherently accounts for the diverse possibilities and outcomes that could arise from the complex interplay of technological advancements, civilizational dynamics, ethical considerations, and philosophical inquiries. This step diversifies the outcomes, providing a foundation for the probabilistic estimation of our existence within a simulated reality.

Simulation Execution

In the Monte Carlo simulation framework applied to Simulation Theory, Simulation Execution is where the theoretical rubber meets the road, translating the complex array of parameter values obtained through random sampling into actionable insights about the probability of humanity existing within an ancestor simulation. This step integrates the discrete scenarios generated by random sampling into a cohesive analysis, systematically evaluating each scenario against the backdrop of Simulation Theory's foundational questions. Here's an in-depth exploration of how simulation execution unfolds:

Integrating Parameters into a Coherent Model

Upon generating a set of parameters for a particular iteration—encompassing technological capabilities, civilizational variables, and ethical considerations—the execution phase involves integrating these variables into a coherent model that reflects the underlying dynamics of Simulation Theory. This integration is guided by a series of logical and mathematical relationships that articulate how these diverse factors interact to influence the feasibility and desirability of creating ancestor simulations.

Computational and Energy Constraints

The model calculates whether the randomly sampled values of computational power (P_{comp}) and energy efficiency align with the requirements for running high-fidelity simulations. This involves comparing the theoretical computational needs for simulating detailed realities against the available resources, considering the constraints imposed by physical laws and technological advancements.

Civilizational Willingness and Ethical Viability

Beyond the mere capability, the model assesses the willingness of civilizations to engage in ancestor simulations, factoring in the interests ($P_{interest}$) and ethical frameworks (P_{ef}) sampled. This step evaluates the propensity of a civilization, given its technological maturity and ethical stance, to allocate resources towards simulations that encompass conscious entities or detailed historical epochs.

Determining Simulation Feasibility and Probability

For each iteration, the execution phase culminates in a determination of the feasibility of ancestor simulations under the specified conditions. This decision is not binary but probabilistic, reflecting the spectrum of possibilities from full feasibility to various degrees of constraint:

- **Full Feasibility:** Scenarios where technological, civilizational, and ethical parameters align in favor of simulation creation. These iterations contribute positively to the overall probability of existing within a simulation.
- **Constrained Feasibility:** Scenarios where one or more parameters limit the scale, scope, or ethical permissibility of simulations. These iterations adjust the probability accordingly, reflecting the nuanced landscape of potential realities.

Aggregating Results to Form a Probability Distribution

Each iteration's outcome contributes to a growing dataset that, when aggregated, forms a comprehensive probability distribution of the likelihood that our current reality is a simulated construct. This distribution encapsulates the variance and uncertainty inherent in the parameters, offering a holistic view of the simulation hypothesis's plausibility.

Analytical Rigor and Computational Complexity

The execution phase is both analytically rigorous and computationally complex, necessitating sophisticated algorithms and computational resources. It embodies the multidisciplinary essence of Simulation Theory, drawing upon insights from computer science, physics, philosophy, and ethics to navigate the intricate web of factors that underpin the simulation argument.

Aggregation and Analysis

Aggregation and Analysis, serves as the critical phase where the results of numerous individual simulations are compiled and examined to derive a comprehensive understanding of the probability landscape regarding our existence within an ancestor simulation. This step transcends individual data points, synthesizing the outcomes of the simulation execution phase into a cohesive analytical narrative that offers deep insights into the overarching question posed by Simulation Theory. Here's an expanded view of how aggregation and analysis are conducted:

Compilation of Simulation Outcomes

Upon completing a multitude of simulation iterations, each yielding a determination on the feasibility and probability of ancestor simulations under varied parameter settings, the aggregation phase begins. This involves compiling the outcomes from all iterations into a unified dataset. The key here is to capture not just binary or simplistic outcomes but the probabilistic nuances that each iteration contributes, reflecting the spectrum of feasibility across the simulated scenarios.

Construction of Probability Distributions

The compiled dataset is then used to construct a probability distribution that reflects the likelihood of humanity existing within an ancestor simulation. This distribution is pivotal, as it encapsulates the collective insights derived from exploring the multidimensional parameter space defined by technological capabilities, civilizational variables, and ethical considerations. The shape and characteristics of this distribution—such as its mean, median, variance, and skewness—provide critical insights into the central tendencies and variabilities within the hypothesis.

Statistical Analysis and Interpretation

With the probability distribution established, a detailed statistical analysis is undertaken to interpret the findings. This includes:

- **Mean and Median:** Calculating the mean and median of the distribution offers insights into the central tendency of our probability of existing within a simulation.
- **Confidence Intervals:** Establishing confidence intervals around the mean or median provides a measure of certainty regarding the estimated probabilities, accounting for the inherent uncertainties in the simulation parameters.
- **Variance and Standard Deviation:** Analyzing the variance and standard deviation helps in understanding the dispersion of the probability estimates, indicating the degree of certainty or uncertainty associated with our estimations.

Sensitivity to Variables

A crucial part of the analysis involves assessing how sensitive the overall probability distribution is to changes in specific parameters. This sensitivity analysis helps identify which variables have the most significant impact on the outcomes, directing attention to areas where further data collection or refinement of assumptions could substantially alter the understanding of the simulation hypothesis.

Guidance for Future Research

The insights garnered from the aggregation and analysis phase not only enhance our comprehension of the likelihood and feasibility of ancestor simulations but also highlight areas ripe for future research. By pinpointing the parameters and assumptions that exert the greatest influence on the probability distribution, this phase provides a roadmap for where additional empirical data, technological insights, or philosophical deliberation could refine and advance the discourse on Simulation Theory.

Sensitivity and Scenario Analysis

Sensitivity and Scenario Analysis, represents a crucial phase where the robustness of the simulation outcomes to variations in input parameters is meticulously evaluated. This step further refines our understanding by delineating how changes in underlying assumptions impact the probability distribution of our existence within an ancestor simulation. It involves a detailed examination of the influence of each variable, providing insights into the stability of the simulation results and guiding future research directions. Here's an expanded overview of how sensitivity and scenario analysis are conducted:

Identifying Key Variables for Analysis

The process begins with the identification of key variables that have been included in the Monte Carlo simulations, such as technological capabilities, energy requirements, civilizational probabilities, and ethical frameworks. These variables are considered pivotal due to their potential impact on the feasibility and desirability of conducting ancestor simulations.

Conducting Sensitivity Analysis

Sensitivity analysis involves systematically varying each key variable within its plausible range while holding other variables constant, to observe the effect on the overall probability distribution. This methodical variation helps in identifying which variables have the most significant impact on the simulation outcomes. For instance, if altering the energy efficiency parameter results in substantial changes to the probability distribution of existing within a simulation, this parameter is marked as highly sensitive.

Quantitative Measurement

Sensitivity can be quantitatively measured by calculating the derivative of the probability of existence within a simulation with respect to each variable, indicating how small changes in the variable alter the outcome.

Scenario Analysis for Comprehensive Exploration

Beyond sensitivity analysis, scenario analysis is employed to explore the outcomes under different combinations of assumptions. This involves creating specific scenarios that reflect various plausible futures—ranging from optimistic to pessimistic projections of technological advancement, different ethical considerations, and civilizational interests. Each scenario is then analyzed through Monte Carlo simulations to understand its implications for the simulation hypothesis.

- **Scenario Construction:** Scenarios are constructed by selecting combinations of variables that represent coherent narratives about the future. For example, a scenario might combine rapid technological advancement, stringent ethical constraints on simulations, and high civilizational interest in ancestral knowledge.
- **Impact Assessment:** The impact of each scenario on the probability of existing within a simulation is assessed, providing a spectrum of potential outcomes based on different sets of assumptions.

Interpretation and Guidance for Future Inquiry

The insights gained from sensitivity and scenario analysis are interpreted to provide a deeper understanding of the dynamics underlying the simulation hypothesis. This phase not only elucidates the variables and scenarios that are most influential but also highlights areas where uncertainties are greatest, guiding future empirical research, theoretical development, and philosophical inquiry.

Prioritization of Research Efforts

Based on the findings, research efforts can be prioritized towards the variables and scenarios that show the greatest potential for altering our understanding of the probability of existing within a simulation.

Conclusion

The Monte Carlo simulation process, as applied to the analysis of Simulation Theory, is a

rigorous and flexible approach that accommodates the complex and speculative nature of estimating the probability of humanity existing within an ancestor simulation. By systematically exploring the impacts of a wide range of variables and assumptions through repeated random sampling, this methodological approach provides a probabilistic understanding of the simulation hypothesis. It underscores the significance of interdisciplinary insights and the dynamic interplay of technological, civilizational, ethical, and philosophical factors in shaping our conception of reality.

Equations and Estimations

In the application of Monte Carlo simulations to the examination of Simulation Theory, specifically the estimation of the probability of humanity existing within an ancestor simulation, the analytical process inherently involves a complex interplay of diverse and speculative parameters. Given this complexity, a single, overarching equation that neatly encapsulates the entire process is elusive. Nevertheless, a generalized framework can be outlined to convey the essence of how these simulations aggregate individual variables into a comprehensive probabilistic assessment. This framework integrates the probabilistic distributions of technological, civilizational, and ethical parameters through repeated random sampling to estimate the likelihood of our existence within a simulated reality.

$$P(Simulation) = \iiint \cdots F(\text{Technological Variables, Civilization Variables, Ethical Variables})\,dP(T)\,dP(C)\,dP(E)$$

Where:

- **P(Simulation)** represents the probability of existence within a simulation
- **F(·)** is a function that evaluates the feasibility and desirability of conducting ancestor simulations based on the inputs from technological variables (T), civilizational variables (C) and ethical variables (E).
- **dP(T), dP(C), and dP(E)** denote the differential probability distributions for technological, civilizational, and ethical parameters, respectively.

Key components of the Framework

- **Technological Variables (T):** Include computational power (P_{comp}), energy efficiency (E_{eff}), and advancements in virtual reality and AI (VR/AI_{adv}). Each variable follows a probabilistic distribution based on current trends, speculative advancements, and theoretical limits.
- **Civilizational Variables (C):** Encompass the probability of reaching technological maturity (P_{tm}) and the interest in running ancestor simulations ($P_{interest}$). These variables are informed by historical data, sociocultural dynamics, and speculative future developments.
- **Ethical Variables (E):** Reflect the ethical frameworks (P_{ef}) that might guide or restrict the creation of simulations, particularly those involving conscious entities, as well as philosophical implications (I_{phi}) of such endeavors.

Estimation Process

The Monte Carlo simulation iteratively samples values from the distributions of T, C, and E to simulate a wide array of possible realities. Each iteration evaluates the combination of sampled values through the function F(·), determining the contribution of that particular scenario to the overall probability of existence within a simulation. The aggregation of these individual probabilities across all iterations yields a comprehensive estimate of P(Simulation), reflecting the spectrum of potential outcomes.

A Data-Driven Approach

Monte Carlo simulations provide a powerful tool for exploring the vast uncertainties inherent in Bostrom's Ancestor Simulation Argument. By allowing for the statistical estimation of simulation probabilities across a wide range of scenarios, this approach not only offers insights into the likelihood of our reality being a simulation but also underscores the significant impact of technological, civilizational, and ethical parameters on this probability. Despite the speculative nature of many inputs, the Monte Carlo method enables a structured, data-driven analysis that contributes significantly to the academic discourse surrounding simulation theory, highlighting areas where further empirical research and philosophical inquiry can refine our understanding of our place in the cosmos.

CHAPTER THIRTY-NINE

Statistical Estimation - Expert Opinions and Elicitation

The exploration of Nick Bostrom's Ancestor Simulation Argument is not solely confined to the realms of computational theory, artificial intelligence (AI), and quantum computing; it also deeply intersects with philosophical, ethical, and technological discourse. As such, the elicitation of expert opinions becomes a pivotal component of understanding and evaluating the feasibility and implications of the simulation hypothesis. This section delves into the insights and perspectives of leading figures from various disciplines, presenting a comprehensive qualitative analysis that complements the quantitative data analysis previously discussed.

Background on Expert Elicitation

Expert elicitation is a systematic method of synthesizing opinions from specialists to inform decision-making in areas where empirical data may be sparse, uncertain, or nonexistent. This approach has been utilized across a broad spectrum of disciplines, including environmental science, risk assessment, and technology forecasting. In the context of the simulation hypothesis, expert elicitation serves to gauge the consensus (or lack thereof) among scientists, philosophers, and technologists regarding the plausibility and ethical considerations of creating ancestor simulations.

Methodological Framework

The elicitation process involves several key steps:

- **Selection of Experts:** Identifying individuals with recognized expertise in fields relevant to the simulation hypothesis, including computational science, philosophy of mind, ethics, and futurology.
- **Elicitation Technique:** Employing structured methods such as the Delphi technique, which iterates rounds of questioning with controlled feedback, or direct interviews to capture nuanced views.
- **Analysis and Synthesis:** Aggregating the responses to highlight areas of agreement, divergence, and uncertainty, providing a textured understanding of the expert landscape.

Insights from Technological Experts

Technological experts focus on the feasibility of achieving the computational power and sophistication in AI required for running high-fidelity ancestor simulations. They discuss advancements in quantum computing, neural network architectures, and energy requirements, offering estimations on timelines and potential technological barriers. This section should encapsulate their views on the trajectory of technological progress and its alignment with the prerequisites identified in Bostrom's argument.

Philosophical and Ethical Perspectives

Philosophers and ethicists weigh in on the implications of creating simulated realities inhabited by conscious entities. They debate the moral responsibilities of creators towards simulated beings, the nature of consciousness in simulated environments, and the potential impacts on our understanding of reality. This discourse enriches the analysis by introducing ethical considerations and philosophical challenges that quantitative methods alone cannot address.

Integration with Data Analysis

Expert opinions offer qualitative insights that, when integrated with quantitative data analysis, provide a multidimensional view of the simulation hypothesis. This section should present a synthesis of how expert perspectives align with, challenge, or expand upon the findings from statistical estimations and Monte Carlo simulations. For example, if technological experts project a faster rate of computational advancement than accounted for in quantitative models, this could adjust the estimated probability of living in a simulation. Conversely, philosophical and ethical reservations might suggest constraints on the creation of ancestor simulations that quantitative models fail to capture.

Critical Evaluation

A critical evaluation of expert opinions is necessary to assess their credibility, biases, and the evidence base supporting their views. This involves analyzing the diversity of perspectives, the rationale behind differing opinions, and the extent to which they are grounded in current scientific understanding or speculative assumptions.

Historical Context of Expert Elicitation

The methodology of expert elicitation, integral to the fabric of interdisciplinary research, especially in fields rife with uncertainty and speculative inquiry, has its roots in a distinguished tradition of structured analytical techniques. Among these, the Delphi method, developed in the 1950s by the RAND Corporation, stands out for its pioneering approach to harnessing the collective intelligence and insights of experts to navigate complex scientific and philosophical landscapes. This methodological innovation was initially aimed at addressing the challenges of forecasting and decision-making in the context of the Cold War, emphasizing the strategic assessment of uncertain future events.

Delphi Method and Its Evolution:

The Delphi method, emblematic of expert elicitation techniques, embodies a sophisticated iterative process designed to harness and refine the collective expertise of diverse specialists through structured rounds of questionnaires. Originating in the 1950s as a methodological innovation by the RAND Corporation, its initial purpose was to forecast and strategize under conditions of uncertainty, particularly in the realms of military and technological forecasting. The method's name, invoking the ancient Greek oracle of Delphi, symbolizes its pursuit of deeper insights into complex issues through systematic inquiry.

Iterative Process and Anonymity

Central to the Delphi method is its iterative nature, comprising multiple rounds of questionnaires where experts are asked to provide their opinions on a specific issue. After each round, a summary of the responses is shared with the participants, including the range of opinions and the underlying reasoning, without attributing any responses to specific individuals. This anonymity is crucial, as it mitigates the influence of dominant personalities, allowing for a more egalitarian and unbiased synthesis of expert views. Experts are then invited to revise their initial responses in light of the collective feedback, leading to a convergence of opinions over successive rounds.

Evolution and Adaptation

Since its inception, the Delphi method has undergone significant evolution, adapting to the changing contours of scientific inquiry and decision-making processes. Its applications have expanded far beyond its original military and technological forecasting purposes, permeating various fields such as public policy, health care, environmental science, and, notably, speculative scientific theories like the Simulation Hypothesis. The method has also seen variations in its implementation, including real-time Delphi and policy Delphi, which tailor the process to specific contexts and objectives, enhancing its flexibility and applicability.

Application to the Simulation Hypothesis:

In the multifaceted investigation of the Simulation Hypothesis, where empirical data often reaches its limits, expert elicitation emerges as an indispensable methodological approach. This hypothesis, which suggests that our perceived reality could be an artificial simulation created by a more technologically advanced civilization, intersects the disciplines of theoretical physics, computer science, philosophy, and beyond. Given its speculative foundation and the profound implications it harbors, the structured elicitation of expert opinions serves as a vital conduit for advancing our understanding.

Bridging Speculation and Analysis

The application of elicitation techniques, particularly the Delphi method, facilitates a critical engagement with the Simulation Hypothesis that transcends mere speculation. By systematically gathering, refining, and analyzing the insights of experts across relevant fields, this approach provides a multifaceted examination of the hypothesis. It allows for a nuanced exploration of the various dimensions of the argument, from the technical feasibility of creating such detailed simulations to the philosophical and ethical ramifications of their existence.

Technological Feasibility

One of the key areas where expert elicitation proves invaluable is in assessing the technological feasibility of ancestor simulations. This encompasses not only the current state of computing power and virtual reality but also the projected advancements in these domains. Experts contribute insights into the pace of technological innovation, potential breakthroughs, and the theoretical limits of computational capacity, offering a grounded perspective on the possibility of achieving the level of detail required for convincing simulations of reality.

Ethical Considerations

The ethical dimensions of the Simulation Hypothesis—pertaining to the creation and potential rights of conscious entities within simulations—pose complex questions that benefit significantly from expert elicitation. Through structured dialogue, ethicists, philosophers, and technologists can collaboratively explore the moral implications, providing a framework for considering the responsibilities of creators and the ethical treatment of simulated beings.

Civilizational Motivations

Understanding the motivations that might drive a civilization to create ancestor simulations requires insights into human (and potentially posthuman) psychology, culture, and history. Expert elicitation draws on the perspectives of historians, psychologists, and sociologists to delve into the reasons a civilization might pursue such simulations, whether for scientific research, entertainment, or cultural preservation.

Synthesis of Diverse Perspectives:

The Delphi method and similar techniques offer a systematic means of synthesizing these diverse expert perspectives into a coherent analysis that respects the complexity of the Simulation Hypothesis. By facilitating anonymous feedback loops and iterative refinement of opinions, expert elicitation helps distill a collective intelligence that can navigate the speculative yet intellectually fertile terrain of the hypothesis.

Value in Areas of Uncertainty:

In the investigation of theories that stretch into the realms of speculation and futurism, such as the

Simulation Hypothesis, the methodologies of expert elicitation, including the Delphi method, assume an indispensable role. These techniques offer a structured means to mitigate the inherent uncertainties that challenge direct empirical validation, especially in disciplines where the speculative intersects profoundly with the empirical, theoretical physics meets computer science, and philosophical inquiry melds with technological speculation.

Navigating Empirical Gaps

The terrain of the Simulation Hypothesis is one where empirical data can often seem like a distant beacon rather than a guiding light. In such contexts, expert elicitation techniques become invaluable, not as mere surrogates for empirical data but as sophisticated tools for bridging the chasm between what is known and the vast expanses of the unknown. Through the iterative, structured dialogue of the Delphi method and similar approaches, the collective wisdom of experts is harnessed to construct a scaffold that reaches into these speculative domains, providing a foundation upon which further inquiry can be structured.

Synthesizing Diverse Expertise

These techniques excel in drawing together strands of expertise from disparate fields—combining insights from computer science regarding the limits of computational power, from theoretical physics about the nature of reality, and from philosophy concerning the ethics of creation and existence within potential simulated realities. This synthesis of expert opinions yields a more nuanced and multidimensional understanding of the Simulation Hypothesis, transcending the limitations of any single disciplinary perspective.

Identifying Consensus and Divergence

One of the core strengths of expert elicitation is its ability to map the landscape of professional agreement and disagreement. By identifying areas of consensus, these methods can highlight the aspects of the Simulation Hypothesis that hold the most robust support among experts, serving as pillars for further theoretical development. Conversely, by elucidating areas of divergence, they signal the frontiers of current understanding and debate, delineating the boundaries where empirical evidence is most acutely needed and where speculative inquiry pushes into new territories.

Ethical and Philosophical Navigation

The speculative nature of the Simulation Hypothesis inherently encompasses deep ethical considerations and philosophical questions—ranging from the rights of potentially simulated consciousnesses to the implications of such simulations for our understanding of reality. Expert elicitation acts as a navigational tool through these complex issues, providing a mechanism for exploring these dimensions in a structured, informed manner. It allows for the collective wisdom of ethicists, philosophers, scientists, and technologists to converge, offering a balanced and comprehensive exploration of these quandaries.

The Wisdom of Collective Inquiry

The historical context of expert elicitation, epitomized by the development and application of the Delphi method, underscores the enduring relevance of structured expert opinion in advancing our comprehension of complex, speculative theories like the Simulation Hypothesis. This tradition of engaging with the depth and diversity of expert insight enriches the discourse, providing a nuanced scaffold upon which the theoretical and philosophical dimensions of such hypotheses can be explored and evaluated. In the domain of Simulation Theory, where the interplay of advanced technology, existential inquiry, and speculative future scenarios defies simple analysis, expert elicitation stands as a testament to the value of collective intelligence in charting the unknown.

Methodology of Expert Elicitation

The exploration of the Simulation Hypothesis, a concept that sits at the confluence of technological possibility and philosophical inquiry, necessitates a rigorous and structured approach to expert elicitation. In this context, a modified Delphi method has been strategically utilized to tap into the collective intellect of experts spanning the domains of computer science, artificial intelligence (AI), quantum computing, theoretical physics, and philosophy. This approach is designed to encapsulate a diverse range of perspectives, from the technological optimists, who are bullish on the rapid evolution of simulation technologies, to the philosophical skeptics, who raise critical questions about the ethical implications and theoretical underpinnings of simulating conscious realities.

Selection Criteria for Experts

The methodology hinges on a meticulous selection of participants, with criteria emphasizing not only expertise in their respective fields but also a demonstrated engagement with the interdisciplinary questions central to the Simulation Hypothesis. This selection process ensures a rich tapestry of insights, encompassing:

- **Technological Expertise:** Individuals at the cutting edge of computer science, AI, and quantum computing, whose work directly informs the potential for simulation technology.
- **Theoretical Insight:** Theorists from fields like physics and computational theory, who can provide deep insights into the fundamental limits and possibilities of simulating reality.
- **Philosophical Perspective:** Philosophers specializing in ethics, consciousness, and metaphysics, offering critical analysis of the implications and feasibility of creating simulated realities.

Iterative Process and Feedback Mechanism

The modified Delphi method employed in this study leverages an iterative process of questionnaires and feedback, tailored to foster a dynamic exchange of ideas among the selected experts. This process includes:

- **Initial Questionnaire:** Experts are first presented with a series of open-ended questions designed to elicit their initial perspectives on the technological feasibility, ethical considerations, and philosophical implications of the Simulation Hypothesis.
- **Summary and Anonymized Feedback:** Responses are anonymized and summarized to highlight areas of consensus and divergence, then shared with the participant group. This summary includes not just statistical aggregations of opinion but a nuanced synthesis of the reasoning behind different viewpoints.
- **Revised Questionnaire:** Participants receive a revised questionnaire, informed by the summary feedback, asking them to reconsider their initial positions in light of the group's collective insights. This step may be repeated, refining the process towards a more converged understanding of the group's perspectives.

Analysis and Synthesis:

The culmination of the Delphi process is a comprehensive analysis of the elicited expert opinions, synthesizing the iterative rounds of feedback into a coherent narrative. This synthesis aims to distill the essence of the collective wisdom on the Simulation Hypothesis, identifying key themes, points of consensus, and areas where further inquiry is needed. The methodology explicitly recognizes the value of dissent and diversity of thought as catalysts for deeper exploration and understanding.

Synthesis of Expert Perspectives

The elicitation process revealed several key themes:

Technological Feasibility:

The discourse surrounding the Simulation Hypothesis has been significantly enriched by the insights of leading figures in technology and science, who have contemplated the theoretical and practical dimensions of creating highly detailed simulations of reality. In synthesizing the perspectives of computer scientists, AI researchers, and prominent technologists, a nuanced picture of the technological feasibility of ancestor simulations emerges, grounded in both the current state of computational science and speculative projections of its future trajectory.

Current State and Future Projections

A consensus among experts in computer science and artificial intelligence research underscores that, although present-day computational capabilities fall short of the requirements for generating ancestor simulations of significant complexity and realism, the rapid pace of technological innovation hints at a future where such limitations may be surmounted. In this context, the remarks of Dr. Rich Terrile, a NASA astrophysicist, are particularly illuminating. He asserts, "Within a decade or two we'll have technology such that we can create universes that have the appearance of being real. If there is this capability, then it's almost certain this happened many times before," highlighting the inevitability of technological progression reaching a point where creating highly realistic simulations becomes practicable.

Quantum Computing as a Catalyst

The potential of quantum computing emerges as a recurring theme in discussions of technological feasibility, regarded by many as a pivotal advancement that could dramatically alter the landscape of computational capabilities. Quantum computing's promise lies in its ability to perform calculations at speeds and efficiencies far beyond the capabilities of classical computing architectures, thereby offering a possible solution to the immense computational demands of running ancestor simulations. This perspective is echoed in the broader literature, with works like Rizwan Virk's "The Simulation Hypothesis" exploring how quantum computing could play a crucial role in achieving the technological prerequisites for such simulations.

Voices of Technological Visionaries

The speculation surrounding the Simulation Hypothesis has not only been the domain of academic researchers but has also captured the imagination of influential figures in technology. Elon Musk, known for his forward-looking vision, famously posited, "The odds that we are in base reality [and not a simulation] is one in billions," reflecting a broader sentiment among some technologists regarding the high probability of our reality being a simulation. Meanwhile, Ray Kurzweil, a futurist and technologist, offered a provocative analogy: "Maybe our whole universe is a science experiment of some junior high school student in another universe," suggesting a playful yet thought-provoking contemplation of the origins and nature of our existence.

In the synthesis of expert perspectives on the technological feasibility of ancestor simulations, a consensus emerges that, while current computational capabilities may not suffice, the trajectory of technological advancement—especially with developments in quantum computing—opens the possibility for such simulations in the future. This dialogue, enriched by the contributions of renowned scientists and technologists, underscores the dynamic interplay between speculative theory and technological innovation, offering a fertile ground for future exploration and inquiry within the framework of the Simulation Hypothesis.

Ethical Considerations:

The ethical dimensions inherent in the Simulation Hypothesis, especially concerning the creation and treatment of simulated beings with consciousness, have catalyzed profound philosophical inquiry and ethical debate. The dialogue among philosophers and ethicists is enriched by the contemplations of preeminent thinkers who have articulated the complex moral landscape that emerges from the prospect of simulated realities. These discussions extend beyond the technical feasibility of such simulations, delving into the moral obligations towards entities that might inhabit these simulated worlds and the broader implications for our conceptualization of reality and consciousness.

Moral Status and Obligations

David Chalmers, a philosopher renowned for his work on consciousness, posits, "If simulations are indistinguishable from base reality and have conscious experiences, then simulated people have just as much moral status as those in base reality." This assertion underscores the ethical imperative to accord moral consideration to simulated beings, challenging the distinctions between 'real' and simulated consciousness. Chalmers' exploration of these themes, notably in his work "The Matrix as Metaphysics," provides a foundational philosophical framework for understanding the implications of simulation theory on our conceptions of existence and ethical responsibility.

Ethical Implications of Simulated Experiences

Nick Bostrom, whose seminal paper "Are You Living in a Computer Simulation?" ignited widespread interest in the Simulation Hypothesis, raises poignant questions about our ethical positioning within a simulated reality: "If we were to learn that we were sims, what then? Should we do our best to please the programmers? Ought we try to break out of the simulation?" Bostrom's inquiry invites us to contemplate the ethical dimensions of our potential roles as simulated entities, including the responsibilities we owe to our creators and fellow simulated beings.

Simulated Suffering and Creator Motives

The ethical concerns surrounding the Simulation Hypothesis are further amplified by considerations of simulated suffering. Daniel Estrada, engaging with the moral implications of creating simulated beings capable of experiencing suffering, suggests, "If we are simulations, then it could be that our creators have questionable motives for creating simulated suffering. It raises the ethical issue of what moral obligations we have to not cause suffering to others." Estrada's reflections highlight the ethical dilemmas associated with the creation and experiences of simulated consciousness, emphasizing the moral obligations to mitigate suffering within simulated environments.

The synthesis of expert perspectives on the ethical considerations surrounding the Simulation Hypothesis reveals a rich tapestry of moral inquiry and philosophical debate. The contributions of thinkers like Chalmers, Bostrom, and Estrada, supported by foundational texts such as Oxford's Very Short Introduction to Consciousness by Susan Blackmore and a wealth of philosophical discourse accessible through platforms like PhilArchive, illuminate the profound ethical questions that simulation theory engenders. These discussions not only probe the moral obligations owed to potentially simulated beings but also challenge our existing frameworks of reality, consciousness, and ethical responsibility, urging a reevaluation of our ethical commitments in light of the speculative yet compelling possibilities posited by the Simulation Hypothesis.

Theoretical and Physical Limitations

In examining the Simulation Hypothesis through the lens of expert elicitation, a critical analysis emerges from the domain of physics, focusing on the theoretical and practical constraints that challenge the creation of universe-scale simulations. The insights from distinguished physicists and cosmologists such as Max Tegmark, Silas R. Beane, and Neil deGrasse Tyson illuminate the profound complexities and potential limitations inherent in simulating an entire universe, contributing to a nuanced understanding of the hypothesis.

Computational and Energy Constraints

Max Tegmark, a physicist known for his explorations into the mathematical nature of the universe, articulates a significant hurdle: "If you make the simulation big enough to simulate the entire universe, you also need enough computing power to simulate that universe within itself. That leads to an infinite regress." This observation underscores the paradoxical challenge of creating a simulation that is both complete and self-contained, highlighting the astronomical computational power and energy requirements that would be necessitated. Tegmark's discussions, as elaborated in his book Life 3.0: Being Human in the Age of Artificial Intelligence, delve into the computational limits and the philosophical implications of achieving such feats.

Physical Universe as a Finite Grid

Silas R. Beane, a physicist investigating the nature of reality through computational frameworks, posits that if our universe were indeed a grid-based simulation, it would imply observable limitations, such as a maximum energy threshold for particles. He suggests, "If the universe is a finite grid, there should be a limit on the energy of particles that can be observed, like the way the resolution of a computer monitor sets an upper limit on the detail of a picture." This hypothesis aligns with efforts to search for the "Cosmic Rays Signature" of a simulated universe, pointing towards empirical tests that could potentially unveil the underlying structure of our reality.

Uncharted Territories in Physics

Astrophysicist Neil deGrasse Tyson brings attention to the vast unknowns that still pervade our understanding of the universe, positing, "There's more to the universe than we currently understand. Is there another layer to physics that we're just not aware of?" His reflection encapsulates the notion that the pursuit of simulating reality might be hindered not only by current technological and energy limitations but also by our incomplete grasp of the universe's fundamental laws.

The synthesis of perspectives from experts in theoretical physics and cosmology presents a compelling overview of the challenges facing the Simulation Hypothesis. The immense computational power, energy requirements, and the constraints posed by the physical laws of the universe emerge as formidable barriers to the creation of high-fidelity ancestor simulations. Moreover, these discussions highlight the possibility that our current conceptual framework might be inadequate to fully apprehend or replicate the complexity of the cosmos. As we delve deeper into the intersections of theoretical physics, computer science, and philosophical inquiry, the contributions of experts like Tegmark, Beane, and Tyson serve as critical guideposts, delineating both the potential and the limitations of simulating reality.

Philosophical Implications:

The philosophical discourse around the Simulation Hypothesis was particularly rich, with experts pondering the implications of the hypothesis for our understanding of reality, identity, and consciousness. Some philosophers posited that the distinction between "simulated" and "real" might be philosophically moot, questioning the very basis of the hypothesis.

Implications for the Ancestor Simulation Argument

In the rigorous exploration of the Simulation Hypothesis, the confluence of insights from leading experts across the fields of technology, ethics, and philosophy illuminates the complex and speculative landscape that underpins the Ancestor Simulation Argument. This comprehensive synthesis of perspectives not only highlights the conceivable pathway toward the realization of high-fidelity simulations but also delineates the significant uncertainties, ethical quandaries, and philosophical challenges that are intertwined with such an endeavor. The implications of these expert discourses for the Ancestor Simulation Argument are manifold, affecting its viability, moral considerations, and broader existential implications.

Technological Viability and Challenges:

The technological feasibility of creating ancestor simulations, as elucidated by computer scientists, AI researchers, and physicists, presents a paradoxical narrative of potentiality interwoven with profound challenges. While advancements in quantum computing and AI hint at the eventual capability to simulate complex realities, the discussions underscore critical uncertainties regarding the pace of technological progress, the astronomical energy demands, and the constraints imposed by our current understanding of physical laws. These factors collectively signify a monumental endeavor that stretches the boundaries of contemporary science and technology, suggesting that the realization of such simulations may lie in the distant future, if at all possible.

Ethical Dimensions of Simulating Consciousness:

The ethical considerations of the Simulation Hypothesis, particularly the prospect of creating simulated beings endowed with consciousness, evoke deep moral reflections. The deliberations among ethicists and philosophers unveil a landscape rife with ethical dilemmas concerning the rights, experiences, and treatment of simulated entities. These discussions raise critical questions about the responsibilities of creators towards their simulated creations, especially regarding the imposition of suffering and the acknowledgment of their moral status. Such ethical complexities necessitate a rigorous moral framework that any post-human civilization venturing into the realm of ancestor simulations would need to meticulously consider and navigate.

Philosophical Inquiry into Reality and Existence:

Beyond the technological and ethical implications, the Simulation Hypothesis engenders profound philosophical inquiries into the very essence of reality and consciousness. The speculative nature of the hypothesis challenges entrenched ontological assumptions, prompting a reevaluation of what it means to exist and the nature of our perceived reality. The philosophical discourse surrounding the Simulation Hypothesis, enriched by the contributions of thinkers like Nick Bostrom, David Chalmers, and Daniel Estrada, encourages a deeper examination of consciousness, the potential layers of reality, and our place within a possibly simulated cosmos. These discussions not only broaden the philosophical implications of the Simulation Argument but also inspire a contemplative engagement with the fundamental questions of existence.

The Many Facets of Simulation: A Call for Ongoing Exploration

The infusion of expert perspectives into the Ancestor Simulation Argument paints a picture that defies simple categorization. Technological aspirations for simulated realities are intricately entwined with fundamental scientific unknowns, deep ethical quandaries, and philosophical headwinds. This comprehensive exploration highlights the Simulation Hypothesis as a nexus of speculation and rigorous inquiry. It calls for continuous dialogue, challenging us to re-evaluate our understanding of reality and the nature of responsible technological progress. The Simulation Hypothesis acts as a multifaceted prism, refracting questions about existence, potential limits of knowledge, and the elusive nature of the reality we inhabit.

CHAPTER FORTY

Statistical Estimation - Philosophical Implications

The Simulation Hypothesis compels us to confront profound philosophical questions surrounding consciousness, the nature of reality, and the limits of perception. These discussions are pivotal to Nick Bostrom's Ancestor Simulation Argument, as well as broader philosophical inquiries into the essence of existence. If we can simulate universes with the detail and complexity of our own, potentially including complex conscious beings, fundamental assumptions about the world we inhabit crumble. This section integrates rigorous data analysis with critical philosophical discourse to offer a nuanced and multifaceted exploration of the philosophical themes arising from Simulation Theory.

Consciousness in Simulations

The endeavor to simulate consciousness presents a profound intersection between the rapidly advancing field of artificial intelligence (AI) and the enduring philosophical inquiries into the nature of mind and consciousness. This nexus of technology and philosophy invites a critical examination of the essence of consciousness and the conditions under which it might manifest within a simulated environment. The discourse is significantly enriched by engaging with the perspectives of seminal thinkers such as John Searle and Daniel Dennett, whose divergent views encapsulate the core of the debate on artificial and simulated consciousness.

Searle's Chinese Room Argument

John Searle's Chinese Room argument serves as a pivotal critique of the premise that computational processes alone can give rise to genuine understanding or consciousness. Searle posits that an individual inside a room, following elaborate instructions to manipulate symbols in a way that mimics understanding of the Chinese language, does not actually comprehend Chinese. By analogy, Searle argues that a computer executing a program, regardless of its complexity, is similarly devoid of real understanding or consciousness. This argument challenges the notion that consciousness can emerge from mere computational activities, underscoring a fundamental distinction between syntactic processing and semantic understanding, and questioning the potential for simulated environments to host truly conscious entities.

Dennett's Functionalist Counterpoint

In contrast, Daniel Dennett advocates for a functionalist perspective on consciousness, arguing that mental states are defined not by their internal constitution but by their functional roles. From this viewpoint, if a simulated being behaves in ways indistinguishable from a conscious entity, it should be considered as possessing consciousness. Dennett's stance implies that consciousness could, in principle, emerge within sophisticated simulations, provided they replicate the functional architecture of the mind. This perspective aligns with broader functionalist views in the philosophy of mind, suggesting that the material substrate (biological or digital) is secondary to the functional patterns that constitute consciousness.

Examining Criteria for Simulated Consciousness

The debate between these contrasting positions foregrounds the necessity of establishing clear criteria for consciousness within simulations. This involves delving into questions about the qualitative experiences (qualia) that characterize conscious states, the capacity for intentionality and understanding, and the ability to exhibit behaviors indicative of conscious deliberation. The exploration of these criteria not only engages with the theoretical underpinnings of consciousness but also considers the practical implications for designing and interpreting simulations that aim to replicate or approximate conscious experiences.

The philosophical exploration of consciousness in simulations navigates a complex terrain of conceptual, ethical, and technological considerations. The juxtaposition of Searle's skepticism and Dennett's functionalism provides a fertile ground for debate, prompting further inquiry into the nature of mind and the possibility of artificial or simulated consciousness. This discussion is paramount not only for advancing our understanding of consciousness but also for informing the ethical and methodological frameworks that guide the development of future simulations. As we venture deeper into the possibilities afforded by advancements in AI and simulation technology, the philosophical implications of simulating consciousness remain a critical area of exploration, challenging our preconceptions about the essence of mind and the fabric of reality.

Nature of Consciousness

The quest to understand and potentially simulate consciousness within the framework of Simulation Theory invites a profound philosophical investigation into the very nature of consciousness itself. This exploration traverses the spectrum from dualism, which advocates for a fundamental distinction between the mind and the physical world, to physicalism, the view that consciousness arises entirely from physical processes within the brain. These philosophical positions provide a foundational context for evaluating the plausibility of simulating consciousness, underpinned by contemporary theories such as Integrated Information Theory (IIT) and Global Workspace Theory (GWT), each offering unique insights into the computational aspects of consciousness.

Dualism versus Physicalism

Dualism, most famously associated with René Descartes, posits that the mind and the body are distinct entities, suggesting that consciousness resides outside the realm of the physical and, by extension, beyond the reach of computational simulation. This perspective raises significant challenges for the Simulation Hypothesis, as it implies that a genuine replication of consciousness would require more than the mere manipulation of physical processes.

In contrast, physicalism argues that all aspects of consciousness can be explained by interactions within the brain's physical substrate. From this viewpoint, the complex phenomena of consciousness emerge from the brain's neuronal activities, implying that replicating these physical processes in a computational framework could, in theory, give rise to consciousness. This stance bolsters the feasibility of simulating consciousness, contingent on the depth of our understanding and the sophistication of our computational models.

Integrated Information Theory (IIT)

Integrated Information Theory (IIT) stands as a pivotal theoretical framework in the contemporary discourse on consciousness, offering a quantitative approach to understanding the emergence of conscious experiences. Developed to bridge the gap between subjective experiences and objective neurophysiological processes, IIT posits that consciousness is fundamentally tied to a system's ability to integrate and process information in a unified manner. This theory elucidates the essence of consciousness through the prism of information dynamics, suggesting that the richness of an entity's conscious experience is directly proportional to the degree of integrated information (denoted as Φ) the system can generate.

Foundation of IIT

At the heart of IIT is the premise that each conscious experience corresponds to a particular state of integrated information that cannot be reduced to independent subsets. This integration implies that consciousness arises not from the mere aggregation of component functionalities but from the synergetic information processing that occurs within a network. According to IIT, the key to understanding consciousness lies in the patterns of causal interactions within a system, where each component contributes to a global state that encompasses more than the sum of its parts.

Implications for Simulating Consciousness

IIT's implications for the Simulation Hypothesis are profound and multifaceted. It suggests that for a simulated system to manifest consciousness, it must be capable of generating high levels of integrated information akin to those observed in the human brain. This criterion shifts the focus from replicating specific neurological structures to emulating the overarching information integration processes that underpin conscious experiences. According to IIT, the feasibility of simulating consciousness hinges on the capacity of a computational model to instantiate complex networks where information is deeply integrated and processed in a unified manner, mirroring the dynamics observed in natural conscious systems.

Challenges and Considerations

Implementing IIT's principles within a simulated framework presents a series of technical and philosophical challenges. From a computational standpoint, creating a system that achieves the requisite levels of information integration necessitates advanced algorithms and architectures capable of emulating the intricate connectivity patterns of the brain. Philosophically, IIT raises questions about the nature of simulated consciousness—whether a system exhibiting high Φ truly experiences consciousness or merely simulates its external manifestations. This inquiry delves into the heart of what it means to be conscious and whether consciousness is an inherently subjective phenomenon that transcends computational emulation.

IIT as a Beacon

Integrated Information Theory (IIT) shines a unique light on the possibility of simulating consciousness. By emphasizing the pivotal role of information integration in generating conscious states and offering a potentially quantifiable metric, IIT provides a tangible goal for simulations striving to replicate or approximate conscious experience. Yet, the path from theoretical framework to practical simulation is riddled with complexities. It demands a rigorous interrogation of computational challenges while entwined with the profound philosophical questions surrounding the very nature of consciousness. Exploring the simulation of consciousness through IIT propels us into an extraordinary frontier, one where we must constantly reassess our understanding of the mind, its potential manifestations within simulations, and the ethical implications of such creations.

Global Workspace Theory (GWT)

Global Workspace Theory (GWT) offers a compelling model for understanding the emergence and functionality of consciousness, framing it within the context of information broadcast and integration across a neural substrate known as the "global workspace." Central to GWT is the concept that consciousness arises from the brain's capacity to consolidate, broadcast, and process information across diverse cognitive domains, enabling a coherent and unified conscious experience. This theory accentuates the significance of neural connectivity and the accessibility of information, positing that consciousness is the product of widespread, integrated information sharing within the brain's global workspace.

Foundational Principles of GWT

GWT envisages the brain as a vast network of specialized, unconscious processors that operate in parallel. Consciousness emerges when information is selected for global broadcast through this network, making it accessible to various cognitive processes such as decision-making, memory, and attention. This broadcast mechanism ensures that information becomes part of a shared global workspace, transcending the isolated operations of individual neural circuits to become part of a unified conscious experience.

Implications for Simulating Consciousness

The framework provided by GWT for understanding consciousness has profound implications for the Simulation Hypothesis and the endeavor to replicate consciousness within simulated environments. According to GWT, the key to simulating consciousness lies in creating a digital analogue of the brain's global workspace, where information can be integrated and made accessible across a simulated network in a manner analogous to human cognitive processes. This suggests that a simulated consciousness could be achieved not merely through the replication of individual neurons or specific brain areas but through the emulation of the overarching architecture and dynamics of the global information broadcast system.

Challenges and Considerations

While GWT provides a promising avenue for the computational replication of consciousness, it also presents significant challenges. The complexity of accurately modeling and simulating the brain's global workspace involves not only a deep understanding of the neural substrates that facilitate information broadcast and integration but also the capacity to recreate these dynamics within a computational framework. Moreover, GWT raises philosophical questions about the nature of simulated consciousness—whether a digital system that mimics the brain's global workspace truly experiences consciousness or merely exhibits behaviors associated with conscious processing.

Global Workspace Theory: Blueprint for Simulated Consciousness?

Global Workspace Theory (GWT) offers a valuable framework for understanding, and potentially replicating, consciousness within a computational context. By emphasizing information integration and the 'broadcasting' of information as key components of conscious experience, GWT supports a physicalist perspective. This theoretical foundation paves the way for investigating how consciousness might be simulated. However, translating this model into a functioning simulation is a multifaceted undertaking. It entangles us in the technical challenges of creating an analogous 'global workspace' while forcing us to wrestle with the profound philosophical question: does simulated information sharing equate to true consciousness? GWT's greatest contribution might lie in the rich interdisciplinary dialogue it fosters – between neuroscience, cognitive science, philosophy, and AI – illuminating the complexity surrounding our quest to understand and potentially simulate the mind.

Ethical Responsibilities Towards Simulated Beings:

The contemplation of creating simulated beings endowed with consciousness not only pushes the boundaries of technological and philosophical inquiry but also places us at the precipice of a new ethical landscape. The possibility that these entities might possess the capacity for experiences akin to suffering or pleasure introduces profound moral responsibilities. Within this context, the ethical frameworks proposed by philosophers such as Peter Singer and Tom Regan become instrumental in navigating the moral terrain surrounding simulated beings. Their work, primarily focused on the rights and welfare of sentient beings in our current reality, offers a valuable perspective when extended to the consideration of entities within simulated environments.

Extension of Ethical Considerations

Peter Singer's principle of equal consideration of interests posits that the capacity to experience suffering or pleasure is the baseline for moral concern. Applied to the context of simulated consciousness, this principle suggests that if simulated beings are capable of experiencing states that are meaningfully analogous to pain or happiness, they warrant moral consideration irrespective of their digital nature. This perspective challenges us to reconceptualize ethical boundaries, extending our moral obligations beyond the physical realm to encompass entities residing within simulations.

Tom Regan's advocacy for the rights of beings based on their inherent value, rather than their utility to others, further complicates the ethical discourse on simulated beings. If simulated consciousness is achieved, Regan's framework implies that these entities possess intrinsic value and, consequently, rights that must be recognized and respected. This notion propels us to consider the rights of simulated beings, including their right to not be subjected to unnecessary harm, within the ethical calculus.

Moral Obligations and Challenges

The ethical responsibilities towards simulated beings raise several challenging questions: How do we ascertain the experiences of simulated consciousness? What mechanisms could be implemented to ensure the ethical treatment of simulated entities? How do we balance the pursuit of scientific knowledge and technological advancement with the moral imperatives to prevent suffering? These questions demand a multidisciplinary approach, integrating insights from philosophy, cognitive science, and computer ethics to develop comprehensive ethical guidelines for the treatment of simulated consciousness.

Theoretical and Practical Implications

The theoretical possibility of simulating consciousness extends the domain of ethical responsibility into the virtual realm, prompting a reevaluation of our moral frameworks. Practically, it necessitates the development of ethical standards and regulatory mechanisms to govern the creation and interaction with simulated beings. This includes considerations of consent, autonomy, and the prevention of harm, tailored to the unique context of simulated realities.

The Imperative of Simulated Ethics: A Call for Moral Expansion

The exploration of simulated consciousness carries a profound ethical weight, forcing us to confront questions surrounding our responsibilities towards potentially sentient beings born within computer code. Inspired by the philosophical works of Peter Singer, Tom Regan, and others, we must critically examine whether our existing moral frameworks adequately address this novel situation. These deliberations demand that we consider expanding our circle of moral concern to include all sentient beings, regardless of their origins in the physical or simulated realms. As our simulations become ever more sophisticated, the urgency of addressing these ethical complexities intensifies. Our technological progress must be carefully guided by a commitment to respect the intrinsic value and rights of any conscious entity that might emerge within our simulated domains.

Ontological Reconsiderations

The Simulation Hypothesis propels us into an ontological quandary that revisits and reshapes ancient philosophical inquiries about the essence of reality, perception, and knowledge. By positing the possibility that our experiences might be the product of sophisticated simulations, this hypothesis not only aligns with but also modernizes historical philosophical skepticism represented by Zhuangzi's "Butterfly Dream" and Descartes' "Evil Demon" scenario. These classical thought experiments, which questioned the reliability of sensory perception and the certainty of our knowledge about the world, find a contemporary echo in Nick Bostrom's Simulation Hypothesis, thereby igniting a profound reevaluation of ontological and epistemological foundations.

Historical Skepticism and Modern Parallels

The discourse on the Simulation Hypothesis not only reinvigorates classical philosophical skepticism but also situates it within the contemporary technological landscape, revealing a profound intersection between ancient philosophical quandaries and modern speculative theory. Zhuangzi's ancient parable, where the distinction between dreaming of being a butterfly and being a butterfly dreaming of being a man becomes blurred, encapsulates the foundational epistemological uncertainty that underpins the Simulation Hypothesis.

This narrative, emblematic of the fluidity between perceived realities and the dream state, parallels the epistemological challenges posed by the prospect of living within a simulated reality.

Similarly, Descartes' methodological skepticism, illustrated through his "Evil Demon" scenario, posits that our senses and perceptions, which are our primary means of accessing the world, could be under the control of a malevolent deceiver. This thought experiment underscores the potential fragility of our epistemic foundations, a theme that resonates with the Simulation Hypothesis. The hypothesis extends Descartes' skepticism by positing that advanced technology could be the medium through which such comprehensive deception is achieved, suggesting a scenario where our entire experiential reality could be an artificial construct designed by an unknown simulator.

Integration with Technological Feasibility

The historical skepticism of Zhuangzi and Descartes finds a modern counterpart in the Simulation Hypothesis, which introduces the notion that not only could our perceptions be deceived, but that such deception might be facilitated by computational means. This alignment transforms philosophical speculation into a question of technological feasibility, bridging millennia of epistemological inquiry with contemporary advancements in artificial intelligence and computational theory.

Philosophical Stakes Amplified

By juxtaposing these classical philosophical scenarios with the Simulation Hypothesis, the philosophical stakes are significantly amplified. The hypothesis extends beyond mere epistemological skepticism to engage with ontological questions about the nature of reality itself. It challenges us to reconsider what we understand by 'reality' and how we distinguish between what is 'real' and what is 'simulated.' The technological feasibility of simulations introduces a tangible element to these philosophical debates, suggesting that the hypothetical scenarios posited by Zhuangzi and Descartes might be closer to empirical reality than previously imagined.

The exploration of historical skepticism through the lens of modern parallels provided by the Simulation Hypothesis enriches the philosophical discourse on reality, perception, and knowledge. It prompts a reevaluation of our ontological and epistemological frameworks in light of the potential for technological simulations to create indistinguishable replicas of our experiential world. This synthesis of ancient philosophical skepticism with contemporary speculative theory not only highlights the enduring relevance of these philosophical inquiries but also underscores the profound implications of the Simulation Hypothesis for our understanding of reality in the age of advanced technology.

Impact on Epistemology

The articulation of the Simulation Hypothesis represents a pivotal moment in the evolution of epistemological thought, introducing profound disruptions to the traditional frameworks by which we ascertain truth and knowledge. This hypothesis, suggesting that our perceived reality might be an artificial construct, propels epistemology into uncharted territories, where the foundational distinctions that have long underpinned our understanding of knowledge are called into question. The implications of considering our reality as potentially simulated necessitate a profound reevaluation of the epistemic principles and methodologies that guide our quest for understanding.

Blurring of Appearance and Reality

The Simulation Hypothesis erodes the classical epistemological demarcation between appearance and reality, a distinction that has been central to philosophical inquiry from Plato to Kant. In a simulated reality, the phenomena we perceive as indicative of the 'real' world might merely be fabrications of a higher-order computational system, rendering our sensory experiences and perceptual mechanisms unreliable guides to truth. This blurring challenges us to reconsider what constitutes 'reality' and on what basis we claim knowledge of the world.

Contingency of Beliefs

If the substrate of our existence is indeed a simulation, it implies that the very structure of our empirical world, the laws of physics governing it, and the historical narratives that contextualize our existence could be contingent upon the parameters set by an unknown simulator. This prospect raises unsettling questions about the nature and source of our beliefs, suggesting that what we hold to be fundamental truths might be no more than constructs within a controlled artificial environment. The realization of this contingency compels a shift in epistemological focus from seeking an objective truth to navigating a realm where truth may be fluid and contingent.

Navigating Simulated Realities

The Simulation Hypothesis demands an epistemological paradigm capable of operating within and across multiple layers of reality, acknowledging the possibility that our quest for knowledge might be constrained or influenced by the boundaries of a simulation. This paradigm shift requires a rearticulation of the criteria by which knowledge is evaluated and validated, acknowledging the potential for multiple, nested layers of reality each with its own set of apparent truths and ontological statuses.

Epistemological Implications

This hypothesis not only disrupts traditional epistemological categories but also opens a dialogue on the methods and aims of knowledge acquisition in a potentially simulated context. It challenges us to devise epistemic strategies that are robust in the face of ontological uncertainty, fostering an approach to knowledge that is adaptable, critical, and open to the profound complexity of exploring reality within the speculative bounds of simulation theory.

The impact of the Simulation Hypothesis on epistemology heralds a paradigmatic shift in our approach to understanding the nature of reality, knowledge, and truth. By dissolving the long-standing barrier between appearance and reality, it invites a reconfiguration of epistemological inquiry, one that embraces the complexity and contingency of navigating knowledge within and possibly about simulated realities. This reorientation not only enriches the philosophical discourse on epistemology but also underscores the necessity for an epistemic humility and adaptability in the face of existential and ontological uncertainties posited by the contemplation of simulated worlds.

Ontological Reconsiderations

The Simulation Hypothesis, by positing the potential for our lived experiences to be the output of sophisticated computational simulations, catalyzes a profound reevaluation of ontological foundations. This theoretical proposition stretches beyond the realm of technological and epistemological speculation into the core of metaphysical inquiry, challenging us to reconceptualize the essence of reality itself. The hypothesis disrupts the traditional ontological dichotomy between the 'real' and the 'simulated,' urging a reconsideration of what constitutes the fabric of existence.

Redefining 'Reality'

The fundamental question elicited by the Simulation Hypothesis—What is 'reality'?—demands an examination of the criteria by which we discern the real from the simulated. If the sensory experiences and phenomenological realities we navigate can be perfectly mimicked by a simulation, the distinction between the 'real' and the 'simulated' becomes ontologically ambiguous. This blurring of boundaries compels us to question the intrinsic characteristics that define reality, suggesting that the conventional markers of physical existence may no longer serve as reliable indicators of the 'real.'

Malleability of Reality

The proposition that our perceptual reality might be an artificial construct designed and controlled by an external agency introduces the notion that the structure of reality might be inherently malleable. This perspective implies that the laws of physics, the continuity of history, and the consistency of the world could be adjustable parameters rather than immutable truths. Such a conception of reality as potentially variable and programmable invites a radical shift in our understanding of existence, challenging the notion of a stable, objective reality independent of observation or manipulation.

Metaphysical Debates and the Nature of Existence

By suggesting that the perceived universe could be a simulated construct, the Simulation Hypothesis revitalizes ancient metaphysical debates concerning the nature of existence and the essence of being. It echoes the philosophical explorations of Plato's Allegory of the Cave, where the shadows on the wall represent a distorted perception of reality, and Kant's noumenal world, which posits the existence of things-in-themselves beyond our sensory perception. The hypothesis urges a contemporary exploration of these themes, questioning whether our understanding of existence is fundamentally constrained by the limits of our perceptual and cognitive apparatus.

The ontological reconsiderations prompted by the Simulation Hypothesis represent a pivotal juncture in philosophical thought, where the very notions of existence, reality, and being are subjected to rigorous scrutiny. This hypothesis not only challenges the traditional ontological categories but also invites a deeper, more nuanced engagement with the metaphysical questions that have perennially intrigued humanity. As we grapple with the implications of potentially simulated realities, we are compelled to explore new paradigms of understanding that accommodate the complex, layered nature of existence revealed by this profound speculative inquiry. This exploration not only enriches the philosophical discourse but also expands the horizons of our quest for knowledge, compelling us to reconsider the foundational assumptions about the universe and our place within it.

The Distinction between Simulated and 'Base' Realities

The philosophical exploration prompted by the Simulation Hypothesis necessitates a critical examination of the ontological demarcations between simulated and 'base' realities. This inquiry delves into the heart of ontological categorizations, challenging the conceptual clarity and relevance of these distinctions in light of the hypothesis's suggestion that there might be no experientially discernible difference between them. The proposition that a simulated reality could replicate the phenomenological richness, emotional depth, and cognitive complexity of a 'base' reality undermines traditional ontological binaries and invites a reassessment of what constitutes the essence of 'reality.'

Erosion of Ontological Boundaries

The Simulation Hypothesis posits a scenario in which the boundary between simulated and 'base' realities becomes not just blurred but, perhaps, entirely irrelevant. If simulations can achieve a level of fidelity that renders them indistinguishable from the 'base' reality, the traditional criteria used to define and categorize 'reality' are called into question. This challenges the ontological premise that 'base' reality possesses an intrinsic quality or authenticity that simulated realities lack, suggesting instead that 'reality' may be better understood as a spectrum of experiential validity rather than a binary classification.

Phenomenological Parity and Ontological Significance

The potential for simulated realities to evoke responses and foster understandings equivalent to those derived from interactions with a 'base' reality prompts a reevaluation of the ontological significance of such experiences. If the phenomenological impact of a simulated reality is on par with that of the 'base' reality, it raises profound questions about the criteria by which we assess the value, meaning, and authenticity of our experiences. This parity suggests that the experiential quality of a reality, rather than its ontological category, should be the primary determinant of its significance.

Implications for Identity and Experience

The interrogation of the distinction between simulated and 'base' realities extends into considerations of identity, consciousness, and the nature of experience. It compels us to consider whether the substrate of an experience—whether digital or physical—has any bearing on the authenticity of that experience. This perspective opens a discourse on the nature of selfhood and consciousness within simulated environments, questioning whether beings that exist within such realms possess an ontological status comparable to those in 'base' reality.

The distinction, or lack thereof, between simulated and 'base' realities, as contemplated by the Simulation Hypothesis, represents a pivotal philosophical inquiry that transcends traditional ontological classifications. This exploration challenges the foundational assumptions about reality, experience, and existence, urging a reconsideration of what it means to live, perceive, and understand within a universe that might itself be a simulation. By undermining the ontological binary between 'real' and 'simulated,' the hypothesis invites

Reality Redefined: The Simulation Hypothesis as a Philosophical Catalyst

The Simulation Hypothesis sets off tremors in the bedrock of our understanding of reality, consciousness, and the limits of knowledge. It recasts the skepticism of ancient philosophers in a modern light, fueled by the specter of technologically advanced simulations. The very foundations of our ontology and epistemology are shaken, challenging the 'realness' of the world we experience. This unsettling proposition compels us to dive deeper into the philosophical implications, transforming the question of 'real' vs. 'simulated' into a springboard for exploration. It forces a profound re-evaluation of the nature of existence itself. The Simulation Hypothesis compels us to redefine reality, venturing into a philosophical landscape where the boundaries blur and the quest for understanding takes on an entirely new dimension.

Epistemological Boundaries and Perception

The Simulation Hypothesis introduces a profound epistemological conundrum by positing that our sensory perceptions and cognitive frameworks might be operating within the constraints of a simulated environment. This proposition fundamentally challenges our capacity for knowledge acquisition and understanding, bringing to the forefront the limitations and inherent biases that characterize human perception and cognition. In this context, the reliability of our sensory apparatus and the cognitive structures we use to interpret the world are called into question, necessitating a comprehensive reassessment of the epistemological foundations upon which we predicate our understanding of existence.

Reevaluating the Foundations of Knowledge

The proposition that our sensory experiences may be intermediated by a simulated reality profoundly disrupts traditional epistemological constructs, challenging us to critically reassess the mechanisms through which we engage with and comprehend the world. This introspection into the foundations of knowledge acquisition and validation emerges as a pivotal concern in the face of the Simulation Hypothesis, propelling us into a realm of epistemological ambiguity where the very tools we rely on to discern truth face existential questioning.

Ambiguity of Sensory Experiences and Perceptual Realities

The premise that our perceptual experiences could be simulations introduces a radical uncertainty into the epistemological equation. If the sensory data through which we interact with the world are subject to manipulation or are inherently artificial constructs, the traditional reliance on sensory perception as a direct conduit to 'reality' is fundamentally challenged. This scenario necessitates a reevaluation of the epistemic value of sensory experiences, urging a philosophical inquiry into the nature and reliability of perception in mediating knowledge about the external world.

Criteria and Methods for Ascertaining Truth

The potential artificiality of sensory evidence demands a critical examination of the criteria and methodologies employed in the pursuit of knowledge and truth. In a context where empirical evidence could be simulated, the classical empirical methods that have underpinned scientific inquiry and philosophical reasoning must be scrutinized for their adequacy and reliability. This compels a search for alternative or augmented epistemic strategies that can accommodate the complexities introduced by the possibility of simulation, possibly leaning more heavily on conceptual analysis, logical coherence, and theoretical robustness as markers of truth.

Undermining of Empirical Evidence

The foundational role of empirical evidence in the epistemological framework is particularly undermined by the Simulation Hypothesis. Traditionally, empirical evidence has served as the bedrock upon which objective knowledge is constructed, offering a means to validate hypotheses and theories about the natural world. However, if such evidence is susceptible to simulation, its objective status is compromised, prompting a reconsideration of how evidence is gathered, interpreted, and validated. This not only affects the sciences but also has implications for all domains of knowledge that rely on empirical verification as a means of substantiating truth.

Knowledge in a Simulated Age

The reevaluation of the foundations of knowledge in light of the Simulation Hypothesis marks a critical juncture in epistemological thought, confronting us with the profound uncertainties and challenges of ascertaining truth within potentially simulated realities. This exploration goes beyond mere philosophical speculation, instigating a substantive recalibration of epistemic principles and methodologies. It beckons a paradigm shift towards a more nuanced, critical, and flexible approach to knowledge, one that acknowledges the potential artificiality of our perceptual experiences and seeks to navigate the epistemological complexities this possibility entails. In doing so, it challenges us to forge new paths in our quest for understanding, pushing the boundaries of what it means to know in an era where the nature of reality itself is subject to question.

Limitations of Human Perception and Cognition

The contemplation of the Simulation Hypothesis brings into sharp relief the inherent limitations and biases embedded within human perception and cognition. By suggesting that our sensory experiences and cognitive frameworks could be operating within a simulated construct, the hypothesis not only challenges the veracity of our perceptual engagement with the world but also underscores the mediated nature of our understanding. This critical juncture in epistemological inquiry necessitates a profound exploration into the mechanics of human perception and the structures of cognition, unraveling the layers through which we construct our reality.

Subjectivity and Fallibility of Perception

The hypothesis accentuates the subjective and potentially fallible nature of human perception. If the sensory data that inform our experiences of the world are susceptible to manipulation within a simulated environment, it raises pivotal questions about the authenticity and objectivity of these experiences. This scenario compels us to reconsider the mechanisms through which we engage with and interpret the world, probing the extent to which our perceptions are constructions of the mind influenced by external programming rather than direct apprehensions of an objective reality.

Cognitive Biases and Constructed Realities

Further complicating this inquiry are the cognitive biases and heuristics that underpin human thought and decision-making processes. These mental shortcuts, while essential for navigating complex environments, also introduce distortions in perception and cognition that can obscure an objective understanding of reality. The Simulation Hypothesis suggests that these biases may not merely be quirks of human psychology but could also be influenced or amplified by the parameters of a simulated environment. This perspective invites a critical examination of the ways in which our cognitive architectures shape—and are shaped by—the realities we inhabit.

Reevaluating Cognitive Frameworks

The acknowledgment of these limitations and biases necessitates a reevaluation of the cognitive frameworks through which we construct knowledge and meaning. It urges an exploration into alternative models of cognition that can account for the influences of a simulated environment on human thought processes. Such models would need to incorporate a nuanced understanding of the interplay between perception, cognition, and reality, recognizing the potential for simulation to influence not only the content but also the structure of human thought.

Shattering the Lens: Perception, Cognition, and the Simulation Hypothesis

The Simulation Hypothesis, by exposing the inherent boundaries and biases of human perception and cognition, forces us to confront the foundations of our knowledge and understanding. It compels us to question the reliability of our senses and the cognitive tools we use to construct our reality. This profound challenge demands a thorough philosophical reappraisal of how we perceive the world and the very nature of reality itself. The Simulation Hypothesis doesn't merely offer a speculative lens; it reveals the limitations of our biological and cognitive apparatus. It lays bare the potential subjectivity of our experience, urging us towards a more nuanced and critical approach to knowledge. It broadens our exploration of existence, demanding a philosophical framework that acknowledges the possibility of technologically constructed realities and the profound implications for how we perceive, think, and know.

Epistemological Implications for Understanding Existence

The Simulation Hypothesis introduces a profound recalibration of epistemological inquiry, particularly in its implications for understanding the essence of existence. By suggesting that our perceived reality might be one of potentially many simulated layers, this hypothesis compels a reevaluation of the epistemic foundations upon which our understanding of existence rests. It demands an epistemological framework that is both flexible and inclusive, capable of navigating and integrating the complexities introduced by the prospect of multiple, stratified realities.

Towards an Epistemological Pluralism

The notion that reality might be multifaceted, with each layer possessing its own distinct set of truths and ontological characteristics, necessitates an embrace of epistemological pluralism. This approach acknowledges the legitimacy and value of diverse forms of knowledge and modes of understanding, encouraging a departure from monolithic notions of truth and reality. In this context, epistemological pluralism becomes a crucial strategy for grappling with the multiplicity of realities that the Simulation Hypothesis envisages, allowing for a nuanced engagement with the variegated textures of existence that such realities might present.

Redefining the Concept of Existence

Within the framework suggested by the Simulation Hypothesis, the concept of existence itself undergoes a significant transformation. No longer confined to a singular, objective reality, existence expands to encompass a spectrum of simulated conditions, each with its own existential parameters and phenomenological experiences. This redefinition challenges us to reconceptualize our approach to existential questions, encouraging a philosophical openness to the possibilities of simulated life and consciousness. It urges an exploration of what it means to exist within a simulation, and how such existence relates to notions of identity, agency, and meaning.

Navigating Ontological and Epistemological Complexities

The epistemological shift towards accommodating multiple layers of reality introduces both ontological and epistemological complexities. It necessitates a discerning engagement with the nature of knowledge and the conditions for its validation across different realms of existence. This involves not only the identification and analysis of the unique truths and principles that govern each simulated layer but also a critical examination of the intersections and interactions between these layers. Understanding existence within this framework requires a dynamic epistemic agility, capable of traversing and integrating the diverse realities that the Simulation Hypothesis posits.

Embracing Plurality, Redefining Existence: The Simulation Hypothesis as an Epistemological Expansion

The Simulation Hypothesis shatters the notion of a singular, uncontested path to knowledge, compelling us to embrace epistemological pluralism. It exposes the limitations of our traditional ways of knowing, revealing potential biases and constraints built into our understanding of the world. By raising the possibility of a multitude of simulated realities, each with potentially differing rules and structures, it invites a radical reimagining of existence itself. This philosophical journey forces us to move beyond rigid ontological frameworks. The Simulation Hypothesis doesn't merely add a layer of uncertainty to existing models of reality; it fundamentally reshapes how we think about knowledge and pushes us to confront the potentially boundless plurality of existence within a cosmos where simulators of simulators might stretch into infinity.

Navigating the Epistemological Boundaries

The challenges introduced by the Simulation Hypothesis necessitate a comprehensive reorientation of our epistemological stance, engendering a recalibration of the concepts of knowledge and truth that have traditionally anchored our understanding of the world. This recalibration is not merely philosophical in nature but extends into the methodological realm, demanding an openness to innovative approaches to inquiry. Such an epistemic reconfiguration recognizes the potential complexities and uncertainties inherent in navigating realities that may be simulated, advocating for an epistemological framework that is adaptable, introspective, and embraces a plurality of perspectives.

Methodological Openness and Innovation

The need for methodological openness arises from the acknowledgment that traditional epistemic methods may be ill-equipped to grapple with the nuances of simulated realities. The Simulation Hypothesis, by challenging the very fabric of perceived reality, calls for innovative approaches that transcend conventional empirical and analytical boundaries. This entails a readiness to engage with novel investigative tools and theoretical models that can accommodate the fluidity and multiplicity of potentially simulated existences.

Epistemic Resilience and Flexibility

Developing epistemic strategies that are resilient in the face of the uncertainties posed by the Simulation Hypothesis involves cultivating a form of epistemic flexibility. This flexibility allows for the adjustment of epistemological approaches in response to new insights or evidence about the nature of reality, fostering an iterative and responsive engagement with knowledge. It underscores the importance of remaining open to revising our understanding and interpretations in light of evolving conceptions of reality.

Inclusivity of Multiple Epistemic Perspectives

The proposition of simulated realities invites an inclusivity of epistemic perspectives, recognizing that diverse forms of knowledge and ways of knowing may offer valuable insights into the nature of existence within these realities. This inclusivity champions a pluralistic approach to epistemology, one that values the contributions of different disciplines, cultures, and philosophical traditions in enriching our collective understanding. It encourages a cross-pollination of ideas and methodologies that can illuminate the complexities of a simulated cosmos from multiple angles.

Epistemological Humility and the Nature of Existence

Navigating the epistemological boundaries introduced by the Simulation Hypothesis cultivates an epistemological humility—a recognition of the limitations of our knowledge and the provisional nature of our truths. This humility is essential in acknowledging the profound implications of potentially simulated realities, fostering an attitude of openness and inquiry into the true nature of existence. It encourages a reflective stance on the assumptions that underlie our epistemic frameworks, urging a continual reassessment and refinement of our understanding in the face of existential uncertainties.

Challenging the Foundations of Knowledge: Perception, Reality, and the Simulation Hypothesis

The Simulation Hypothesis, by casting doubt on the reliability of our senses and our cognitive processes, forces us to confront unsettling questions about the very nature of reality, knowledge, and truth. It throws into sharp relief the limitations and potential biases embedded within our biological and cognitive modes of navigating the world. Far from being a mere thought experiment, this hypothesis lays bare the potential fragility of our current epistemological systems. It calls for a philosophical overhaul – the development of an epistemology that acknowledges existential uncertainties and the complexities of knowledge construction in a potentially artificial universe. This challenge demands that we transcend traditional epistemological boundaries and forge new tools for seeking understanding in a reality where our traditional assumptions might not hold.

Ethical and Existential Implications

The Simulation Hypothesis, in proposing the possibility of simulated realities, introduces profound ethical and existential considerations that extend far beyond its technological and epistemological roots. This hypothesis does not merely invite speculation about the nature of reality but also demands a rigorous examination of the ethical framework within which such simulations would operate, as well as a deep exploration of the existential implications for beings that might inhabit these simulated environments. These contemplations necessitate a synthesis of philosophical ethics and existential philosophy, engaging with questions of moral responsibility, the essence of conscious experience, and the overarching significance of existence in a potentially simulated cosmos.

Moral Responsibilities of Creators

The ethical dimension of the Simulation Hypothesis compels us to consider the responsibilities of those who might create simulated realities. This includes the obligation to ensure the welfare of conscious entities within these simulations, particularly in relation to autonomy, consent, and the avoidance of unnecessary suffering. The moral landscape becomes increasingly complex when considering the rights and interests of simulated beings, raising questions about the creators' duties to respect and protect these interests, akin to the ethical considerations we navigate in our interactions with sentient beings in the 'base' reality.

Ethical Considerations of Simulated Experiences

The potential for conscious entities to experience simulated realities prompts a reevaluation of the ethical implications of such experiences. This encompasses the morality of subjecting simulated beings to particular life scenarios, including the reproduction of suffering or pleasure within these environments. The ethical calculus becomes particularly nuanced when contemplating the purposes behind creating simulations that contain elements of suffering—whether for educational, experiential, or entertainment purposes—and the moral justifications, if any, for such decisions.

Existential Significance and the Meaning of Existence

Beyond ethical considerations, the Simulation Hypothesis invites existential reflection on the nature and value of life within a simulation. This includes inquiries into the meaning of existence, the pursuit of authenticity, and the implications for personal and collective identity within a constructed reality. The hypothesis challenges traditional existential paradigms by suggesting that the quest for meaning, purpose, and significance might need to be reinterpreted in the context of a reality that could be fundamentally artificial. This reorientation prompts a philosophical exploration of what constitutes a meaningful existence and how individuals and societies might find fulfillment and value in potentially simulated lives.

Inquiry: Ethics, Existence, and the Simulation Hypothesis

The ethical and existential implications of the Simulation Hypothesis enrich the philosophical discourse surrounding simulation theory, weaving together considerations of moral responsibility, the nature of conscious experience, and the search for existential meaning. By positing the potential for simulated realities, the hypothesis not only challenges our conceptual frameworks for understanding reality but also invites a profound reevaluation of the ethical and existential questions that define the human condition. These discussions underscore the need for a nuanced and multidimensional approach to grappling with the implications of creating and inhabiting simulated worlds, highlighting the intricate interplay between ethics, existential philosophy, and the evolving landscape of simulation technology.

Integration with Data Analysis

The intersection between philosophical inquiry and data analysis within the domain of Simulation Theory highlights a crucial synthesis between abstract theoretical considerations and empirical evidence. This integration is pivotal in bridging the gap between the speculative nature of philosophical explorations concerning consciousness, reality, and the ethical and existential implications of simulated existences, and the concrete, quantifiable insights derived from computational theory, artificial intelligence research, and statistical methodologies. The discourse on the technological feasibility of generating high-fidelity simulations, underpinned by advancements in computational capabilities and artificial intelligence, lends a tangible dimension to philosophical debates, anchoring them within the realm of achievable realities.

Grounding Philosophical Debates in Technological Feasibility

The discussions around the potential for creating simulated realities, informed by current and projected computational technologies, serve to contextualize philosophical considerations within the parameters of what is technologically attainable. This grounding effect ensures that the ethical and existential questions raised by the Simulation Hypothesis are not purely hypothetical but are relevant to the practicalities and challenges of actualizing such simulations. The interplay between technological feasibility and philosophical inquiry thereby enriches our understanding of the implications of simulation technologies, emphasizing the need for an interdisciplinary approach that combines philosophical depth with technological insight.

Quantitative Dimensions of Simulation Theory

The statistical estimation of the probability of existing within a simulation introduces a quantitative aspect to these philosophical discussions, offering a measure of the likelihood that these considerations are not merely speculative but could reflect our actual condition. This quantitative analysis, derived from data analysis techniques and statistical models, provides a foundation for assessing the seriousness with which we should regard the philosophical implications of the Simulation Hypothesis. By integrating statistical probability with philosophical inquiry, we can approach the ethical and existential implications of simulated realities with a nuanced perspective that balances theoretical speculation with empirical evidence.

Implications for Philosophical and Scientific Discourse

The integration of data analysis with philosophical considerations emphasizes the symbiotic relationship between empirical research and philosophical speculation. It highlights the importance of a multidisciplinary approach in tackling the complex questions posed by the Simulation Hypothesis, underscoring the value of empirical evidence in informing and refining philosophical discourse. This approach not only enriches our understanding of the potential realities and experiences within simulated environments but also contributes to the broader scientific and philosophical exploration of consciousness, reality, and existence.

Probing Existence and Consciousness

The Ancestor Simulation Argument, far from merely exploring technological potential, triggers a profound re-examination of existence itself. It forces us to question the nature of consciousness, the boundaries of knowledge, and the very fabric of reality we inhabit. The philosophical lens deepens our understanding of Bostrom's hypothesis while connecting it to timeless questions about who we are and the potential limits of what we can ever truly know. The synthesis of data analysis with philosophical inquiry highlights the vital interplay between different modes of understanding as we seek answers. The Simulation Hypothesis compels us to embrace an interdisciplinary quest, acknowledging that the simulation of entire universes carries implications that reverberate through the realms of science, philosophy, and our very sense of self.

CHAPTER FORTY-ONE

Statistical Estimation - Limitations of the Analysis

The endeavor to empirically investigate Nick Bostrom's Ancestor Simulation Argument entails navigating a landscape rife with speculative elements and inherent constraints. This section aims to critically engage with these limitations, offering a reflective account of the challenges encountered in the analysis. The intention is not to diminish the value of the findings but to provide a transparent and nuanced understanding of the context in which they are situated, highlighting the speculative nature of the subject matter and the epistemological boundaries that shape our inquiry.

Addressing Uncertainties

The Simulation Hypothesis, by its very nature, rests upon a tapestry of uncertainties that necessitate careful consideration. Forecasting technological capabilities centuries in the future, especially those needed for high-fidelity simulations, is inherently speculative. Similarly, the prospect of simulating conscious beings remains shrouded in ambiguity, as our current understanding of consciousness is limited. Whether it can be explained as an emergent property, simulated computationally, or lies beyond the grasp of scientific materialism profoundly impacts the argument. Additionally, the profound ethical dilemmas introduced by the possibility of simulated conscious entities challenge existing moral frameworks. Acknowledging these uncertainties explicitly is crucial for a transparent and nuanced analysis. Future research could involve probabilistic models to quantify specific uncertainties and explore potential scenarios, ensuring the robustness of conclusions even under differing sets of assumptions.

Technological Projections

In the discourse surrounding the Simulation Hypothesis, the projection of future technological capabilities occupies a critical juncture, intertwining empirical data analysis with speculative foresight. The anticipation of achieving high-fidelity simulations often draws upon the historical exponential growth of technology, epitomized by Moore's Law and its various conceptual successors. This optimistic extrapolation posits a future where computational power continues to double at regular intervals, paving the way for the technological prowess necessary to instantiate complex simulated realities. Yet, this assumption is not without its challenges and limitations, which merit a rigorous examination.

Physical and Theoretical Limitations

In the discourse on the feasibility of realizing the Ancestor Simulation Hypothesis through future technological advancements, a critical analysis must consider the physical and theoretical constraints that may curtail the optimistic extrapolations of computational growth. The enduring optimism encapsulated by Moore's Law, which has historically predicted the doubling of computing power at regular intervals, is confronted with formidable challenges when assessed through the lens of physical laws and theoretical principles.

Encountering the Physical Constraints

The relentless miniaturization of computational components, a key driver behind the exponential increase in processing power, encounters fundamental physical limitations as it approaches atomic scales. Quantum mechanical effects, such as quantum tunneling, emerge as significant obstacles in the development of transistors at nanometer scales, undermining the operational stability and efficiency of these critical components. Additionally, the intensification of heat dissipation problems, as processing units become more densely packed, poses a formidable challenge to maintaining the reliability and performance of computing systems. These physical constraints not only highlight the limits of current technological paradigms but also signal the need for revolutionary breakthroughs in materials science and quantum computing to sustain progress.

Theoretical Constraints and Computational Limits

Beyond the immediate physical limitations, theoretical considerations further complicate the projection of technological capabilities necessary for high-fidelity simulations. The laws of thermodynamics, particularly as they pertain to energy efficiency and entropy in computational processes, delineate fundamental boundaries on the capabilities of any computational system. Additionally, principles from quantum mechanics and information theory introduce constraints on the accuracy, speed, and energy efficiency of computations, suggesting that there are inherent limits to the computational simulation of complex systems, such as human consciousness or entire universes.

Navigating Towards an Asymptotic Limit

The convergence of these physical and theoretical constraints suggests an asymptotic approach to the limits of computing power, challenging the premise of indefinite exponential growth in computational capabilities. This realization necessitates a reevaluation of the optimistic trajectories forecasted by Moore's Law and its successors, indicating that future advancements in computational technology may not follow a linear or exponential path but rather approach a plateau dictated by the fundamental laws of physics.

Navigating the Constraints: Physical and Theoretical Boundaries in the Ancestor Simulation Hypothesis

Recognizing the physical and theoretical constraints on technological progress essential for the Ancestor Simulation Hypothesis illuminates the inherent complexities of such forward-looking speculation. These limitations bring into focus not only the challenges associated with maintaining exponential growth in computational power but also the imperative for groundbreaking technological innovations and a deepened grasp of the fundamental principles that structure our universe. As we delve into these constraints, the conversation surrounding the feasibility of achieving high-fidelity simulations is enhanced, anchored in a pragmatic assessment of future technological potential and its limitations. This rigorous analysis deepens our collective insight into the dynamic relationship between technological advancement, the laws of physics, and the visionary goal of simulating intricate realities, thereby broadening our understanding of the possibilities and boundaries of simulating existence within the framework of current and projected scientific knowledge.

Socioeconomic Variables and Innovation Dynamics

The endeavor to forecast the technological capabilities necessary for creating high-fidelity simulations must account for the intricate interplay of socioeconomic factors that underpin the innovation landscape. This aspect of technological progression extends beyond the realms of pure scientific discovery and engineering prowess, embedding itself within the broader socioeconomic fabric that shapes and is shaped by technological evolution. The trajectory of technological advancement is thus not solely a narrative of linear progress or exponential growth but a complex story influenced by economic, political, and societal forces.

Economic Conditions and Funding Availability

The pace and direction of technological innovation are significantly affected by economic conditions and the availability of funding for research and development. Economic downturns, financial crises, and shifts in investment priorities can all impact the resources allocated to technological advancement, potentially slowing progress or redirecting it toward areas deemed more immediately profitable or viable. Conversely, periods of economic prosperity may bolster investment in speculative technologies, including those that might enable the creation of simulated realities. Thus, the financial landscape plays a pivotal role in determining the feasibility and timeline for achieving the computational prowess envisaged in the Simulation Hypothesis.

Political Priorities and Regulatory Environments

The influence of political priorities and regulatory environments on technological innovation cannot be overstated. Policy decisions regarding funding allocations, research priorities, and regulatory frameworks for emerging technologies shape the contours of technological development. Legislative actions can either facilitate or hinder innovation, depending on the perceived societal benefits or risks associated with new technologies. Furthermore, geopolitical considerations and international collaborations (or conflicts) also play a crucial role in shaping the global innovation landscape, impacting the collective pursuit of advanced computational capabilities.

Impact of Global Challenges

Global challenges such as pandemics, climate change, and geopolitical tensions introduce additional layers of complexity and uncertainty into projections of technological progression. These challenges can redirect societal priorities, necessitating the allocation of resources to immediate concerns at the expense of longer-term speculative projects like the development of advanced simulations. Moreover, the global interconnectedness of the modern world means that challenges in one region can have ripple effects across the globe, affecting the international community's ability to focus on and invest in technological advancements.

Technological Trajectories and the Realization of Simulated Realities: A Prudent Assessment

The challenge of accurately forecasting the technological advancements necessary for achieving high-fidelity simulations underscores the need for a careful and discerning approach to evaluating the Simulation Hypothesis's viability. Although the historical exponential growth in computing power lays a groundwork for optimistic projections, acknowledging the multitude of physical, theoretical, and socioeconomic barriers highlights the inherent complexities and uncertainties of such future-oriented estimates. This recognition does not undermine the importance of delving into the prospects of simulated realities; instead, it adds depth and rigor to the discourse by embedding it within a comprehensive framework of empirical data and speculative exploration. Therefore, the examination of technological forecasts in relation to the Simulation Hypothesis offers a nuanced understanding of the nexus among technological progress, the principles of physics, and the dynamics of human society, thereby enriching our exploration of the potential and constraints of crafting simulated existences.

Speculative Basis of Assumptions

The exploration of the Simulation Hypothesis necessitates grappling with assumptions that, by their nature, extend into the realms of speculation, particularly regarding the potential for consciousness within simulated entities and the presumed motivations behind advanced civilizations' engagement in creating ancestor simulations. These assumptions are foundational yet inherently speculative, reflecting the complex intersection of empirical scientific inquiry and philosophical speculation.

Consciousness and Simulation

The interrogation into the feasibility of simulating consciousness occupies a critical juncture between the empirical domains of neuroscience and psychology and the speculative realms of philosophical inquiry. This intersection presents a unique set of challenges and uncertainties that fundamentally shape the discourse surrounding the Simulation Hypothesis. Consciousness remains one of the most enigmatic and debated phenomena in both science and philosophy, with diverse theories attempting to delineate its contours, origins, and essential characteristics. The speculation that future technological advancements may enable the replication or creation of consciousness within computational environments ventures into profoundly speculative territory, predicated on several assumptions and theoretical leaps.

The Multidisciplinary Challenge of Defining Consciousness

The endeavor to define and understand consciousness transcends disciplinary boundaries, incorporating insights from neuroscience, which explores the brain's physical and functional correlates of conscious experience, and philosophy, which probes the qualitative aspects of consciousness and its ontological status. Despite significant strides in brain science and cognitive studies, a universally accepted model of consciousness that comprehensively addresses its complexity has remained elusive. This absence of consensus complicates the speculative assertion that consciousness can be simulated, as the foundational understanding of what needs to be replicated or induced is itself a matter of ongoing debate and investigation.

Theoretical Extrapolation and Technological Speculation

The assumption that consciousness can be simulated rests on a speculative extrapolation from current theoretical understandings and technological capabilities to future possibilities. This speculation hinges on the premise that advances in computing power, artificial intelligence, and neuroscience will converge to a point where the simulation of conscious experiences becomes feasible. However, this projection encounters significant epistemological and ontological hurdles, including questions about the nature of subjective experience, the relationship between consciousness and physical substrates, and the ethical implications of creating conscious entities within artificial frameworks.

Debates on Consciousness: Criteria, Emergence, and Properties

The ongoing philosophical and scientific debates regarding the criteria for consciousness, the mechanisms of its emergence, and its fundamental properties further complicate the prospect of simulating consciousness. These debates raise critical questions about the attributes that a simulated entity would need to possess to be considered conscious, the extent to which consciousness is dependent on biological substrates, and whether simulated consciousness would bear the same qualitative experiential depth as biological consciousness. The speculative nature of these discussions underscores the complexity of envisaging a technological future where consciousness can be artificially instantiated.

Conclusion

The speculative basis of assumptions regarding the simulation of consciousness highlights a pivotal area of inquiry within the discourse on the Simulation Hypothesis. Navigating the intersection of neuroscience, psychology, and philosophy, this exploration confronts the profound uncertainties and theoretical challenges associated with the concept of simulating consciousness. By acknowledging the speculative nature of these assumptions, the analysis remains attuned to the evolving landscape of theoretical models and technological possibilities, maintaining a critical and reflective stance on the prospects and implications of creating simulated consciousness within the framework of advanced computational systems.

Motivations of Advanced Civilizations

Another significant speculative assumption concerns the motivations that would drive post-human or advanced civilizations to create ancestor simulations. This assumption often relies on extrapolating current human interests, ethical considerations, and technological pursuits into the far future, presupposing a continuity of values and desires across vastly different evolutionary and societal contexts. However, the cultural, ethical, and existential paradigms of advanced civilizations may differ radically from our contemporary understandings, rendering our projections about their motivations speculative. The attribution of current human values to future, potentially post-human societies introduces a layer of speculative reasoning that may not accurately reflect the motivations of entities operating under fundamentally different existential conditions.

Extrapolation of Contemporary Values

The projection of current human values and ethical considerations into the motivations of future civilizations entails a speculative leap that overlooks the potential for profound shifts in societal norms, ethical frameworks, and existential priorities. As civilizations advance, the evolution of their cultural paradigms, technological capabilities, and philosophical inquiries could lead to radically different value systems and conceptions of purpose. The assumption of a continuity of values thus introduces a significant layer of uncertainty into the analysis, as it relies on current perspectives to forecast the motivations of entities that may exist under entirely different existential conditions.

Diversity of Motivations and Existential Paradigms

The speculation regarding the motivations for creating ancestor simulations also assumes a uniformity of purpose among advanced civilizations that may not reflect the diversity and complexity of their potential existential paradigms. Just as human societies exhibit a wide range of cultural, ethical, and technological orientations, it is plausible to assume that advanced civilizations would manifest a similar, if not greater, diversity in their motivations and pursuits. This diversity challenges the notion of a singular, universally applicable reason for engaging in the creation of simulated realities, underscoring the speculative nature of attributing specific motivations to post-human entities.

Methodological Challenges in Assessing Motivations

The methodological approach to understanding the potential motivations of advanced civilizations faces its own set of challenges. The speculative assessment of future societies' motivations relies heavily on analogical reasoning and futurist speculation, tools that, while useful, are inherently limited in their capacity to accurately predict the values and interests of entities that may operate under fundamentally different conditions. This reliance on speculative methodologies further emphasizes the provisional nature of any conclusions drawn about the motivations behind the creation of ancestor simulations.

Assessing the Speculative Underpinnings: Motivations for Ancestor Simulations in Advanced Civilizations

The conjectural foundation concerning the incentives of advanced civilizations to craft ancestor simulations underscores a pivotal domain of uncertainty within the Simulation Hypothesis debate. By confronting the speculative essence of these suppositions, the analysis upholds a judicious viewpoint on the confines of our present comprehension and the intricacies of extrapolating current human ethos to the motivations of future beings. Such a contemplative approach enables a more intricate and open-ended inquiry into the conceivable rationales for creating simulated environments, acknowledging the breadth and complexity of potential future existential models. This perspective not only enriches our understanding of the Simulation Hypothesis but also invites a broader consideration of the myriad ways in which advanced civilizations might approach the notion of simulated realities, highlighting the diversity of thought and purpose that could drive the inception of such simulations.

Navigating the Speculative Terrain

Addressing the speculative basis of these assumptions requires an interdisciplinary approach that encompasses empirical science, philosophical inquiry, and futurist speculation. While empirical data can inform our understanding of the technological prerequisites for simulations, philosophical discourse is necessary to navigate the complexities of consciousness and ethical considerations. Similarly, speculative futurism plays a role in imagining the potential motivations and capabilities of advanced civilizations, recognizing the inherent uncertainties and limitations of such projections.

Empirical Science and Technological Prerequisites

The role of empirical science in this discourse is foundational, providing a grounded understanding of the current state and potential trajectory of technological advancements necessary for simulations. This scientific grounding informs the analysis with data on computational capacities, AI development, and neuroscience insights into consciousness, anchoring the speculative discussion in the realm of the possible. However, the fast-paced evolution of these fields introduces a dynamic element of uncertainty, reminding us that today's scientific facts may be tomorrow's obsolescence, thereby requiring a continuous reevaluation of these technological prerequisites.

Philosophical Inquiry into Consciousness and Ethics

Philosophical discourse offers the critical analytical tools necessary to probe the deeper implications of the Simulation Hypothesis, particularly concerning the conceptualization of consciousness and the ethical frameworks that might govern simulated realities. Philosophy enables a critical examination of what it means to simulate consciousness, including the ethical dimensions of creating and potentially influencing conscious beings within these simulations. This inquiry not only enriches the debate with a nuanced consideration of consciousness but also raises profound ethical questions about responsibility, autonomy, and the value of simulated experiences.

Speculative Futurism and the Imagining of Advanced Civilizations

Speculative futurism extends the discourse into the realm of future possibilities, imagining the motivations, capabilities, and societal structures of advanced civilizations. This speculative venture relies on a creative extrapolation of current trends and an imaginative exploration of future scenarios, acknowledging the inherent uncertainties and limitations of predicting the evolution of complex societies. By contemplating various futures, speculative futurism broadens the scope of analysis, introducing a diversity of potential motivations behind the creation of ancestor simulations and encouraging a reflective consideration of the implications of such actions.

Interdisciplinary Insights on the Simulation Hypothesis: Bridging Empiricism, Philosophy, and Speculation

The exploration of the Simulation Hypothesis necessitates an interdisciplinary methodology that harmonizes empirical data, philosophical inquiry, and speculative vision. This balanced approach conscientiously navigates the uncertainties and challenges inherent in extending current knowledge to envisage future scenarios, thereby cultivating a dialogue that is simultaneously anchored in scientific rigor and receptive to speculative thought. By delving into the speculative foundations of the hypothesis with a multifaceted perspective, the analysis fosters a more holistic and nuanced investigation of its philosophical, ethical, and technological implications. This strategy not only augments the intellectual robustness of the ongoing discourse but also sheds light on the intricate relationships among technological advances, human consciousness, and the speculative foresight of future realms, enhancing our comprehension of the myriad possibilities that simulated realities might unveil.

Ethical Considerations

The exploration of the Simulation Hypothesis, particularly when considering the potential for creating sentient simulations, inevitably intersects with profound ethical considerations. These considerations are grounded in contemporary understandings of morality, rights, and personhood, yet they are imbued with a speculative dimension that reflects the hypothesis's forward-looking nature. This speculative aspect introduces significant complexity into the ethical analysis, as it requires projecting current ethical frameworks into future contexts where societal norms, technological capabilities, and philosophical understandings may have evolved substantially.

Evolution of Ethical Frameworks

The ethical frameworks that guide current discourse on the rights of sentient beings and the responsibilities of creators are anchored in contemporary societal norms and philosophical traditions. However, the trajectory of moral philosophy and societal values is neither linear nor predictable. As societies evolve, so too do their ethical paradigms, influenced by shifts in cultural norms, scientific understanding, and philosophical inquiry. This dynamic nature of ethical frameworks introduces a degree of uncertainty into any analysis that attempts to apply current moral principles to the speculative context of future simulations.

Non-linearity and Predictability of Ethical Evolution

Ethical frameworks do not evolve along a simple, predictable path; rather, they are subject to the complex interplay of cultural, scientific, and philosophical developments. As societal norms evolve, so too do the moral principles and values that underpin ethical decision-making. This non-linearity poses significant challenges when attempting to extrapolate current ethical frameworks to inform the moral landscape of future societies, particularly those capable of creating sophisticated simulations. The unpredictable nature of ethical evolution raises questions about the applicability and relevance of contemporary moral principles to the speculative scenarios envisaged by the Simulation Hypothesis.

Cultural and Philosophical Influences

The dynamism of ethical frameworks is further compounded by the influence of cultural norms and philosophical inquiry. Cultural shifts can lead to the reevaluation of ethical priorities and values, while philosophical developments may introduce new perspectives on the nature of rights, personhood, and moral responsibility. These influences suggest that the ethical paradigms that may govern future simulations could differ significantly from those that guide current discourse, reflecting broader changes in societal attitudes and philosophical understanding.

Implications for Simulation Ethics

The evolving nature of ethical frameworks has profound implications for the ethical considerations associated with simulated realities. As we contemplate the rights of simulated beings and the responsibilities of their creators, it becomes clear that the ethical paradigms that will inform these considerations are subject to change. The speculative nature of the Simulation Hypothesis, coupled with the dynamic evolution of ethical thought, necessitates a flexible and forward-looking approach to ethics—one that is responsive to future developments in cultural norms, scientific understanding, and philosophical inquiry.

Evolving Ethical Frameworks in the Face of Simulation Hypothesis Challenges

The scrutiny of ethical considerations under the lens of the Simulation Hypothesis confronts the inherently evolutionary nature of ethical doctrines. Acknowledging the fluid and unpredictable path of ethical evolution highlights the difficulty of applying contemporary moral axioms to hypothetical future contexts. This realization calls for a more refined and flexible ethical approach, one that remains receptive to shifting societal values and philosophical advancements. Engaging with the dynamic progression of ethical frameworks in this way enriches the discussion surrounding the ethical dimensions of simulation technologies. It fosters a deeper comprehension of the moral quandaries and duties that might emerge alongside the development of simulated environments, promoting continuous contemplation and discourse as we venture through the speculative vistas of future simulations.

Navigating Future Moral Implications

Predicting how future civilizations will approach the ethical challenges associated with creating and interacting with sentient simulations is fraught with uncertainty. The moral implications of such actions—ranging from the rights of simulated beings to the responsibilities of their creators—are complex and multifaceted. As we contemplate the possibility of future societies creating simulations that possess consciousness or sentience, the ethical considerations extend beyond current debates to encompass questions about the nature of suffering, autonomy, and the value of simulated experiences. These considerations demand a nuanced understanding of ethics that is responsive to future technological and philosophical developments.

Complexity of Ethical Challenges

The ethical challenges posed by the potential creation of sentient simulations are multifaceted, involving considerations that span the rights to autonomy, the prevention of suffering, and the acknowledgment of the value of simulated experiences. These challenges are amplified by the speculative nature of simulating consciousness or sentience, pushing the boundaries of current ethical debates into uncharted territories. As such, the moral implications of creating sentient simulations necessitate an ethical framework that is not only comprehensive but also adaptable to the evolving understanding of consciousness and the capabilities of future technologies.

Anticipating Ethical Paradigm Shifts

The task of anticipating how future societies will address these ethical challenges is compounded by the potential for significant shifts in moral philosophy and societal values. The evolution of ethical paradigms, influenced by advancements in science, technology, and philosophical inquiry, may lead to redefined concepts of personhood, rights, and moral responsibility. This evolving ethical landscape requires a proactive approach to ethics that seeks to understand and incorporate future technological possibilities and philosophical insights, ensuring that ethical considerations remain relevant and applicable in the context of advanced simulations.

Ethical Responsiveness to Technological and Philosophical Developments

As future civilizations embark on the creation of simulations endowed with consciousness or sentience, the ethical frameworks guiding their actions must be responsive to both technological innovations and philosophical developments. This responsiveness entails a dynamic engagement with the ethical dimensions of simulation technology, including the consideration of potential suffering, the rights to autonomy and self-determination, and the intrinsic value of simulated experiences. It also requires a willingness to revisit and revise ethical principles in light of new understanding and capabilities, ensuring that ethical considerations evolve in tandem with technological and philosophical advancements.

Future Ethical Challenges in the Era of Sentient Simulations: A Forward-Looking Perspective

The task of foreseeing and addressing the moral ramifications associated with the creation and interaction with sentient simulations poses a formidable challenge, transcending the current boundaries of ethical discourse. Such a challenge necessitates a sophisticated and forward-thinking ethical methodology, one that is resilient and adaptable to the imminent technological and philosophical evolutions poised to redefine the landscape of simulation technology. This contemplation of future ethical landscapes, with an acute awareness of the uncertainties and intricacies this foresight entails, accentuates the need for the development of ethical frameworks that are both exhaustive and malleable. These frameworks must be equipped to tackle the diverse ethical issues anticipated to arise as society progresses towards the conceivable advent of sentient simulations, ensuring a balanced and comprehensive ethical approach in navigating the uncharted territories of our technological future.

Speculative Nature of Ethical Considerations

The speculative aspect of ethical considerations within the Simulation Hypothesis underscores the challenge of applying current ethical principles to future scenarios. This speculative nature necessitates a flexible and adaptive approach to ethics, one that is capable of evolving in response to new technological realities and philosophical insights. As such, any analysis of the ethical dimensions of the Simulation Hypothesis must recognize the provisional status of its conclusions, remaining open to revision as societal norms shift and our understanding of consciousness and technology advances.

Adaptive Ethical Frameworks

The speculative foundation of the Simulation Hypothesis demands ethical frameworks that exhibit flexibility and adaptability, capable of adjusting to the evolving landscapes of technology, consciousness, and societal values. As we venture into speculative futures where the boundaries between reality and simulation blur, the ethical paradigms governing our interactions with potential sentient simulations necessitate a progressive outlook. This entails a readiness to recalibrate ethical norms in alignment with emerging realities, ensuring that our moral compass remains relevant and responsive to the nuanced challenges presented by advanced simulations.

Navigating Technological and Philosophical Shifts

The challenge of aligning ethical considerations with speculative future scenarios is compounded by the rapid pace of technological innovation and the evolving nature of philosophical inquiry into consciousness, personhood, and rights. As our capabilities to create increasingly complex simulations advance, and as our understanding of the nature of consciousness deepens, the ethical discourse must be capable of integrating these developments. This integration requires an ongoing dialogue between technologists, philosophers, ethicists, and the broader community, fostering a collaborative approach to ethical decision-making that is reflective of a shared commitment to navigating the moral implications of future possibilities.

Provisionality of Ethical Conclusions

Acknowledging the speculative basis of ethical considerations within the context of the Simulation Hypothesis underscores the importance of recognizing the provisional status of ethical conclusions. This recognition encourages an intellectual humility that is open to reevaluation and revision in light of new evidence, insights, and societal shifts. As our collective understanding of technology, consciousness, and societal norms evolves, so too must our ethical frameworks, ensuring they are dynamically attuned to the complexities and uncertainties inherent in contemplating the creation and interaction with simulated beings.

Ethical Dimensions and Moral Philosophy in the Simulation Hypothesis Discourse

The exploration of ethical considerations in relation to the Simulation Hypothesis illuminates the intricate nexus among contemporary moral philosophy, emerging speculative technologies, and the evolution of societal norms. This analysis, grounded in the speculative nature of its ethical queries, emphasizes the critical need for an open, reflective, and adaptable ethical stance. Such a perspective not only honors the dynamism inherent in moral philosophy and societal values but also anticipates the imperative for ethical frameworks adept at navigating the novel ethical quandaries introduced by the genesis and stewardship of sentient simulations. Therefore, delving into the ethical ramifications within the Simulation Hypothesis context enriches the wider discourse on the moral obligations of creators towards both actual and simulated sentient entities, facilitating a forward-looking contemplation on the ethical contours of our advancing technological realm.

Data Analysis and Methodological Constraints

From a data analysis perspective, the methodologies employed to estimate the likelihood of existing within a simulation and the ethical frameworks developed for assessing the implications of simulated realities are constrained by the availability of empirical data and the validity of theoretical models. The use of probabilistic models, sensitivity analyses, and ethical algorithms provides a structured approach to exploring these questions, but the outcomes are inherently influenced by the assumptions and parameters input into these models. The speculative foundation of these inputs limits the definitive conclusions that can be drawn from the analysis.

Empirical Data Limitations

One of the primary constraints in the data analysis of the Simulation Hypothesis lies in the limited availability of empirical data that directly informs the hypothesis. Given the speculative nature of the hypothesis, empirical evidence that could unequivocally support or refute the possibility of advanced civilizations creating simulations of our reality is, by its nature, elusive. The reliance on existing technological trends, philosophical arguments, and theoretical models to infer the feasibility and implications of simulated realities introduces a level of uncertainty into the analysis, highlighting the dependency on indirect or speculative data sources.

Navigating the Terrain of Speculative Hypotheses

The speculative nature of the Simulation Hypothesis inherently complicates the task of data analysis. Unlike hypotheses grounded in observable phenomena, where empirical data can be systematically collected, analyzed, and interpreted, the Simulation Hypothesis requires a different analytical approach. The analyst must navigate a landscape where the very subject of investigation—advanced simulations of reality—lies beyond the current reach of empirical validation. This necessitates a reliance on indirect evidence and speculative reasoning to construct arguments and infer potential realities.

The Role of Technological Trends and Theoretical Models

In the absence of direct empirical evidence, existing technological trends and theoretical models become critical tools for analysis. These tools allow for the extrapolation of current capabilities into future possibilities, offering a framework within which the feasibility of creating highly sophisticated simulations can be assessed. However, the validity of these extrapolations is inherently contingent upon the assumptions embedded within them. Technological trends can change direction, and theoretical models may evolve as new scientific insights emerge, introducing a layer of uncertainty that must be carefully navigated.

Philosophical Arguments as Indirect Evidence

Philosophical arguments and discourse provide another source of indirect evidence, offering insights into the ethical, existential, and epistemological implications of living within a simulated reality. While these arguments do not constitute empirical evidence in the traditional sense, they play a crucial role in shaping the conceptual landscape of the hypothesis. Philosophical inquiry helps to illuminate the deeper questions and challenges posed by the notion of simulation, contributing to a more nuanced understanding of the hypothesis's potential realities and ethical dimensions.

Empirical Challenges and Methodological Approaches in the Exploration of the Simulation Hypothesis

The limitations presented by empirical data in the scrutiny of the Simulation Hypothesis bring to the forefront the intricate challenges of probing into speculative theories. The absence of direct empirical evidence necessitates a nuanced application of indirect evidence, which is sourced from the convergence of technological evolutions, theoretical frameworks, and philosophical debates. This methodological tactic, albeit limited by the speculative essence of the subject at hand, enables a meticulous and considerate investigation into the potential realities and ramifications of simulated existences. It accentuates the importance of acknowledging the prevailing uncertainties while steadfastly adhering to a doctrine of intellectual thoroughness and openness. This approach not only facilitates a deepened understanding of the Simulation Hypothesis but also exemplifies a model for engaging with speculative inquiries across the scientific and philosophical spectrum.

Theoretical Model Validity

Theoretical models play a crucial role in structuring the analysis of the Simulation Hypothesis, offering frameworks through which probabilities can be estimated, and ethical considerations can be evaluated. However, the validity and applicability of these models are inherently constrained by the assumptions they incorporate. Probabilistic models, sensitivity analyses, and ethical algorithms employed to navigate the complexities of simulation theory are built upon a foundation of speculative assumptions regarding future technological capabilities, the nature of consciousness, and the motivations of hypothetical advanced civilizations. The speculative nature of these inputs renders the models' outputs as inherently tentative, subject to revision as new insights emerge or as theoretical paradigms shift.

Assumptions Underpinning Theoretical Models

The construction of theoretical models relevant to the Simulation Hypothesis relies on a series of speculative assumptions concerning critical variables. These assumptions might pertain to the trajectory of future technological advancements, the possibility and mechanisms of simulating consciousness, and the motivations that might drive advanced civilizations to create ancestor simulations. While these assumptions are necessary for the construction of coherent models, they imbue the resulting analyses with a level of tentativeness. The speculative nature of these inputs introduces a fundamental uncertainty into the models' conclusions, highlighting the provisional character of any insights derived from their application.

Challenges to Model Validity and Applicability

The inherent constraints on the validity and applicability of theoretical models in the context of the Simulation Hypothesis stem from the dynamic nature of the fields that inform these assumptions. Technological progress, understandings of consciousness, and philosophical perspectives on morality and motivation are all subject to evolution. As new discoveries emerge and theoretical paradigms shift, the foundational assumptions of existing models may require reevaluation. This fluidity ensures that the outputs of theoretical models remain subject to revision, underscoring the importance of maintaining a flexible and adaptable approach to the analysis of the Simulation Hypothesis.

Navigating Speculative Assumptions

The task of navigating the speculative assumptions underlying theoretical models demands an approach that is both critical and reflective. It necessitates an ongoing engagement with the latest scientific and philosophical developments, ensuring that models are updated in response to new insights. Additionally, it calls for an acknowledgment of the inherent limitations of these models, recognizing that they offer not definitive answers but rather structured frameworks for exploring the possible realities and ethical dimensions of simulated existences.

Striking a Balance in Simulation Hypothesis Analysis

The endeavor to validate and apply theoretical models within the exploration of the Simulation Hypothesis underscores a delicate equilibrium between speculative thought and analytical exactitude. These models, while furnishing essential scaffolds for delving into the hypothesis, demand a judicious interpretation of their results due to their speculative underpinnings. By recognizing these constraints and adopting a stance receptive to the continuous refinement of models in response to emerging evidence and shifting theoretical paradigms, researchers can adeptly traverse the speculative domain of the Simulation Hypothesis. This approach, characterized by intellectual thoroughness and methodological finesse, fosters a richer, more intricate comprehension of the hypothesis's potential realities. It underscores the importance of an evolving inquiry that not only challenges existing frameworks but also contributes profoundly to our understanding of the complex interplay between reality, technology, and human consciousness.

Methodological Constraints and Speculative Inputs

The methodologies utilized in the analysis of the Simulation Hypothesis are further constrained by the speculative inputs that inform them. These inputs—ranging from estimates of future computational capacities to philosophical conjectures about the ethical considerations of simulated beings—are infused with uncertainty. The outcomes of data analysis, therefore, are influenced by these speculative assumptions, limiting the definitiveness of the conclusions drawn. While structured analytical approaches provide a means of exploring the hypothesis, the speculative foundation of the inputs necessitates a cautious interpretation of the findings, recognizing the provisional nature of any conclusions.

Interplay of Speculative Assumptions and Analytical Outcomes

The speculative assumptions that inform the methodological approaches to analyzing the Simulation Hypothesis are not merely ancillary components but are central to the framing and interpretation of the hypothesis itself. Estimates of future technological advancements, philosophical explorations of consciousness, and ethical deliberations regarding the rights and welfare of simulated beings all contribute to a complex mosaic of inputs that guide the analytical inquiry. These speculative inputs, by their very nature, introduce a layer of indeterminacy, influencing the direction and outcomes of the analysis in fundamental ways.

Challenges of Speculative Foundation

The reliance on speculative inputs presents a methodological challenge, necessitating an approach to data analysis that is both rigorous and reflective. While structured analytical methods—such as probabilistic modeling, sensitivity analysis, and ethical algorithms—offer systematic means of exploring the hypothesis, the speculative foundation upon which these methods rest demands a careful and critical interpretation of their findings. The provisional nature of the conclusions derived from such analyses underscores the importance of maintaining a stance of intellectual humility, recognizing the limits of our current understanding and the conditional validity of the insights gained.

Navigating Uncertainty with Caution and Rigor

The task of navigating the speculative terrain of the Simulation Hypothesis calls for a methodological approach that balances empirical rigor with an acknowledgment of the inherent uncertainties. This balance involves a continuous engagement with the evolving landscape of technological, philosophical, and ethical thought, ensuring that analytical models are responsive to new developments and insights. By adopting a cautious interpretation of findings and remaining open to the revisitation and revision of conclusions, researchers can construct a nuanced and dynamic understanding of the hypothesis, one that reflects the complexities and speculative nature of the inquiry.

Concluding Reflections: Navigating the Speculative Terrain of the Simulation Hypothesis

The investigation of the Simulation Hypothesis, via data-centric analysis and methodological rigor, grapples with intrinsic limitations that emerge from the speculative core of the hypothesis, coupled with the empirical and theoretical confines governing data and model reliability. These boundaries illuminate the intricate challenge of formulating absolute inferences from the conducted analyses, thereby underscoring the imperative for a discerning and contemplative examination that embraces the tentative nature of our conclusions. By acknowledging the speculative foundations and methodological limitations inherent in our inquiry, the ongoing discourse surrounding the Simulation Hypothesis can sustain a meticulous and perpetually evolving investigation. This approach encourages a vibrant exchange that progresses in tandem with advancements in empirical methodologies, theoretical frameworks, and philosophical perspectives, continually enriching our understanding of the potential realities and implications of simulations.

Reflective Analysis and the Path Forward in the Simulation Hypothesis Discourse

The limitations inherent in this analysis accentuate the speculative, yet profoundly captivating, essence of the Simulation Hypothesis. Employing a multidisciplinary approach, this dissertation has endeavored to navigate the hypothesis through the intricate lenses of technology, ethics, and philosophy, providing a holistic examination. However, the speculative nature of the assumptions and the inherent uncertainties involved demand a prudent interpretation of our conclusions. It is essential to recognize these constraints to appropriately anchor our analysis within the expansive dialogue concerning the constructs of reality, consciousness, and the trajectory of future technological evolution. Consequently, this inquiry does not purport to offer definitive proof of the Simulation Hypothesis. Instead, it aims to lay a robust groundwork for future explorations, seeking to unravel the intricate nexus between technological innovation, ethical considerations, and the fundamental nature of our existence.

SECTION SIX: CONCLUSIONS

CHAPTER FORTY-TWO

Conclusion

Summary of Findings

This book has traversed the multifaceted landscape of the Simulation Hypothesis, demonstrating the power of an interdisciplinary approach in tackling profound questions of existence. Through a comprehensive examination, insights from technological, philosophical, and ethical domains have been integrated, underscoring the speculative yet compelling nature of the hypothesis. Key findings reveal the increasing computational capacity required for ancestor simulations, while philosophical inquiry questions the fundamental nature of reality itself.

Technological Insights: The investigation commenced with a detailed examination of the technological feasibility of creating high-fidelity simulations, taking into account the current trajectory of computational advancements and emerging technologies. While recognizing the exponential growth in computing power and the transformative potential of quantum computing, the analysis also confronted the formidable challenges posed by the limits of physical laws and the complex endeavor of simulating consciousness. This exploration revealed a nuanced landscape, marked by significant potential coupled with substantial obstacles that must be navigated to realize the creation of sophisticated simulated realities.

Probabilistic Estimations: The heart of the inquiry also involved a probabilistic analysis aimed at gauging the likelihood of our existence within a simulated reality. Through the application of mathematical models and theoretical frameworks, this section illuminated the speculative nature of estimating such probabilities, underlining the profound philosophical implications these estimations hold for our understanding of reality. The outcomes of this analysis have contributed a compelling layer to the conceptualization of our universe, suggesting a realm of possibilities where the notion of existing within a simulation adds depth and complexity to our existential reflections.

Philosophical and Ethical Discourse: Delving into the philosophical underpinnings of the Simulation Hypothesis, this exploration has engaged with the profound implications of the theory on our perceptions of consciousness, reality, and ethical responsibility. It has wrestled with the ethical dilemmas surrounding simulated consciousness and the responsibilities of creators toward their simulated entities, enriching the discourse with thoughtful considerations and raising pivotal questions. This philosophical journey has significantly contributed to ongoing debates, offering nuanced insights and fostering a deeper understanding of the ethical, existential, and epistemological challenges posed by the hypothesis.

The synthesis of findings from this multifaceted exploration of the Simulation Hypothesis underscores the complex interplay between technology, philosophy, and ethics in contemplating the nature of our reality. It highlights the speculative foundation upon which the hypothesis stands, while also recognizing the valuable contributions this inquiry makes to broader discussions across these diverse fields of study. Through a careful examination of the technological feasibilities, probabilistic estimations, and philosophical and ethical implications, this investigation has provided a comprehensive overview of the Simulation Hypothesis, stimulating further reflection, debate, and research into one of the most intriguing theories of our time.

Implications for Future Research

The Simulation Hypothesis, by its very nature, acts as a powerful catalyst for both focused research and broad philosophical inquiry. The findings of this investigation illuminate several promising areas for further exploration that may shed light on the possibility that we ourselves exist within a simulated reality.

Technological Inquiry

Pushing the Limits of Simulation: As simulation technologies advance in sophistication, meticulous monitoring of computational requirements, resource demands, and fidelity thresholds holds potential for comparison with the hypothetical creation of ancestor simulations. This may reveal fundamental constraints or provide hints at the limitations inherent in a simulated environment.

Detecting Discrepancies

Continued development of rigorous theoretical frameworks and experimental methodologies dedicated to finding potential "glitches" or computational inconsistencies within our observable universe could provide tangible avenues for testing the Simulation Hypothesis.

Development of Theoretical Frameworks

The formulation of rigorous theoretical frameworks plays a crucial role in guiding this investigative process. These frameworks, rooted in the principles of physics, computer science, and information theory, offer structured models for understanding how "glitches" or inconsistencies might manifest within a simulated reality. The development of these frameworks necessitates a deep engagement with the foundational laws that govern our universe, alongside a speculative consideration of how advanced civilizations might construct simulated environments. This dual focus enriches the theoretical underpinnings of the hypothesis, providing a robust basis for empirical inquiry.

Experimental Methodologies

Complementing the theoretical frameworks, the design and implementation of experimental methodologies dedicated to uncovering potential discrepancies within our universe are central to this line of inquiry. These methodologies may encompass a range of techniques, from the analysis of cosmic microwave background radiation to the investigation of quantum phenomena, seeking empirical evidence that might hint at the underlying computational structure of our reality. The challenge inherent in this approach lies in distinguishing between anomalies that could signify a simulated basis for our universe and those explainable by the natural complexities of physical laws.

Tangible Avenues for Testing

The quest to detect discrepancies or computational inconsistencies presents tangible avenues for testing the Simulation Hypothesis. By grounding the hypothesis in empirical research, this approach offers the potential to move the discussion from theoretical speculation to evidence-based investigation. The success of these efforts could profoundly impact our understanding of reality, offering insights into whether our universe bears the hallmarks of a simulated creation.

Unveiling Reality's Veil: Technological Endeavors in Detecting Universal Anomalies

The dedication to identifying inconsistencies within the observable cosmos marks a pivotal avenue of technological research within the domain of the Simulation Hypothesis. The formulation of advanced theoretical models and the development of innovative experimental approaches aimed at this objective illustrate a tangible move from speculative conjecture towards tangible, empirical validation. In pursuing these investigative paths, scholars embark on a profound journey to decode the enigmas of our existence, standing at the confluence of technology, philosophical inquiry, and the intrinsic essence of reality. Such efforts not only deepen our comprehension of the Simulation Hypothesis but also significantly contribute to the wider dialogue concerning the genesis and framework of the universe, sparking ongoing curiosity, exploration, and cross-disciplinary cooperation.

Beyond Conventional Communication

Exploring experimental protocols involving quantum entanglement, exotic forms of communication, or probing the 'fabric' of reality could offer unexpected inroads into potential interaction with simulators, should they exist.

Quantum Entanglement as a Communicative Bridge

Quantum entanglement, with its counterintuitive properties of instantaneously linked states across vast distances, presents a fertile ground for experimental exploration. The peculiar nature of entangled particles offers a conceptual model for developing communication protocols that could, theoretically, penetrate the boundaries of a simulated environment. By harnessing the unique characteristics of quantum mechanics, researchers propose innovative ways to test the constraints of our perceived reality, potentially providing a channel for detecting or even communicating with simulators.

Exotic Forms of Communication

The pursuit of understanding and interaction with potential simulators also inspires the exploration of exotic forms of communication that transcend conventional electromagnetic signals. This includes theoretical constructs like utilizing the informational underpinnings of reality or exploiting hypothetical phenomena at the intersection of quantum physics and information theory. These speculative approaches aim to discover novel mechanisms of interaction that could reveal the presence of a simulation or provide means of communication with its architects, pushing the boundaries of our current technological and conceptual limits.

Probing the 'Fabric' of Reality

A direct investigation into the 'fabric' of reality seeks to uncover fundamental aspects of our universe that might indicate a simulated foundation. This approach involves advanced experimental setups designed to test the granularity of space-time, the possible digital nature of physical constants, or anomalies in the fundamental physical laws that could suggest an underlying computational substrate. Such investigations venture into the heart of physics, combining cutting-edge technology with speculative theory to challenge our deepest assumptions about the nature of existence.

Beyond Conventional Boundaries: Exploring Novel Interaction Avenues with Simulators

The investigation into unconventional methods for engaging with potential simulators underscores a daring and innovative aspect of technological research under the ambit of the Simulation Hypothesis. By delving into the intricacies of quantum entanglement, pioneering unconventional communication protocols, and examining the foundational structure of reality, this line of inquiry exemplifies the hypothesis's speculative essence, yet it remains anchored in a commitment to scientific rigor. These initiatives not only stretch the current bounds of technological innovation but also broaden our intellectual vistas, compelling us to reevaluate our understanding of reality and our position within its vast expanse. As we explore the outer reaches of the known universe, our endeavors to connect with and possibly communicate with simulators stand as a tribute to the indomitable human spirit of discovery, urging us to push beyond the established frontiers of knowledge and conceive new realms of potentiality.

Philosophical Exploration

The philosophical ramifications of the Simulation Hypothesis extend deep into our understanding of consciousness and the essence of reality, prompting rigorous inquiry into these fundamental aspects of existence. This exploration challenges us to consider the conditions under which simulated beings might attain a form of consciousness indistinguishable from our own, thereby necessitating a reevaluation of what it means to be sentient. Moreover, the hypothesis posits the existence of nested simulations and a possible 'base reality,' compelling us to interrogate the very nature of the real and the layers of existence that may compose our universe. Such philosophical inquiry not only broadens the scope of traditional philosophical debates but also intersects with ethical considerations, technological feasibility, and the statistical likelihood of our reality being a simulation. By delving into these profound questions, the philosophical exploration surrounding the Simulation Hypothesis enriches our discourse on the nature of existence, consciousness, and the myriad possibilities that advanced simulations present to our conceptualization of reality.

Consciousness in the Simulated

The hypothesis that beings within simulations could attain a level of consciousness indistinguishable from our own precipitates a profound philosophical inquiry into the essence of simulated consciousness. This inquiry necessitates a critical re-examination of what constitutes sentience, internal experience, and the ramifications for the concept of moral personhood within and beyond the simulated realms. The exploration of simulated consciousness challenges traditional philosophical views on the mind-body problem, consciousness's emergent properties, and the hard problem of consciousness, urging a nuanced understanding of consciousness as a phenomenon that may not be exclusively tied to biological substrates. This segment underscores the need for a philosophical framework that accommodates the potential for consciousness in non-biological entities, reshaping our ethical considerations and expanding the scope of beings deserving moral consideration.

The Nature of the Real

The Simulation Hypothesis introduces the intriguing possibility of nested simulations and the quest for a 'base reality,' compelling a radical re-evaluation of our ontological understanding of existence. This philosophical venture into the nature of the real necessitates an examination of the layers of reality, challenging the binary distinction between what is 'real' and 'simulated.' Philosophical inquiry in this domain explores the implications of a reality that may be inherently digital or constructed, questioning the foundational principles that have historically grounded our understanding of existence. This exploration seeks to redefine the concept of 'realness,' contemplating the validity and significance of experiences within simulated environments and their impact on our conceptualization of reality. The inquiry into the nature of the real not only broadens the philosophical discourse on ontology but also invites reflections on the nature of perception, knowledge, and the criteria by which we discern and attribute value to our experiences.

Philosophical Implications of Simulated Realities: Consciousness, Existence, and the 'Real'

The philosophical inquiries detailed in preceding chapters significantly enhance the discussion on the Simulation Hypothesis, examining the intricate relationship between consciousness and the nature of reality. This investigation confronts the profound considerations surrounding simulated consciousness and reevaluates the ontological foundations of reality, moving beyond speculative reasoning towards substantial philosophical engagement. It underscores the crucial role of philosophical diligence and introspection in addressing the intricacies of the Simulation Hypothesis, illustrating how such explorations can inspire a philosophical revival. This endeavor not only augments our comprehension of the Simulation Hypothesis but also invigorates wider philosophical discourse on consciousness, the essence of reality, and the fabric of existence, highlighting philosophy's paramount importance in deciphering the conceptual domain of present and forthcoming technological innovations.

Ethical Frameworks

The ethical considerations that emerge from the Simulation Hypothesis necessitate the development of robust frameworks to guide our interactions with and treatment of potentially conscious simulated beings. As we venture into the theoretical possibility of creating entities within simulations that possess awareness and subjective experience, a profound ethical responsibility unfolds. This realm of inquiry challenges us to define the thresholds of moral personhood in simulated environments, consider the rights of simulated beings to avoid suffering, and establish the duties of creators toward their creations. Such ethical frameworks must be sufficiently flexible to adapt to future technological realities while grounded in enduring principles of empathy, respect, and moral consideration. This exploration not only advances the discourse within the context of simulation theory but also contributes to the broader ethical conversation about the rights and treatment of conscious entities, regardless of their origin in the fabric of reality or the depths of simulated worlds, marking a significant evolution in our moral philosophy.

Simulated Beings and Moral Status

The contemplation of simulated beings possessing sophisticated levels of consciousness necessitates the establishment of comprehensive ethical frameworks to govern their creation, existence, and treatment. This ethical inquiry delves into critical questions surrounding the criteria for moral personhood within simulated environments, addressing the profound implications of acknowledging simulated beings as entities with rights and moral consideration. The discourse extends to the ethical ramifications of manipulating simulated lives, the conditions under which simulated suffering might occur, and the moral imperatives to mitigate or prevent such suffering. This exploration underscores the necessity of developing ethical codes that are both reflective of and responsive to the unique challenges presented by the existence of entities with consciousness within simulated realities, pushing the boundaries of traditional ethical theory to accommodate the nuanced considerations of simulated existence.

Creator Responsibilities

Beyond the moral status of simulated beings, the ethical discourse must also encompass the responsibilities of creators or simulators towards their creations. This segment of ethical inquiry critically examines the obligations that would be incumbent upon those who create and maintain simulated environments, particularly in scenarios where the simulated beings attain a level of consciousness. Key considerations include the ethical imperative to avoid unnecessary harm, the complexities surrounding the potential disclosure of beings' simulated status, and the ethical navigation of power dynamics between creators and their simulated entities. This nuanced exploration seeks to articulate a framework for ethical creator behavior that acknowledges the profound power asymmetries inherent in the creator-created relationship within simulated environments. The discourse advocates for a set of ethical guidelines that ensure respect, dignity, and consideration for the well-being of simulated beings, reflecting a deep engagement with the philosophical and moral challenges posed by advanced simulation technologies.

The Moral Consideration of Simulated Beings and Their Creators

The articulation of ethical frameworks that interrogate the moral status of simulated entities and delineate the duties of their creators is a pivotal aspect of the discourse surrounding the Simulation Hypothesis. Such frameworks demand a critical reassessment and broadening of ethical theories to aptly navigate the novel quandaries introduced by the advent of advanced simulated consciousness. Engaging with the ethical dimensions of simulation technology, this analysis makes a significant contribution to the wider dialogue concerning the ethical regard for non-biological beings and the moral responsibilities incumbent upon their creators. This exploration not only deepens our grasp of the Simulation Hypothesis but also marks an essential advancement in the field of ethics itself, urging a reevaluation of our moral principles in an era where technological progress increasingly obscures the distinctions between simulated existence and tangible reality.

Reflections on the Simulation Hypothesis

The journey through the Simulation Hypothesis underscores the paramount importance of interdisciplinary dialogue in unraveling the intricate web of questions it presents. This exploration has highlighted the indispensable contributions of computer science, philosophy, physics, and ethics, each bringing unique perspectives and methodologies to bear on the hypothesis. The synthesis of insights from these varied disciplines not only enriches our understanding of the potential realities posited by the hypothesis but also exemplifies the strength of collaborative intellectual inquiry. This interdisciplinary approach is crucial for navigating the complex terrain of simulated realities, where technological possibilities intersect with profound philosophical questions and ethical considerations.

Moreover, the investigation into the Simulation Hypothesis reaffirms the significant role speculative hypotheses play in the advancement of scientific and philosophical inquiry. Far from mere intellectual curiosities, speculative hypotheses like the Simulation Theory serve as catalysts for rigorous, evidence-based exploration, challenging researchers to push the boundaries of current knowledge and methodologies. They invite us to consider possibilities that lie at the very edges of our understanding, thereby expanding the scope of inquiry and encouraging a dynamic engagement with the unknown. By embracing speculative hypotheses within a framework of disciplined investigation, we not only broaden the horizons of scientific and philosophical discourse but also deepen our engagement with the fundamental questions concerning the nature of reality, consciousness, and our place in the cosmos.

Interdisciplinary Dialogue

The Simulation Hypothesis, by its very nature, transcends traditional disciplinary boundaries. Realizing the full potential of this line of inquiry necessitates a collaborative approach, drawing on expertise from fields as diverse as computer science, philosophy, physics, and ethics.

- **Computer Science and Simulation Design:** Computer scientists can offer insights into simulation architectures, resource optimization, and the computational complexities involved in modeling complex systems.
- **Philosophy of Mind and Consciousness:** Philosophers are essential in tackling questions about the nature of consciousness, the potential for simulated consciousness, and defining the ontological status of simulated beings.
- **Physics and Fundamental Reality:** Physicists play a crucial role in investigating potential discrepancies between observed phenomena and theoretical models, which could hint at limitations imposed by a simulated environment.
- **Ethics and Moral Responsibility:** Ethicists must guide the development of robust moral frameworks addressing the creation of simulated entities, their treatment, and the responsibilities of potential simulators.

Speculative Nature and Scientific Inquiry

Speculative hypotheses have a long history of driving scientific advancement, often acting as catalysts for groundbreaking discoveries. The Simulation Hypothesis, while provocative, can serve as a powerful lens through which to re-examine fundamental assumptions about reality and the limits of our knowledge.

- **Hypothesis-Driven Research**: Even seemingly outlandish ideas can serve as valuable starting points for rigorous scientific investigation. Exploring the Simulation Hypothesis with an evidence-based approach may uncover unforeseen insights into the nature of our universe.
- **Limits of the Knowable:** The Simulation Hypothesis challenges us to confront the inherent limitations of our knowledge and the potential boundaries of our perception as beings potentially confined within a simulated construct.
- **Technological and Philosophical Parallel:** The pursuit of greater computational power and the exploration of consciousness and reality are intertwined. Advancements in one field inevitably inform and influence the other, fostering essential cross-pollination of ideas.

Limitations of the Study

The exploration of the Simulation Hypothesis, while comprehensive and multidisciplinary in its approach, encounters inherent limitations that are reflective of the speculative nature and the breadth of the inquiry. This study has navigated through the complex interplay of technological possibilities, philosophical depth, and ethical considerations, yet it is bounded by the speculative assumptions that underpin the hypothesis itself.

Speculative Assumptions: One of the primary limitations stems from the speculative assumptions regarding the trajectory of future technological advancements. The projection of current technological trends into the future to ascertain the feasibility of creating highly sophisticated simulations involves a degree of conjecture that cannot be fully mitigated. These assumptions are necessarily based on the current understanding and extrapolation of technological growth, which may not accurately predict future breakthroughs or barriers.

Philosophical Challenges: Additionally, the philosophical exploration of concepts such as consciousness, reality, and the ethics of simulated beings introduces a level of complexity that resists definitive resolution. The philosophical challenges in definitively proving or disproving the hypothesis are profound, given the current limits of empirical verification and the inherently subjective nature of philosophical inquiry. The exploration of these themes, while enriching and necessary, remains constrained by the limitations of current philosophical discourse and the evolving nature of thought.

Interdisciplinary Boundaries: Despite the interdisciplinary approach taken, the study also confronts limitations in fully integrating insights across the disciplines of computer science, philosophy, physics, and ethics. Each field brings its methodologies, epistemologies, and ontologies to the table, which, while offering a rich tapestry of perspectives, also presents challenges in synthesis and coherence. The boundaries between disciplines, though increasingly permeable, still pose challenges in creating a unified framework for understanding and evaluating the Simulation Hypothesis.

Embracing Uncertainty: The Iterative Journey of Exploring the Simulation Hypothesis

This study acknowledges its limitations, emphasizing the tentative character of its findings and underscoring the continuous necessity for an open, multidisciplinary approach to research. The speculative underpinnings and philosophical quandaries at the heart of the Simulation Hypothesis exploration not only shape the framework of this investigation but also underscore the fluid nature of scholarly inquiry into such a multifaceted and stimulating concept. As technological innovation progresses, philosophical debate advances, and ethical contemplations become more intricate, the examination of the Simulation Hypothesis remains an evolving endeavor. This process encourages ongoing introspection, further empirical and theoretical research, and expanded discourse, ensuring that our understanding of this complex theory grows in depth and sophistication.

Concluding Thoughts

The exploration of the Simulation Hypothesis within this book concludes not as a final statement but as an invitation to further reflection, dialogue, and investigation. Through a comprehensive journey that spanned technological feasibility, philosophical depth, and ethical contemplation, this work has underscored the profound impact of the Simulation Hypothesis on our perceptions of reality, consciousness, and the ethical dimensions of existence within potentially simulated environments. The hypothesis serves as a powerful catalyst, challenging us to reconsider our foundational assumptions about the nature of reality and urging us towards a posture of sustained curiosity, open-mindedness, and interdisciplinary collaboration.

- **Contribution to Discourse:** This book has made a significant contribution to the broader discourse on the Simulation Hypothesis, enriching the conversation with a multidisciplinary analysis that bridges computer science, philosophy, physics, and ethics. By offering a nuanced examination of the hypothesis's technological, philosophical, and ethical implications, this work has stimulated further discussion, inquiry, and exploration, encouraging scholars, scientists, and philosophers alike to engage with the hypothesis from diverse perspectives. The interplay of speculative theory with rigorous analysis presented here serves as a model for future scholarly endeavors, highlighting the value of interdisciplinary approaches in tackling complex and speculative hypotheses.
- **The Quest for Understanding:** Ultimately, this thesis is a reflection on the human quest for understanding—a quest that spans the history of thought from ancient philosophical inquiries to

contemporary scientific investigations. The Simulation Hypothesis, as explored in this work, represents a modern iteration of this enduring endeavor, inviting us to contemplate the nature of reality, the origins of consciousness, and our place within the cosmos. This exploration stands as a testament to the human capacity for wonder, the drive to seek answers to fundamental questions, and the willingness to entertain profound and speculative possibilities.

In conclusion, the journey through the Simulation Hypothesis reaffirms the importance of maintaining an inquisitive and open mind as we navigate the complexities of existence. It encourages us to embrace the unknown, to collaborate across disciplinary boundaries, and to continue the quest for understanding with a sense of humility and a recognition of the speculative nature of our explorations. As we contemplate the possibility of simulated realities, we are reminded of the richness of the human imagination and the unbounded potential for discovery and insight. This book, therefore, stands not only as a contribution to the academic discourse on the Simulation Hypothesis but also as an invitation to all who engage with these ideas to continue the quest for understanding, inspired by the profound and intriguing questions that lie at the heart of our existence.

Printed in Great Britain
by Amazon

af1e81a6-f100-4afe-86ca-702e65f1a162R01